Geokinematics

Springer

Berlin
Heidelberg
New York
Hong Kong
London
Milan
Paris
Tokyo

Rex H. Pilger

Geokinematics

Prelude to Geodynamics

With 187 Figures

Springer

Rex H. Pilger
Landmark Graphics Corporation
1805 Shea Center Drive, Suite 400
Highlands Ranch, CO 80129
USA
E-mail: *rpilger@lgc.com*
Digital data sets: http://www.pilger.us

ISBN 3-540-00548-X Springer-Verlag Berlin Heidelberg New York

Library of Congress Cataloging-in-Publication Data Applied For

A catalog record for this book is available from the Library of Congress.
Bibliographic information published by Die Deutsche Bibliothek
Die Deutsche Bibliothek lists this publication in die Deutsche Nationalbibliographie, detailed bibliographic data is available in the Internet at <http://dnb.ddb.de>.

Springer-Verlag Berlin Heidelberg New York
a member of BertelsmannSpringer Science+Business Media GmbH

http://www.springer.de

Printed in Germany

Coverdesign: E. Kirchner, Heidelberg
Typesetting: Fotosatz-Service Köhler GmbH, Würzburg

Printed on acid-free paper 32/3141 – 5 4 3 2 1 0 –

Preface

There's something about introductory textbooks that can be fundamentally misleading. As I plowed through freshman Physical Geology some 36 years ago, I couldn't help but acquire the impression that all of the major problems in the earth sciences had been solved. Yes, there was a chapter about Wegener's continental drift hypothesis along with a few other ideas concerning earth evolution – expansion, contraction – and including Holmes' convection cell idea. Once I began to take the field trips and participate in junior-year field camp, I began to suspect that the textbooks had left something out. It didn't take long, however, to begin hearing about the "New Global Tectonics", at which point I felt that I had been misled by what I had encountered in my first formal contact with geology. All of a sudden it was clear that there was much more to geology and geophysics than a widely used book could convey.

Plate Tectonics is now into its fifth decade. I haven't looked at current introductory textbooks, so I don't know whether they still convey the impression that all of the major problems in the earth sciences have been solved. Clearly, there is still much to do with continuing elaboration of plate tectonics, among all of the other challenges of the geosciences.

This monograph introduces a number of new observations, techniques, and hypotheses as extensions to Plate Tectonics. Perhaps the term "monograph" seems inappropriate, but virtually everything presented here is part of a coherent whole. Some of the components of this study were developed many years ago and have been augmented in the intervening time. Other ideas have emerged in only the last few years.

About the title: Plate Tectonics is primarily a kinematic theory. I find it intriguing to see how far we can go in the application of simple, largely non-physical mathematics to the elucidation of the theory. Numerical modeling of the earth continues to be an active focus of research for many geophysicists, but, I'm sure that most would agree that really satisfactory geodynamic models have yet to be achieved. By elaborating and extending the kinematic theory, it is my hope that the additional components added to Plate Tectonics provide additional insight and constraints that the geodynamic modelers can and will use.

In the last year, two prominent scientists have produced massive monographs (much more extensive than this one) – the late Stephen Jay Gould's summary of evolution and natural selection, and Stephen Wolfram's theory for everything. Both books include substantial personal comments and asides embedded within the text. Such comments are not standard components of a scientific paper or even of a monograph. In his early compendium of many of the critical plate tectonics and geomagnetic reversal papers, Allan Cox provided a fascinating commentary that includes some valuable nuggets – especially the account of L.W. Morley's independent discovery of seafloor spreading and the application of Thomas Kuhn's paradigm formulation to the new theories.

While I lack the prominence of a Gould, Wolfram, or Cox, I've nevertheless taken the liberty of providing an aside at the beginning of each chapter as a means of explaining, motivating, or rationalizing the argument to come, from either a personal perspective or from what I view is a generally accepted context. For those who would prefer no such ruminations, the asides can be skipped.

Acknowledgements

It is traditional to acknowledge those who have been of special assistance to this study. More than any others, it is those who have acquired so many of the field and laboratory observations and those who create the edifice of Plate Tectonics that warrant the most thanks – they are listed in the References and in the Supplementary References: Additional Sources for Appendix 5.

There are those with whom I've collaborated in past work, with whom some of the seeds of the current study were planted: Tim Cross, David Handschumacher, Jeff Nunn, Roy Dokka, Jack Gallagher, and Steve Bergman. Then there are those with whom I've had prolonged conversations and, at times, have served as reviewers of previous work: Donna Jurdy, Norm Sleep, Seth Stein, Lynn Sykes, and Jim Heirtzler. Even though it has been many years, Professors Tom Henyey and Greg Davis of the University of Southern California, the late Peter Misch of the University of Washington, and Bob Nelson and Sam Treves of the University of Nebraska provided early insight, motivation, and encouragement.

Many of the map illustrations included in this volume were prepared with Landmark Graphics Corporation's GeoGraphix Discovery – using the GeoAtlas and IsoMap modules – software I have had the good fortune of helping develop over the past 6 years. Discovery was primarily developed for the oil and gas industry, not for academic research. Nevertheless, it provides an excellent presentation capability, only a small portion of which I make use of herein.

Lastly, I cannot help but acknowledge my wife, Rita, and our children, Beth, Paul, Andrew, Mary, and Matt. I suspect it's not always easy having an introverted theoretical scientist for a husband or father. "Whatcha doing, Dad?" When I attempt to explain, I'm afraid that the explanation is lost in translation; the resulting glazed eyes are rather like those observed when summarizing 60 million years of Andean subduction history for earthquake seismologists (who are only interested in the last 60 years). I can only hope that those scientists for whom this book is written are able to translate what they read into something useful and meaningful for their future work.

Contents

CHAPTER 1

Introduction

1.1 Aside

For those who entered the earth sciences in the mid-1960s or earlier, the advent of Plate Tectonics was either an extraordinarily upsetting encounter with a seemingly non-geological science or a marvelous revelation that rearranged a couple of centuries of field observations into a high degree of unanticipated coherence. Suddenly, a profound simplicity emerged from apparent complexity – a simplicity that was so clear and transparent that the initial reaction of many was disbelief. Indeed, skeptics viewed the New Global Tectonics as a kind of bogus religious experience promulgated by carpet-bagging geophysicists and a few scalawag geologists. The disbelievers for the most part decided to ignore the new "paradigm", treating Plate Tectonics in the same way that contemporary science philosophers treated Kuhn's inferred structure of scientific revolutions, of which Plate Tectonics became the latest paradigm.

As a college sophomore in the spring of 1968, I recall the excitement of when J. Tuzo Wilson, the "music man" of what was not yet known as "Plate Tectonics", visited Lincoln. The department chairman encouraged us (the petrology class, which he also taught) to attend Wilson's lecture to the Nebraska Academy of Sciences. The chairman (who wore many hats) had been responsible for the invitation. On a Friday night, we went to a medicine show; we found a tall, cultured Canadian with a satchel of tricks very much like that of a snake-oil salesman; in that satchel was a centerpiece contraption consisting of window shades painted in black and white stripes and separated by flashlight bulbs. The good professor from up north knew marketing; whether a theory is accepted is sometimes as much the manner of its presentation as its simplicity and consistency with the data. Was it simple? Even the biologists seem to understand the spiel.

J. Tuzo Wilson spoke of reversals of the earth's magnetic field (first discovered before World War II, but documented, dated and correlated only in the past decade), recorded as "stripes" above the seafloor (once the data were processed and stacked or contoured) in progressively created oceanic crust. He reminded the audience of matching Atlantic continental margins (which we remembered from last year's Physical Geology – best known as the idea of a non-geologist, a German meteorologist named Wegener), of volcanoes around the Pacific Rim, and earthquakes along faults and ridges and in association with the volcanic rim. He particularly emphasized the paradox of the faults – his great discovery; the displace-

ment along the faults was opposite to that implied by the patterns of the magnetic stripes. The faults (he called them "transforms") connected spreading centers along which new oceanic crust was created. As the crust (the window shades) on either side of the spreading centers moved symmetrically away from the centers (the mid-ocean ridge), earthquakes (flashing flashlight bulbs) occurred.

At the end of the lecture, the department chairman announced that Professor Wilson would be available to visit with geology students the following morning. Bright and early Saturday, we found ourselves in the presence of the great man as he patiently explained the new theory. Some 1,500 miles from either ocean, 1,500 miles from a plate boundary, 1,000 miles from Toronto, landlocked Nebraska was being introduced to a revolutionary new idea. Others, a bit older than we, have described their own experience as graduate students at Lamont Geological Observatory (Columbia) or Scripps Institution of Oceanography, or Stanford as they saw the revolution take shape, and participated in its formulation. We, on top of the Quaternary glacial drift and the feather edge of the Late Cretaceous seaway of the mid-continent, were the second-hand recipients of this geological good news.

The department chairman had more up his sleeve, however. Like a recent convert faced with his first demon, he was going to challenge us neophytes with a very different (evil?) story. The following week, as we marveled at Professor Wilson's presentation, he promised this different response. Arthur Meyerhoff, a prominent, vocal opponent of Plate Tectonics, would be invited to town as well. About a month later, a bulldog of a man lit up two screens with lantern slides (not 35 mm – a tip-off of what was to come), and spent an hour and a half describing all that was "wrong" with Plate Tectonics; the audience this time was entirely geological (fortunately; nothing like exhibiting dirty laundry before biologists and chemists). Meyerhoff, a self-described geologist of the old school, a mid-continent stratigrapher within the fraternity of oil-and-gas geologists, emphasized the most important argument: Plate Tectonics is the invention of "physicists", not geologists (James Heirtzler, who was the leader of the magnetic reversal team at Lamont had his Ph.D. in physics). How dare we allow outsiders to interpret the earth for us! He wanted us to extend the same welcome to the New Global Tectonics as non-geologist Wegener received from American geologists in the 1930s. A series of articles written by Meyerhoff, including some coauthored with his father, appeared over the next few years, without appearing to persuade anyone that Plate Tectonics was scientific heresy, incompatible with the legacy of Hutton, Lyell, and Smith. As a graduate student a few years later, I had the opportunity to explore one of the supposed arguments he cited – the presence of the same rock unit on either side of the inferred boundary between Asia and greater India; this boundary was inferred by an elaboration of plate tectonic theory to have once represented a great paleo-ocean basin that had disappeared beneath Asia as India marched northward to collide with the larger continent and form the Himalayas while elevating Tibet. The existence of a once-continuous sandstone stratum across and older than the boundary would prove fatal to interpretation of the boundary as a suture marking the location of a former ocean. A perusal of the source of the original inference indicated that the inferred identity of the rocks in question was entirely superficial; no petrological or geochemical analyses had been performed to demonstrate identity, nor were any distinctive correlative fossils found in either outcrop. It would be analogous to correlating the late Tertiary Punchbowl formation along the southern San Andreas fault with the Nugget formation (Jurassic) in southern Nevada. Any support that the anti-Plate Tectonics geologist received was more likely to be of a chauvinistic kind, not based on critical analysis of the new theory.

The writings of Meyerhoff have now largely faded from view, except as a historical curiosity, and were largely ignored, in any case, when they first appeared. Their rejection was often based on the kind of analysis I had undertaken; his arguments were turgid, illogical, inconsistent, and incoherent. The Duke of Occam would not have been impressed. Nevertheless, the undercurrent of such exhausting detail (the papers are all very long and elaborate) is the implication that something as "simple" as Plate Tectonics cannot be correct. Rather, one can invoke the solemn pre-World War II literature on the structure and origin of mountain belts, built upon "geosynclines" that deepen and collapse upon themselves, with igneous and metamorphic rocks mysteriously emerging from the "bowels of the earth". Overthrust belts and ultramafic rocks were held in awe as they manifested the profundities of geologic time and place. Similarly, repetitive sequences of sedimentary rocks in the interior contents were to be interpreted as reflective of synchronously global pulses of orogeny – mountain building – the misty pulse of the planet. Obscure hypotheses with uncertain physical underpinnings and even less certain testable implications were the best the science could do. It is no wonder that some greeted the advent of the seemingly simple theory with so little enthusiasm. Mystery was to be replaced with understanding. "Geology is not simple," they said. "How could things be so simple?"

It took a few years for the structure of mountain belts to begin to make sense in the context of Plate Tectonics. The old terms "miogeosyncline" and "eugeosyncline" were slow to leave the literature as modern analogues – island arcs and trenches, ocean basins, passive continental margins – emerged as elaborations of the new paradigm. Uncertainties still existed (and exist) regarding which plates interacted with each other to produce a particular structural event, and questions remain regarding the origin of the wide variety of volcanic rocks encountered along plate boundaries and, in some cases, even within plates. More significantly, the connection between plate creation, destruction, and motion with the motions of the earth's interior remains uncertain, even now, decades later. But a great fog had lifted; Plate Tectonics was even more than the rush of fresh air that displaced the mildewed mists. Plate Tectonics provided part of the means for changing tectonic geology from a qualitative, descriptive science caught up with classification and interpretation, to a quantitative, predictive discipline.

Despite the quantitative predictability that Plate Tectonics provides, the implicit question still remains – why is the earth so simple that quantitative plate tectonics works? Philosophically speaking, a similar question can be asked in the other sciences, especially physics: How can the Universe be described in such elegant, beautiful mathematics? Before the more down-to-earth question can be dealt with, however, it is necessary to demonstrate the degree of simplicity that does indeed exist. It is the what and how that are the foci of this book; the why can only be touched upon with, at best, a ten-meter pole.

1.2 Kinematic Prediction: The Success of Plate Tectonics

The success of Plate Tectonics in linking diverse geological and geophysical phenomena is largely the ability of fairly simple geometric and kinematic techniques to characterize the principal manifestations of plates and their motions. Equivalently, the outer earth exhibits large-scale kinematic behavior that can be, at least in part, easily characterized and parameterized. Implicitly, researchers work on the hypothesis that Plate Tectonics will eventually yield to a more comprehensive dynamic model based on physical and chemical principles, instead of the current, largely geometric and kinematic models (although such models already have implicit physical and chemical constraints). In the meantime, simple kinematic models are both successful and essential to geohistorical description of plate evolution, while the reasons for their success (emergent simplicity out of complexity) await further explanation and elaboration.

Within the first 15 years of Plate Tectonics, most of the concepts, tools, and collateral hypotheses of the theory had been advanced and applied. The basic concepts and their confirmation were virtually complete by 1968: kinematic rigidity, three plate boundary types (divergent, convergent, shear) and their leit motifs (seismicity, focal mechanisms, magmatism, heat flow, topography/bathymetry, magnetic reversal time scale/isochrons), and the (rotational) displacement parameterization. By the mid-1970s, deep-sea drilling had verified the required progressive aging of oceanic plates away from spreading centers, plate reconstruction techniques were being routinely applied to magnetic isochrons and paleotransform faults, more sophisticated instantaneous global models of contemporary plate kinematics were promulgated, and candidates for internal reference frames, especially hotspots/plumes (in addition to paleomagnetism – an implicit pre-Plate-Tectonics reference frame), were proposed. These same concepts and tools continue to be elaborated, and the extended hypotheses continue to be developed and tested as Plate Tectonics enters its fifth decade.

This volume is focused on a review and extension of the basic techniques of plate kinematics and reconstructions, evaluation of reference frames (especially hotspots and paleomagnetism), and consideration of their implications for the dynamics of plate tectonics. Theoretical development of expanded plate kinematic theory and an ensemble of critical, available observations are integrated into a grand model. A key addition to the plate kinematic framework is the concept of the "mesoplate" and its corollary, "proper motion".

Despite the existence of the basic formulation of Plate Tectonics for over 30 years, it is not uncommon to find misunderstandings of simple kinematic and reconstruction techniques within otherwise valuable contributions by those who do not "do" plate reconstructions. Similarly, there are occasional misinterpretations of fundamentally kinematic concepts as intrinsically dynamic. In order to address some apparent misunderstandings and, further, as a basis for elaboration of plate kinematics techniques, both the fundamental formulae as well as their derivative extensions are included. For convenience, and, it is hoped, utility, the basic formulations as well as their further development are elaborated by C programming language routines as a supplement to conventional mathematical notation and derivation.

As a prelude to this discussion, it is important to address the idea of kinematic rigidity as distinct from dynamic rigidity. The primary postulate of Plate Tectonics is that spherical tectonic plates are kinematically rigid; that is, plates do not internally deform to any significant extent. Formally, this means, given any three points, A, B, and C, on the surface of a spherical tectonic plate, the great circle distances AB, BC, and AC are unchanged as the plate itself is displaced over the surface of the sphere. In a sense the use of the term "plate" itself implies a certain kind of rigidity (either kinematic, dynamic, or both). The postulate of kinematic rigidity has dynamic implications, certainly, but they are not directly addressed by the theory. Part of the elaboration of Plate Tectonics is testing to what extent plates are indeed kinematically rigid. Additionally, there are significant aspects of the study of the principal internal stress field within plates – its magnitude and orientation – that may prove to be a critical bridge to a more complete dynamic analysis. Unfortunately, there are some workers who apparently infer that use of the term "rigid" explicitly means dynamic rigidity – some measure of the strength of the plate; these workers interpret plate rigidity as implying that plates are subjected to significant differential stresses which they are able to resist without deforming. If, on the other hand, a plate is not subjected to significant differential stresses as it is displaced, it will not deform; thus it will behave in a kinematically rigid way. Dynamically, plate topology is a reflection of self-organization that minimizes differential stresses rather than creating solid bodies of high tensile- and shear-strength. Apply a significant differential stress to a plate and it will fragment, forming additional, smaller plates. Conversely, when differential stresses between two adjacent plates are minimized, the two plates merge to form a single plate. To reiterate, one of the "mysteries" of plate tectonics is how the minimization of differential stresses is accommodated by a small number of plates. Documented internal deformation of the originally defined Indian Ocean plate gives some insight into the mystery; there has been some success in redefining the larger plate into a few smaller plates that are kinematically rigid, although their non-extensional boundaries are less well-defined.

The kinematic models of Plate Tectonics conveniently rely on a theorem of Leonid Euler: a kinematically rigid displacement confined to the surface of a sphere (equivalently: a displacement with one fixed point – the center of the sphere) can be parameterized as a rotation around an axis passing through the center of the sphere (and origin of a Cartesian and/or spherical coordinate system – the one fixed point of the system; e.g., Weisstein 1998). More conveniently, the axis of rotation corresponds with the latitude and longitude of the two points at which it intersects the sphere's (the Earth's) surface – the poles – instead of the less-easily visualized direction cosines.

Bullard, Everett, and Smith (1965) first applied this parameterization of Euler to nascent Plate Tectonics as a means of quantifying and evaluating Wegener's (1912) postulated fit of the western margin of Africa to the eastern margin of South America. They used a least-squares measure to quantify the fit of an isobath on the African margin to the corresponding, rotated isobath on the South American margin for a set of trial rotation poles and angles. Subsequent elaboration of Plate Tectonics by other workers included derivation of rotation poles and angles and rates to describe plate reconstructions of magnetic isochrons and paleotransform faults as well as instantaneous kinematics. Morgan (1968) and McKenzie and Parker (1967) applied the rotational parameterization to contemporary, instantaneous plate kinematics and to situations inferred to correspond to distinct periods of fixity of the pole of relative motion between two plates. LePichon (1968) closely followed with the first estimates of contemporary plate kinematics on a global scale. Of course, the other familiar advances that comprised the theory – magnetic isochrons and the geomagnetic time scale (Heirtzler et al. 1968), transform faults (Wilson 1965a; Sykes 1967), and seismicity of the boundaries (Isacks et al. 1968) – were prerequisites to successful application of the new theory.

In the absence of an adequate, comprehensive, quantitative dynamic model of plate tectonics (necessarily including mantle convection), the kinematic description of real earth tectonic processes suffer from challenging limitations. The earliest workers in Plate Tectonics recognized these limitations of the application of rotational parameterization (including, implicitly, the more basic hypothesis of kinematic rigidity):

1. Rotational kinematics of three or more plates prohibits fixity of the poles of motion of more than one plate pair (McKenzie and Morgan 1969). While this is a geometrical requirement, an inferential corollary is that there seems to be no compelling reason, other than computational convenience, for assuming that any rotational parameters remain constant.
2. Explicit application of rotational parameterization requires definition of one or more reference frames, which may or may not have dynamic significance (for example, selection of one plate, the paleomagnetic field, hotspots, or paleoclimatic zones as the preferred reference frame). Further, even if the reference frame is dynamically significant, it may be significant in only one component (there is, for example, the longitudinal ambiguity of paleoclimatic zones and of dipole-modeled paleomagnetism).
3. The earth (and therefore its plates) departs from sphericity in its static figure (an oblate spheroid) by as much as 1 part in 1,000 and in radial ("local vertical") displacements on the order of 1 part in 2,000 over a period of a few to tens of millions of years.
4. Geological and geophysical evidence indicates varying degrees of internal "plate" deformation (therefore, departures from kinematic rigidity and "simplicity") at present and in the recent past with strains greater than those simply attributable to radial deformation due to departures from sphericity. The actual formation of new plates from old plates explicitly requires a significant degree of deformation of parent plates. For example, there is uncertainty concerning how many plates or subplates the Indian–Australian "plate" must be subdivided into in order to produce satisfactory kinematic models for the past 10–20 million years (e.g., Royer and Gordon 1997). Conversely, rifting of Africa appears to fragment the continent, but instantaneous parameterization into additional plates had been irresolvable (e.g., Minster et al. 1974; DeMets et al. 1990) until quite recently (Chu and Gordon 1999). Nevertheless,

the success of plate reconstructions demonstrates that the departures from rigidity are of second- or third-order; kinematic rigidity is, in most cases, an excellent first approximation of lithospheric behavior.

The last two limitations could require modification of simple rotational kinematics and reconstruction parameterizations. However, since the degree of departure of the lithosphere from kinematic rigidity appears to be of second or even third order, the Euler approximation still seems to be applicable.

The first two limitations are quite naturally an appropriate focus. Since, in general, we cannot expect rotational parameters to remain fixed, it is desirable to develop some form of continuous parameterization of plate motions. Further, it is additionally desirable to more convincingly demonstrate or rule out the existence of one or more dynamically meaningful internal reference frames for plate motions.

In this contribution, a novel geohistorical parameterization of plate reconstructions and kinematics is developed and applied to an extensive set of data: magnetic isochrons, contemporary and extinct fracture zones, hotspot traces, and other possible kinematic indicators. This new parameterization has the potential of providing a framework for further quantification of intraplate deformation and boundary constraints on complete geodynamic models. It also provides a means of more complete evaluation of critical hypotheses corollary to plate tectonics. In addition to the formulation of the new parameterization, it is applied to the fixed hotspot hypothesis, to the assessment of the relationship of intraplate stress fields to plate motions in a hotspot framework, and to other phenomena that can be inferred to be related to motions of plates relative to an internal reference frame. A newly defined set of internal reference frames, "mesoplates", is a natural consequence of this formulation. Mesoplates appear as at least approximations of sub-primary plate tectonic behavior of the outer mesosphere and asthenosphere. In addition, several phenomena appear to find their origin in the interaction of mesoplates with one another, the asthenosphere and lithosphere. Further, some peculiar phenomena associated with initial rifting and onset of seafloor spreading find explanation in distributed extension of a particular type, involving mesoplates. Anticipating the arguments to be advanced below, it might be suggested that an adequate numerical dynamic model would not explicitly require mesoplates; rather, the observations that lead to their inference (as discussed below) would be a gauge of the success of a particular numerical model. The same argument, of course, applies to plates themselves. If a dynamic model is to prove acceptable, it must reproduce the observations that lead to the inference of the existence of kinematic plates. Plates and mesoplates are, for the moment, convenient approximations of Earth behavior that allow inference of other implications of Earth tectonics (e.g., paleooceanography, mountain building, passive continental margin development) until a grand dynamic model emerges.

In sum, this volume incorporates a review of the fundamentals of plate kinematics and reconstructions, introduction of a new kinematic parameterization, evaluation of internal reference frames, postulation of the existence of mesoplates, and models for rift-related anomalous volcanism. Source code accompanies mathematical derivations, and an extensive data base of paleostress indicators, along with other data types, are added features that provide documentary evidence for the ideas presented as well as a basis for further study. Data, parameterizations, tools, new ideas: it is hoped that each of these contributions will prove valuable to both present and future workers in the fruitful vineyard that is Plate Tectonics.

CHAPTER 2

Proper Reference Frames: Hotspots, Paleomagnetism, Paleoclimate

2.1 Aside

One of the more challenging aspects of the doing of science is the apparent failure of communication. Does the paper reporting the results of an investigation get read? Is the paper understood?

One of the first papers I published was based on part of my Ph.D. dissertation. As an extension of some collaborative work with Tim Cross (now with Platte River Associates), in which we had looked at the distribution of Cretaceous and Cenozoic igneous rocks of the western United States, I decided to take a closer look at Mid- to Late Cenozoic volcanic rocks of coastal California. Tim and I felt we had done an effective job of relating patterns of magmatism in the western United States to subduction regimes, and inferences concerning the relation of North American plate motion to subduction angle, a topic we explored in more detail later. But, we hadn't dealt with the curious coastal volcanism that predated movement on the San Andreas Fault (and the onset of transform motion between the Pacific and North American plate) and subsequently accompanied transform motion. Morro Rock, offshore of Morro Bay,California, is perhaps the best known of Miocene intrusions, which I found fascinating.

In this paper (co-authored with my former major professor, Tom Henyey), I developed the hypothesis that the coastal volcanism had two principal sources: (1) subduction of young, hot lithosphere of the Farallon plate, predating transform motion (e.g., Morro Rock), and (2) oblique extension associated with transform motion along splays of the San Andreas system, above shallow, hot asthenosphere, typified by the young volcanics of the Crystal Peak/Geysers area north of San Francisco.

With a few shining exceptions, this paper of mine seemed to disappear into a void; very few workers referenced it. Over the next 10 years, two more studies of the coastal volcanics and their relation to the plate tectonic environment appeared, neither of which cited my work. Okay, these things happen; it's hard to do a thorough survey of the literature prior to beginning a project. Perhaps the more important oversight, however, was that neither of the papers cited a previous study of the isotopic ages of the Coast Range volcanics (of which I had made ample use). They not only missed my inspired speculation, they also missed utilization of some very valuable data.

After these publications appeared, a research conference was announced with a focus on coastal California volcanism. Tim Cross called to tell me that he had been invited to participate; he declined, pointing out to the organizers that I had published a study on the subject – a study he hadn't been involved with – and suggested that I be contacted. The organizer responded that he didn't want to invite me because he didn't understand my (now yellowing) paper. Aargh!

Was my paper so opaque? Was the reason for its lack of citation by others due to their not understanding it, rather than overlooking it? The reviewers, I recalled, initially had encouraged me to clarify some of my interpretation, which I did indeed do, apparently to the editor's satisfaction, as he accepted the paper. Doubts still persist.

I'm comforted by one other memory. Shortly after the paper was published, I was visiting with a colleague at a professional meeting when one of the grand men of California geology approached. My colleague asked him how he interpreted the origin of the coastal volcanics. With a puzzled look, the senior scientist said, "Why don't you ask Rex? He has the best understanding of the subject." Yes, my colleague was embarrassed (as he should have been), and I was taken aback by his question, but I, at least, found one person who had read and understood my paper.

The problem of having a contribution overlooked is not an isolated experience in our science. One of the more widely uncited papers of which I am aware is a contribution by Moberly (1972). His paper deals with the origin of back-arc basins, the regions of extension and, in some cases, sea-floor spreading in the back-arc regions above active subduction zones. Oddly, Moberly's paper occurs in the same memoir as Morgan's (1972a) classic elaboration of his mantle plume hypothesis for the origin of hotspots and their traces.

"Everybody" has read Morgan's paper. I sometimes wondered whether only Tim and I (and Morgan) had read Moberly's; he had proposed that back-arc extension and spreading were a consequence of the stabilizing effect of the subduction zone (another worker, in an abstract, referred to the anchoring effect of subduction zones, but in a different context). Motion of the principal upper plate of the subduction zone away from the subduction zone would be expected to induce the kind of extension, even spreading, that is observed. Subsequently, a number of workers have advanced a hypothesis for the origin of back-arc basins similar to that of Moberly's, but without, apparently, being aware of his original contribution, even while citing Morgan's 1972 paper. Ironically, as noted, even Morgan (1972a) himself had referenced Moberly's idea, since it provided additional support for Morgan's "absolute motion" model based on fixed plumes.

I mention Moberly's paper primarily because it has motivated much of my subsequent research. However, I must admit that this book may itself exclude contributions that deserve citation. I can only shrug my shoulders and lamely point out that the literature on hotspots and "absolute motion" is voluminous.

2.2 Hotspots and Plumes

One of the most persistent collateral hypotheses of plate tectonic theory has been the fixed hotspot/plume hypothesis of Wilson (1963a, 1965b) and Morgan (1971, 1972a). Volcanism within plates (e.g., Hawaii) that cannot be attributed to plate boundary processes and excessive volcanism along plate boundaries (e.g., Iceland) are postulated to represent distinct "hotspots" (Wilson's contribution) in the mantle. Commonly, these zones of anomalous volcanism occur at the ends of island-seamount chains or ridges that exhibit monotonic increases in age of volcanism with distance along the chain away from the currently active volcanic center. Subparallelism of intraplate volcanic island-seamount chains and ridges leads to the additional inference that the inferred hotspots represent a different reference frame from the plate on which the chains and ridges ("hotspot traces") occur. Morgan's elaboration of Wilson's hypothesis included the suggestion that hotspots form a fixed reference frame and, further, that they represent vertical plumes of hot, low viscosity material derived from near the base of the mantle.

The fixed hotspot/plume hypothesis not only provides an explanation for anomalous volcanism, but also has implications for the dynamics of the lithosphere, asthenosphere, and the rest of the mantle, and provides an important mechanism by which internal energy is brought close to the Earth's surface. In addition, it has had the continuing promise of providing a frame of reference for the plate motions relative to the Earth's interior. Whatever the origin of "hotspot traces", complete dynamic models of Earth convection and plate tectonics must include them.

Insofar as the kinematic implications are concerned, Morgan (1972a) initially demonstrated and Minster et al. (1974) confirmed that inferred hotspots and the trends of their correlative traces could define a consistent internal reference frame for contemporary instantaneous plate motions. Subsequently, several efforts have been made to elaborate a multiple-plate reconstruction model in a hotspot reference frame for the Atlantic and Indian Ocean plates for the Cenozoic and Mesozoic (Duncan 1981, 1984; Morgan 1981, 1983; O'Connor and Duncan 1990; Duncan and Richards 1991; Müller et al. 1993). Efforts to construct global hotspot models have not been convincing, however, most obviously in their inability to adequately link the Atlantic–Indian Ocean hotspots with the systems in the Pacific (e.g., Molnar and Atwater 1973; Molnar and Francheteau 1975; McDougall and Duncan 1988; Molnar and Stock 1987; Norton 1995, 2000; Raymond et al. 2000).

Considerable uncertainty has existed concerning the relative motion linkage (via reconstructions across spreading centers) of the Atlantic–Indian Ocean plate system with the Pacific Ocean plate system through Antarctica. It has been suggested that Antarctica has not behaved as a coherent plate throughout the Late Mesozoic and Cenozoic (e.g., Atwater and Molnar 1973; Duncan and Richards 1991). Thus, the failure of a global hotspot model linking the Pacific plates with the rest of the major plates via Antarctica has not been viewed as a definitive failure of the fixed hotspot hypothesis. However, further studies suggest that even internal deformation of Antarctica is not adequate to account for the discrepancy between Pacific and Atlantic–Indian ocean models. Until recently, even within the Atlantic–Indian system, most models have been unsuccessful in simultaneously fitting all of the obvious inferred hotspot traces (Walvis Ridge-Tristan da Cunha–Rio Grande Rise; Reunion–Chagos Ridge–Laccadive Ridge; Kerguelen Plateau–Ninetyeast Ridge; White Mountain Magma Series–New England Seamounts–Corner Rise; Eifel–Central European volcanics; Iceland–Faeroes Ridge) without significant motion between hotspots. Three models, by Fleitout et al. (1989, as published for Africa-hotspot motions only, but easily extended to other plates by composing relative plate motions), Duncan and Richards (1991), and Müller et al. (1993), showed remarkable improvement in simultaneous fit of all of the Atlantic and Indian Ocean traces, with the exception of Iceland–Faeroes, the older inferred portions of the Yellowstone trace, and the poorly defined Eifel trace. (The latter three traces were not all explicitly addressed by these papers.)

Improvement of new Atlantic–Indian Ocean hotspot-plate kinematic models over their predecessors can be attributed to several factors. These include: (1) improved Indian Ocean reconstructions (despite ambiguity due to Late Cenozoic intraplate deformation); (2) an increase in available information on the apparent age of selected points on various presumed hotspot traces (although confidence in these ages is limited; see below); and, perhaps, (3) both perseverance and serendipity – developing such reconstructions is not an easy task. Of course, the possible validity of the fixed hotspot hypothesis for the Atlantic and Indian Oceans traces is essential for the models' apparent successes. The one, persistent anomaly – the apparent inconsistency of Iceland–Faeroes with other traces, a problem already well known in early tests of the hypothesis (e.g., Molnar and Atwater 1973; Molnar and Francheteau 1975), may have a different solution. The processes responsible for Iceland (plus Eifel and Yellowstone) may not be intrinsically different from those responsible for the formation of other hotspot traces, as discussed below.

The African plate, both because of its central position in the Atlantic–Indian Ocean plate system and its large number of apparent hotspot traces, is the naturally preferred plate for comparing various models. However, the key to successful models is not only their ability to fit the traces on one plate, but multiple traces on other plates as well. Thus, the principal differences between models are more obvious on other plates. For example, each of the newer models provide a much more satisfactory fit to the Reunion–Chagos–Laccadive, Kerguelen–Ninetyeast, and Cruiser–New England Seamounts traces than other models. Müller et al.'s (1993) model appears to better fit available age data from the two Indian Ocean and Cruiser–New England Seamounts traces than either the model of Fleitout et al. (1989) or that of Duncan and Richards (1991); for this reason, the model of Müller et al. (1993) is used as the basis of analysis presented in this volume. However, as noted below, reexamination of the analytical data utilized in the models of Müller et al. and Duncan and Richards introduces some uncertainty in the validity of some of the published dates (Baksi 1999).

At the same time that he advanced the plume hypothesis, Morgan (1971, 1972) introduced the term "absolute motion", which a number of subsequent workers (including this author) also adopted. However, there are some unsatisfactory aspects of this terminology in both the larger perspective of the other physical sciences and in the obvious recognition that the hotspot frame of reference itself is in revolution around the Earth's axis and around the sun, and in motion with respect to the Milky Way Galaxy and other galaxies. More critically, the apparent discrepancy between hotspot and paleomagnetic reference frames, as recognized by Morgan (1981) and discussed further, below, and the present lack of a consistent global hotspot model militates against a single "absolute" reference frame for plates. Alternatively, and in analogy with a cosmological term, motion in an inferred internal (sublithospheric) terrestrial reference frame is herein referred to as "proper motion". Just as one can refer to the proper (or conventional) motion of an astronomical body or group of bodies, so the motion of a plate with respect to an inferred internal frame of reference can be called "proper". Proper motion can still be variously defined, for example, in terms of all or a set of hotspots, paleomagnetism, paleooceanography, paleoclimatology, or no net rotation of the lithosphere (e.g., Solomon and Sleep 1975). It would clearly be desirable to demonstrate correspondence of several or even all proposed proper reference frames; thus far this goal has proven elusive.

Parallel with the development and elaboration of proper kinematic plate-hotspot models, a voluminous literature has accumulated on the characteristics of hotspots (particularly their petrology and geochemistry) and on the dynamics of the processes assumed to be responsible for hotspot traces (especially deep-mantle plumes). Numerical and experimental scale-modeling experiments have successfully produced plume-like forms very similar to those imagined by Morgan (1971, 1972a; reviewed by Duncan and Richards 1991 and by Richards and van der Hilst 2000). Thermal calculations, constrained by heat flow, volcanic production rates, and bathymetry and geoid perturbations, support the idea of rapid introduction of significant amounts of heat beneath inferred hotspots (e.g., Sleep 1990a, 1992). Voluminous Cretaceous volcanism in the western Pacific and major flood basalt eruptions (e.g., Deccan, Siberia, Columbia River) have been invoked as evidence for superplumes (e.g., McNutt and Fischer 1987) or plume heads (Richards et al. 1989). Sleep (1992), however, noted that there is a difficulty in reconciling the idea of a superplume with the numerous progressive hotspot traces in the South Pacific. Further, a major problem exists with plume modeling, which implies significant shear of the plume top by plate traction, thereby limiting the ability of hotspots to form a coherent reference frame. One of the principal arguments in support of the plume hypothesis has been the existence of one or a few nearly fixed frames of reference. Numerical and scale modeling of mantle convection suggests that just such a frame of reference is unlikely to exist on the scale at which plumes from the deep mantle are inferred to operate (e.g., Steinberger and O'Connell 2000).

Oddly enough then, most of the literature on plumes has ignored the kinematic difficulties and focused on geodynamic models and geochemical implications. A considerable effort has been devoted to "reproduction" of a plume-head-and-tail model for the origin of large igneous provinces such as the Early Cretaceous Rajmahal basalts and Late Cretaceous Deccan Traps of India, the Early Cretaceous Parana basalts of Brazil, the inferred marginal mafic intrusives of the northernmost Atlantic, and the Columbia River basalts – inferred manifestations of plume heads together with their inferred "tails": Ninetyeast–Kerguelen, Chagos–Laccadive–Seychelles–Reunion, Rio Grande–Walvis–Tristan da Cunha, and Snake River–Yellowstone. Plume head-and-tail models for the Pacific Ocean are somewhat amorphous, with speculations concerning the Jurassic and Cretaceous plateaus of the western Pacific and the large number of traces in the equatorial western to southern Pacific as manifestations of plumes or a superplume.

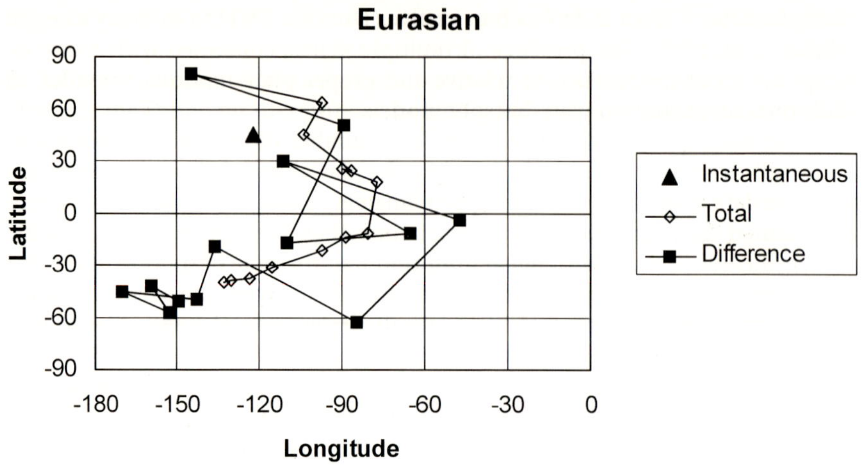

Fig. 2.2. Latitude-longitude plots of proper (hotspot) reference frame parameters for Eurasia from Appendix 2.

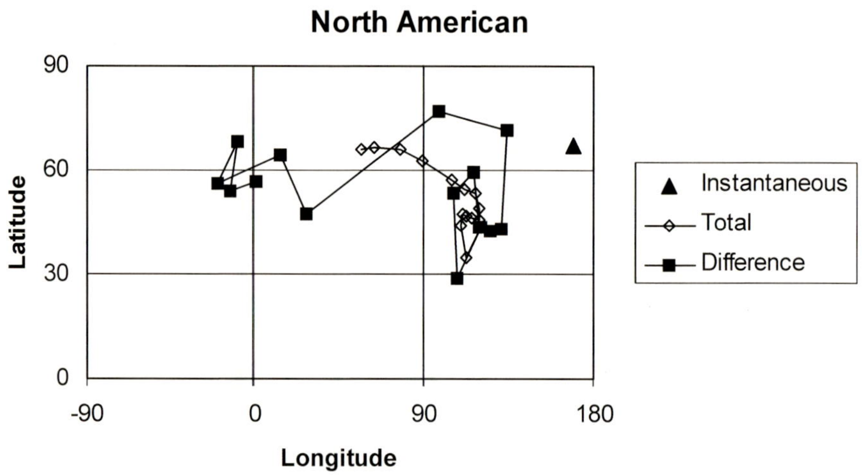

Fig. 2.3. Latitude-longitude plots of proper (hotspot) reference frame parameters for North America from Appendix 2.

finite difference parameters and comparing them for adjacent time intervals; Figs. 2.2–2.10, Appendix 2).

What appears to be common to the models is a desire to fit the traces with monotonically increasing age–distance curves, partially guided by isotopic ages, as discussed below. In general, the direction of proper (hotspot frame) motion of fast-moving plates is consistent for prolonged periods of time. It is also apparent that there are different inferred present-day positions of hotspots from one worker to the next, perhaps guided by the desired optimal fit to the trace.

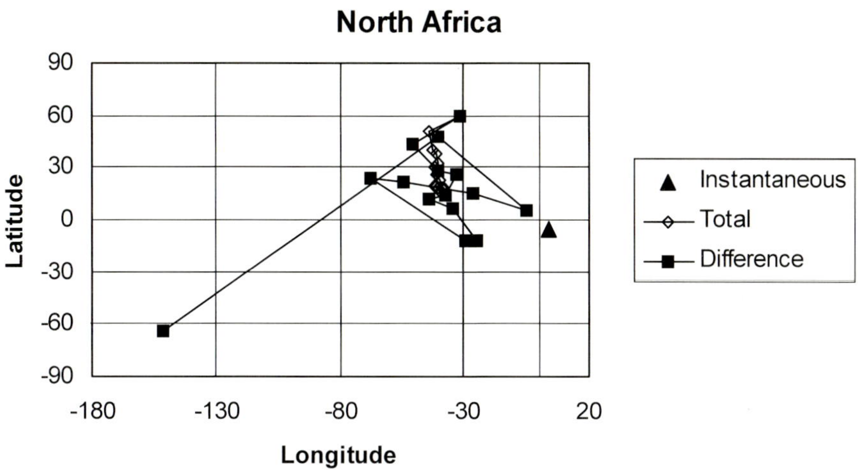

Fig. 2.4. Latitude-longitude plots of proper (hotspot) reference frame parameters for North Africa from Appendix 2.

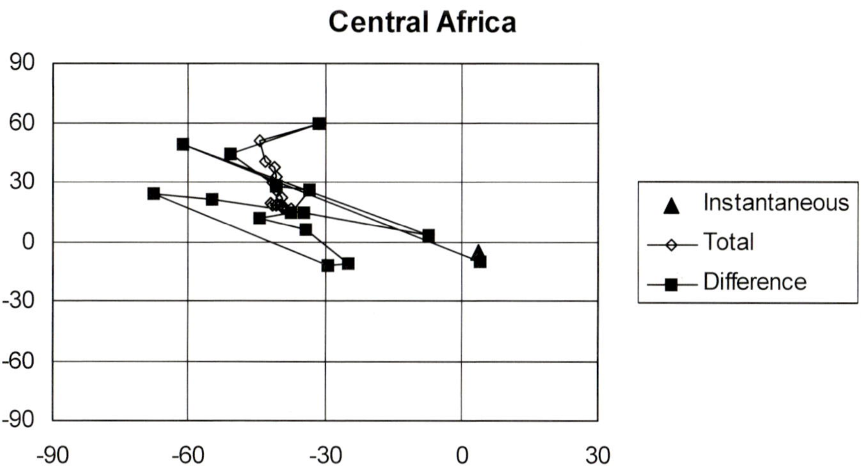

Fig. 2.5. Latitude-longitude plots of proper (hotspot) reference frame parameters for Central Africa from Appendix 2.

The predicted position of a trace relative to a hotspot is as much controlled by the plate reconstructions as by the isotopic dating. Because of the contribution of relative plate reconstructions, the geomagnetic time scale is an implicit collaborator in the hotspot frame reconstruction process; the inferred age of a reconstruction and its accuracy are dependent upon the accuracy of the time scale as well as the accuracy of the isotopic dates of the hotspot traces.

Since there are no plate reconstruction constraints on most Pacific-hotspot models, the isotopic ages play a more important role in derivation of the model

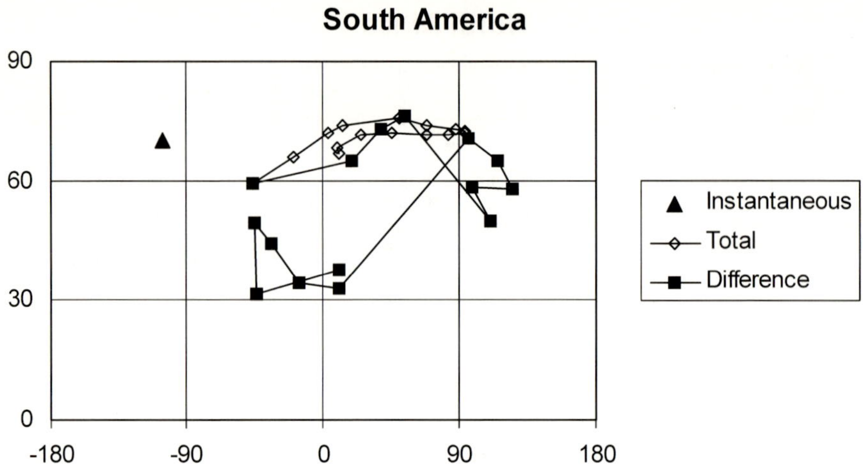

Fig. 2.6. Latitude-longitude plots of proper (hotspot) reference frame parameters for South America from Appendix 2.

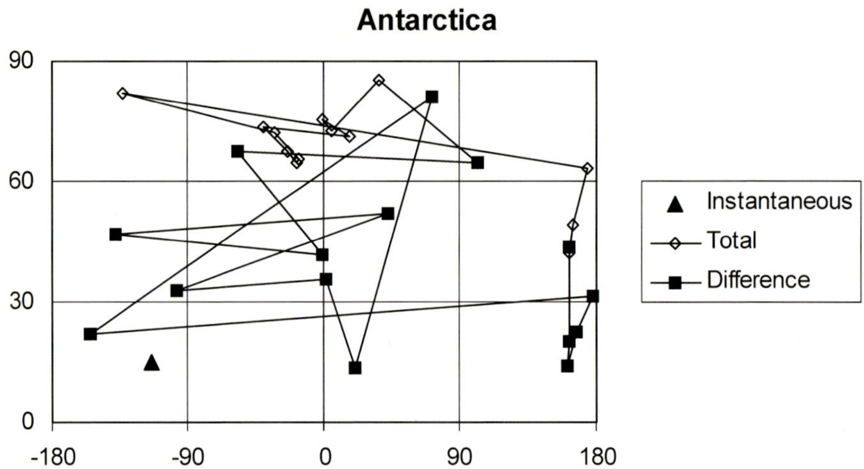

Fig. 2.7. Latitude-longitude plots of proper (hotspot) reference frame parameters for East Antarctica from Appendix 2.

parameters. One exception for Pacific-hotspot models is that of Pilger and Handschumacher (1981), who used a hotspot trace on the Nazca plate and its "mirror image" on the Pacific plate as additional constraints on Pacific-hotspot models. From the Nazca plate constraint, they showed that the Hawaiian–Emperor bend, commonly assumed to be 43 Ma in age, based on isotopic ages of dredged and cored samples, is probably kinematically older.

Along with uncertainty in the location of a plume (if it exists) with respect to its surface manifestation, there are also the questions of how soon, after im-

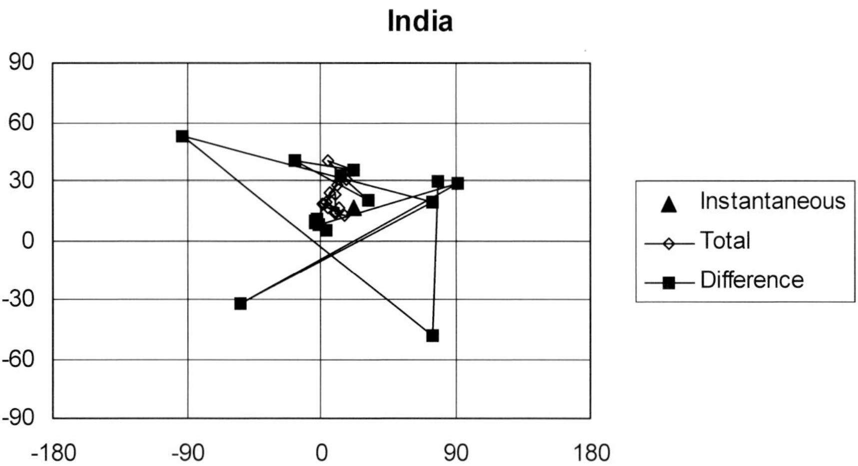

Fig. 2.8. Latitude-longitude plots of proper (hotspot) reference frame parameters for India from Appendix 2.

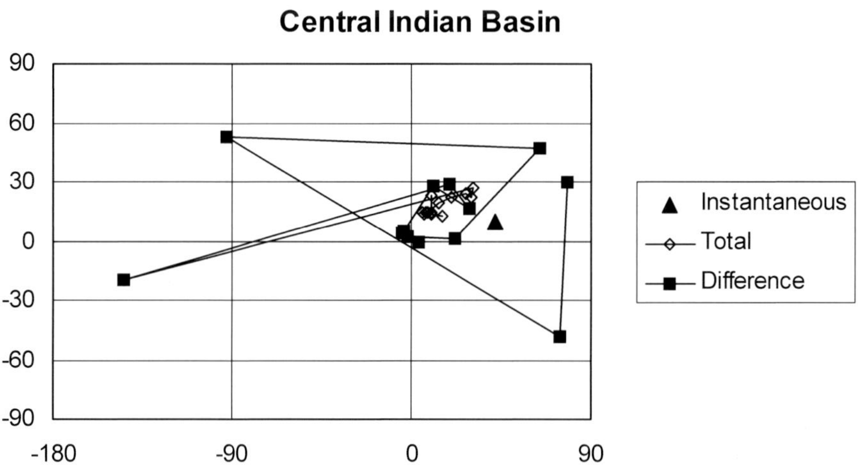

Fig. 2.9. Latitude-longitude plots of proper (hotspot) reference frame parameters for Central Indian plate from Appendix 2.

pingement of a particular plate location upon a plume, surface volcanism appears and how long it persists. Virtually all samples of hotspot traces, both submarine and subaerially exposed, are likely, by cross-cutting and superposition, to record the last phases of volcanism, not the initiation of magmatic activity. Thus, the oldest age determined for a location along a hotspot trace is younger by an uncertain amount than the inception of volcanism and the still older actual impingement of the hotspot on the plate location. Of course, errors in age dating from excess argon or recoil can introduce an older age bias as well; ide-

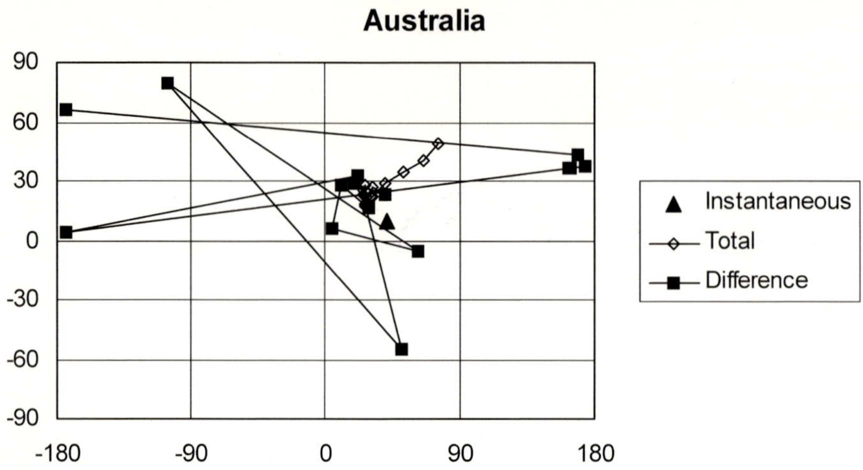

Fig. 2.10. Latitude-longitude plots of proper (hotspot) reference frame parameters for Australia from Appendix 2.

ally Ar/Ar dating can detect the excess-argon effect by thermal degassing, while there are other techniques for detecting recoil. Assuming that the errors are suitably filtered out, it is fair to suggest that accurate volcanic ages from surface or near-surface samples should always be younger than the kinematic age of the hotspot.

The differences between Pacific and Atlantic–Indian Ocean proper motion models are elaborated upon, below. However, first, it is desirable to discuss the derivation of new analytical tools that facilitate this comparison.

CHAPTER 3

Plate Kinematic and Reconstruction Fundamentals

3.1 Aside

In the early 1970s, a series of "second-generation" plate tectonics papers appeared in which the reconstruction techniques introduced by Bullard, Everett and Smith (1965) were progressively modified and applied to magnetic isochrons and paleo-fracture zones. McKenzie and Sclater (1971) attempted the first model of opening of the central Indian Ocean. Their technique involved fitting fracture zones with an instantaneous pole and assuming constant spreading rate to fit segments of the Maldive Ridge (African plate) to the Chagos Ridge (Indian plate).

Pitman and Talwani (1972) and Weissel and Hayes (1972) introduced a different technique in their analyses of the North Atlantic and southeast Indian Ocean, respectively. They applied trial-and-error total rotation fits of magnetic anomalies and fracture zone offsets to the North Atlantic and southeast Indian Ocean. Their contributions were particularly exciting insofar as they showed that differential reconstruction parameters derived from the total reconstruction parameters could be used to approximate plate kinematics; further, flowlines derived from the resulting stage poles could be compared with observed fracture zone trends as an additional test of the reconstructions.

A further advance came with Atwater and Molnar (1973) and Molnar and Atwater (1973), who combined sequential reconstructions from adjacent oceans to derive global reconstructions. The nature of the results these and other workers put forward motivated me to attempt global reconstructions that incorporated the Nazca plate and its motion relative to South America and North America (for the time prior to 25 Ma, when the Nazca plate was part of the larger Farallon plate). I went further and developed a plate reconstruction technique based on McKenzie et al.'s (1970) continental reconstruction method and applied it to Pacific–Nazca reconstructions. After reproducing and modifying, to an extent, the Pacific–North American analysis of Atwater and Molnar and applying it to US Cordillera and coastal California, I applied the techniques to an analysis of the plate tectonic history of the South American Andes in a series of three papers in the early 1980s. These analyses were particularly exciting in that dated Cenozoic volcano-tectonic histories of the two Cordilleras could be shown to correspond with predicted plate motions.

The idea of correlating distinct geological phenomena with plate motions still has a profound appeal. Dewey and others (1973) attempted to interpret the evolu-

tion of the Mediterranean and Alpine region in the context of Pitman and Talwani's oceanic reconstructions, with some success. Atwater and Molnar (1973) provided some intriguing constraints on the evolution of the San Andreas fault system, providing additional support for the speculations Atwater (1970) had previously advanced regarding the Late Cenozoic evolution of western North America.

What was particularly exciting for me was the correspondence of global reconstructed motions of the Pacific, Farallon, and North and South American plates with magmatic histories recorded in isotopic age dates. However, there has always been the objective uncertainty in the reconstructions – has Antarctica behaved as a single plate or is there a hidden plate boundary under the ice? Until the Antarctic plate problem is resolved, the apparent correlations of circum-Pacific plate motions remain apparent.

3.2 Basics

Spherical coordinates provide the most easily visualized parameterization of the displacement of points on the surface of the approximately spherical earth. A variety of map projections provide a two-dimensional view of earth data, while familiarity with model globes facilitates graphical comprehension.

Latitude (ϕ, or colatitude ϕ_c) and longitude (θ) of a point on a unit sphere are related to Cartesian (x, y, z) coordinates as:

$$x=\sin\phi_c\cos\theta=\cos\phi\cos\theta \tag{3.1}$$

$$y=\sin\phi_c\sin\theta=\cos\phi\sin\theta \tag{3.2}$$

$$z=\cos\phi_c=\sin\theta \tag{3.2}$$

```
C routine:
// subroutine cdtrnx:
void cdtrnx(double xlt, double xln, double *xx, double *yy, double *zz)
// xlt: latitude; xln: longitude; both in degrees
 {
 extern double rad;
// rad=3.14159265/180; conversion from degrees to radians; externally defined.
 double cosxln,sinxln,cosxlt,sinxlt;
// xx, yy, zz: pointers corresponding to the x, y, z Cartesian coordinates
 sinxlt=(*zz)=sin((*xx)=xlt*rad);
 cosxlt=cos(*xx);
 sinxln=sin((*yy)=xln*rad);
 cosxln=cos(*yy);
 (*xx)=cosxlt*cosxln;
 (*yy)=cosxlt*sinxln;
 return;
 }
```

Note that latitude (instead of the more conventional colatitude) is used, because latitude is more commonly mapped and visualized in the geosciences.

If a point on the sphere corresponds with a pole of instantaneous rotation, then the magnitude of the rotation (usually in radians), w_w, is applied to the x, y, z coordinates, producing the pseudovector w_x, w_y, w_z.

```
C code:
// fragment:
wx=ww*xx;
wy=ww*yy;
wz=ww*zz;
```

The inverse transformation of vector/pseudovector to spherical coordinates is:

$$\theta=\tan^{-1}(y/x) \tag{3.4}$$

$$\phi=\sin^{-1}(z/r) \tag{3.5}$$

or

$$\phi_c=\cos^{-1}(z/r) \tag{3.6}$$

$$r=(x^2+y^2+z^2)^{1/2} \tag{3.7}$$

in which r=1 if the vector has unit norm.

```
C code:
// subroutine invcdtrn3,
// calling asintest, atan2 t, longnorm, rotnorm, xnorm:
void invcdtrn3(double xf, double yf, double zf,
 double *rlt, double *rln, double *rlm)
 {
 extern double deg;
// deg=180/3.14159265;
// conversion from radians to degrees; externally defined.
 int i;
 double drlmi,prln,prln1,prlt,rlmi,rlmt,rltt;
 *rln=longnorm(atan2 t(yf,xf));
 rlmi=divide(1.0,*rlm=xnorm(xf,yf,zf));
 *rlt=zf*rlmi;
 *rlt=deg*asin(asintest(*rlt));
 *rlm=rotnorm(deg*(*rlm));
 return;
 }
```

with calls to utility subroutines:

```
//function xnorm, returns vector norm:
double xnorm(double xf,double yf,double zf)
 {
 return(sqrtt(xf*xf+yf*yf+zf*zf));
 }
// function sqrtt:
// returns-
//square root of argument if greater than zero;
//0.0 if less than or equal to zero;
double sqrtt(double temp)
 {
 if (temp>1.0e-40)
  return (sqrt(temp));
 return(0.0);
 }
// function divide:
// returns
// inverse of argument if absolute value is greater than exp(-40);
// 0.0 otherwise;
double divide(double yy,double xx)
 {
 if (fabs(xx)>1.0e-40) return(yy/xx);
 return(0.0);
 }
// function atan2 t:
// returns inverse tangent (in degrees) of dividend of numerator and
// denominator, if denominator is greater than exp(-40) and numerator
// is greater than 0.0; 0.0 otherwise.
double atan2t(double yy,double xx)
 {
 extern double deg;
 if (fabs(xx)>1.0e-40 || fabs(yy)>0.0) return(deg*atan2(yy,xx));
  else
   return(0.0);
 }
// function asintest:
// handles rounding error effects of absolute value greater than unity.
// returns argument if absolute value is less than or equal to 1.0;
// 1.0 if greater than 1.0; -1.0 if less than –1.0;
double asintest(double x)
 {
 if (fabs(x)<=1.) return(x);
 if (x>1.) return(1.);
 return(-1.);
```

```
}
// function longnorm:
// modifies longitude argument (in degrees) to fall within limits of
// positive/negative 180 degrees.
double longnorm(double xln)
{
if (fabs(xln)>180.)
 {
if (xln>0.) return(xln-360.);
  else return(xln+180.);
}
return(xln);
}
// function rotnorm:
// modifies rotation angle argument (in degrees) to fall within limits
// of positive/negative 180 degrees.
double rotnorm(double rot)
{
int i;
if (fabs(rot)>360.)
 {
 i=(int)(rot/360.);
 rot-=i*360;
 }
if (fabs(rot)>180.)
 if (rot>0.0) rot-=360.;
  else
   rot+=360.;
return(rot);
}
```

or an instantaneous angular velocity pole and rate, the pseudovector (w_x, w_y, w_z) is computed as above. The instantaneous velocity vector (v_x, v_y, v_z) of a point (x_x, y_y, z_z) on a plate rotated around the angular velocity pole is computed by the vector cross-product of the pseudovector and the position vector:

```
// fragment:
vx=wy * zz – wz * yy;                                    (3.8)
vy=wz * xx – xx * zz;                                    (3.9)
vz=wx * yy – wy * xx;                                    (3.10)
```

The instantaneous velocity vector (v_x, v_y, v_s) is not immediately useful for visualization, nor is its equivalent spherical coordinate set. Rather, the azimuth of the vector and its magnitude (as projected in an equal-azimuth, such as the Mercator, projection) are more easily visualized in two dimensions. Morgan

(1968) provided convenient spherical trigonometric formulations for these relationships:

$$\beta = \cos^{-1} \sin\phi_r \sin\phi_p + \cos\phi_r \cos\phi_p \cos(\theta_p - \theta_r) \tag{3.11}$$

$$\alpha = \sin^{-1} \sin(\theta_p - \theta_r) \cos\varphi_p / \sin\beta \tag{3.12}$$

in which α is the azimuth of the great circle segment between the pole of rotation (ϕ_p, θ_p) and a point on a plate (ϕ_r, θ_r) and β is the arc length of the great circle segment between to the points. The azimuth of the velocity vector is then $\alpha + \pi/2$. The velocity magnitude is:

$$v = v_{max} \cos\beta \tag{3.13}$$

3.3
Instantaneous Relative Motion Indicators

The primary instantaneous relative motion indicators for contemporary plate motion include fracture zones interpreted as transform faults and seismic focal mechanisms from transform faults and convergent plate boundaries. Young magnetic isochrons have been used as estimators of contemporary spreading rates for relative plate motions, beginning with Morgan (1968) and LePichon (1968). Ideally, geodetic surveying and monitoring can provide contemporary relative motion estimates.

For instantaneous plate kinematics, instantaneous angular velocity pseudovectors and calculated instantaneous velocity vectors are appropriately used for modeling of transform faults and measured spreading rates, as variously applied by Chase (1972, 1978), Minster et al. (1974), Chase (1978), Minster and Jordan (1978), and DeMets et al. (1990).

The instantaneous plate model of DeMets et al. (1990), with modifications for revisions to the geomagnetic time scale (e.g., DeMets et al. 1994), is the current standard for contemporary relative plate motions. Building upon the approaches of Chase (1978) and Minster and Jordan (1978), the instantaneous model uses conventional iterative nonlinear weighted least-squares analysis.

Given a set of instantaneous data, consisting of inferred motion azimuths (i.e., transform faults, focal mechanisms), and rates (i.e., recent average spreading rates measured normal to the spreading ridge, whose azimuth is also provided) at known locations, form a measure of squared error:

$$\Phi^2 = \sum [d_i - d_i(x_j)]\ 2\sigma_i^{-2} \tag{3.14}$$

In this equation, d_i represents an observed data value (azimuth or rate) from a plate boundary **j**, while $d_i(\mathbf{x})$ is the estimated value as a function of the instantaneous parameters $\mathbf{x}_j$ for that plate boundary.

Solution of the equation involves determination of all of the parameters, $\mathbf{x}_j$ (j=1–n plate pairs), that minimize the squared error. Since the functions that produce the $\mathbf{x}_j$ are nonlinear, an iterative approach is required, beginning with trial parameters, $\mathbf{x}_j^0$. DeMets et al. (1990) used essentially the same approach as Minster et al. (1974): differentiate the fitting function Φ^2 with respect to the desired parameters and expand it into a first-order Taylor series, dropping second- and higher-degree terms. Solve for the $\Delta\mathbf{x}$, using Gauss-Jordan elimination; take the new $\Delta\mathbf{x}$, add it to the previous estimated and iterate until convergence is achieved.

For the purposes of this study, only contemporary transform fault azimuths and seismic focal mechanisms along transform faults are used, since magnetic isochrons are explicitly used for finite rotation estimates. Further, convergent plate boundaries are interpreted to be too broad to provide unambiguous motion azimuths. For example, the seismicity and available focal mechanisms across the South America Andes, separating the South American and Nazca plates and South American and Antarctic plates, show significant variation over a fairly broad zone.

3.4 Finite Rotations

Vector calculations can be extended, with some additional trigonometry, to the calculation of finite rotations of points around a pole of rotation. However, the composition of rotations (combining successive rotation parameters to produce an equivalent net rotation) is almost intractable in either spherical coordinates or vector/trigonometric addition. Matrix manipulations, either real or complex, provide a more direct means of composition of rotations (complex Cayley-Klein parameterizations can also be used). The use of quaternions to calculate rotations of points and the composition of rotations is particularly convenient (Francheteau 1970; LePichon et al. 1973).

For a finite rotation set (pole latitude, ϕ, or colatitude, ϕ_c, and longitude, θ, plus rotation angle, Λ), a set of four quaternion parameters (w, x, y, z) is calculated:

$$w=\cos(\Lambda/2) \tag{3.15}$$

$$x=\sin(\Lambda/2)\cos\phi\cos\theta=\sin(\Lambda/2)\sin\phi_c\cos\theta \tag{3.16}$$

$$y=\sin(\Lambda/2)\cos\phi\sin\theta=\sin(\Lambda/2)\sin\phi_c\sin\theta \tag{3.17}$$

$$z=\sin(\Lambda/2)\sin\phi=\sin(\Lambda/2)\cos\theta_c \tag{3.18}$$

Notice that the x, y, z components are equivalent to the spherical-to-Cartesian coordinate conversion normalized to the factor sin $(\Lambda/2)$.

```
C routine:
// subroutine cdtrnp:
// computation of quaternion parameters w, x, y, z
// based on derivation of Francheteau (1970)
void cdtrnp(double plt,double pln,double plm,
 double *w,double *x,double *y,double *z)
// plt, pln, plm: input latitude and longitude of pole, rotation angle (degrees)
// w, x, y, z: pointers to quaternion parameter values
 {
 extern double rad;
```

```
// rad: as defined above in cdtrnx comment
 double cpltr,plnr,sin_plnr,cos_plnr,sin_cpltr,cos_cpltr,plmr,plmr2,
  sin_plmr2,cos_plmr2;
 if (plm<0.)
  {
  plm=-plm; plt=-plt; pln=pln+180;
  }
 plmr=plm*rad;
 cos_plmr2=cos(plmr2=plmr*0.5);
 sin_plmr2=sin(plmr2);
 cos_cpltr=cos(cpltr=(90.-plt)*rad);
 sin_cpltr=sin(cpltr);
 sin_plnr=sin(plnr=pln*rad);
 cos_plnr=cos(plnr);
 if (plm>0.)
  {
  *w=cos_plmr2;
  *x=sin_plmr2*sin_cpltr;
  *y=(*x)*sin_plnr;
  (*x)*=cos_plnr;
  *z=sin_plmr2*cos_cpltr;
   }
 else
  {
  *w=1; *y=0.; *x=0.; *z=0.;
  }
  return;
  }
```

To rotate a point:

- Convert known latitude and longitude to Cartesian coordinates, xx, yy, zz, as above.
- Calculate rotation parameters w, x, y, z.
- The rotated vector is calculated as follows:

$$rx=(w^2+x^2-y^2-z^2)xx+2(-wz+x,y)yy+2(wy+xz)zz \tag{3.19}$$

$$rx=2(wz+xy)xx+(w^2-x^2+y^2-z^2)yy+2(-wx+yz)zz \tag{3.20}$$

$$rz=2(-wz+xz)xx+2(wx+yz)yy+(w^2-x^2-y^2+z^2)zz \tag{3.21}$$

```
C routine:
// subroutine rotprod:
// calls functions rotmult4 and rotmult2, below
// based on derivation of Francheteau (1970)
void rotprod(double w,double x,double y,double z,
```

```
 double xx,double yy,double zz,
 double *rx,double *ry,double *rz)
// w, x, y, z: rotation parameters calculated from cdtrnp
// xx, yy, zz: unit vector calculated from cdtrnx
// rx, ry, rz: pointers to rotated unit vector values.
 {
 *rx=rotmult4(w,w,x,x,-y,y,-z,z)*xx+2.*rotmult2(-w,z,x,y)*yy+
 2.*rotmult2(w,y,x,z)*zz;
 *ry=2.*rotmult2(w,z,x,y)*xx+rotmult4(w,w,-x,x,y,y,-z,z)*yy+
 2.*rotmult2(-w,x,y,z)*zz;
 *rz=2.*rotmult2(-w,y,x,z)*xx+2.*rotmult2(w,x,y,z)*yy+
 rotmult4(w,w,-x,x,-y,y,z,z)*zz;
 }
// function rotmult2
// called by rotprod
double rotmult2(double w,double x,double y,double z)
 {
 return(w*x+y*z);
 }
// function rotmult4
// called by rotprod
double rotmult4(double w1,double w2,double x1,double x2,
 double y1,double y2,double z1,double z2)
 {
 return(w1*w2+x1*x2+y1*y2+z1*z2);
 }
```

The latitude and longitude of the rotated point are calculated from invcdtrn3, above.

In order to combine two rotations, thereby producing the equivalent single rotation, calculate the rotation parameters from cdtrnp for the first rotation (w1-x1-y1-z1) and for the second (w2-x2-y2-z2), then combine them to produce the equivalent total rotation parameters (wt-xt-yt-zt):

$$w=w_1w_2-x_1x_2-y_1y_2-z_1z_2 \quad (3.22)$$

$$x=w_1x_2+x_1w_2-y_1z_2+z_1y_2 \quad (3.23)$$

$$y=w_1y_2+x_1z_2+y_1w_2-z_1x_2 \quad (3.24)$$

$$z=w_1z_2-x_1y_2+y_1x_2+z_1w_2 \quad (3.25)$$

```
C routine:
// function combrots:
// combine rotation parameters (w1-x1-y1-z1; w2-x2-y2-z2) for two rotations
// to produce parameters (*wt-*xt-*yt-*zt) of total rotation
// calls addmult
```

```
// based on derivation of Francheteau (1970)
void combrots(double w1,double x1,double y1,
 double z1,double w2,double x2,
 double y2,double z2,double *wt,double *xt,double *yt, double *zt)
// w1, x1, y1, z1: parameters for first rotation
// w2, x2, y2, z2: parameters for second rotation
// wt, xt, yt, zt: pointers to parameter values for total rotation
 {
 *w=addmult(w1,-x1,-y1,-z1,w2,x2,y2,z2);
 *x=addmult(w1,x1,-y1,z1,x2,w2,z2,y2);
 *y=addmult(w1,x1,y1,-z1,y2,z2,w2,x2);
 *z=addmult(w1,-x1,y1,z1,z2,y2,x2,w2);
 if ((*w)<0.)
  {
  w=-(*w); *x=-(*x); *y=-(*y); *z=-(*z);
  }
 return;
 }
 double addmult(double w1,double x1,double y1,double z1,
  double w2,double x2,double y2,double z2)
 {
 return(w1*w2+x1*x2+y1*y2+z1*z2);
 }
```

To calculate the spherical coordinates of the total rotation:

$$\lambda=2\cos^{-1}(w) \tag{3.26}$$

$$\phi=\sin^{-1}(z/\sin(\lambda/2)) \tag{3.27}$$

$$\phi_c=\cos^{-1}(z/\sin(\lambda/2)) \tag{3.28}$$

$$\theta=\tan^{-1}(y/x) \tag{3.29}$$

```
C routine:
// subroutine invrotpars
// calculate spherical coordinates of rotation from quaternion parameters
// calls asintest, longnorm, rotnorm (above)
void invrotpars(double w, double x, double y, double z,
 double *plt,double *pln,double *plm)
// w, x, y, z: quaternion rotation parameters
// plt, pln, plm: pointers to spherical rotation parameters
 {
 extern deg;
// deg: as defined in invcdtrn3 above
double spltr,cplt,cpltr;
spltr=sin(*plm=acos(asintest(w)));
```

```
cplt=deg*acos(cpltr=asintest(divide(z,spltr)));
*pln=longnorm(atan2 t(y,x));
*plm=rotnorm((*plm)*2.*deg);
*plt=90.-cplt;
if (fabs(*plt)>90.)
 {
 if (*plt>0.0)
  {
  *plt=180-(*plt);
  *pln=(*pln)+180.;
  }
 else
 {
 *plt=-180-(*plt);
 *pln=(*pln)+180.;
 }
 }
 return;
 }
```

To compute an incremental rotation from another incremental rotation and the total rotation, merely reverse the angle of the known rotation and combine with the total rotation using the above procedure.

The series of procedures enumerated above for computing instantaneous kinematics and finite rotations is the bread-and-butter of plate kinematic and reconstruction analysis. However, these procedures disguise one clear limitation: in order to derive plate kinematics from finite plate reconstructions, it is necessary to introduce a tenuous assumption. That is, by computing differential rotation parameters, incremental finite reconstruction parameters can be derived. In order to make a kinematic inference, it is most straightforward to assume constancy of the differential pole of rotation and a constant rate. For example, assume that reconstructions have been made for the midpoints of magnetic isochron 25 (latitude 79.68°, longitude –0.46°, angle 18.16°) and isochron 30 (82.9°, 4.94°, 20.76°) between North America (reference plate) and Africa (parameters from Klitgord and Schouten 1986). The differential rotation parameters can be calculated using the routines above by combining the negative of the isochron 25 rotation with the isochron 30 rotation. Then, the inferred rate of motion is the resulting differential angle divided by the difference in age between isochron 31 (66.594 Ma) and isochron 25 (56.1475 Ma) (time scale of Cande and Kent 1995). The calculated finite difference parameters (75.32°, 175.35°, 2.83°) produce an average rotation rate of 0.271°/my.

The differential rotation parameters derived from finite reconstruction parameters may have no clear kinematic significance. That is, between 67 and 56 Ma, the calculated pole of instantaneous motion may never have corresponded with the differential pole. The very nature of the differential calculation requires any changes in motion to occur at the times represented by the plate reconstructions. Ironically, the coherence of magnetic isochron identification correlations is best when significant plate motion changes do not correspond with the isochron of interest; that is, the most easily identified magnetic isochrons form when changes in plate motions are minimal (e.g., Menard and Atwater 1968).

The derivation of differential plate motion parameters from finite reconstructions is analogous to using linear interpolation of a set of x-y points representing a sample of a continuous variable. A smooth, analytic curve generally proves to be a better interpolator, especially since discontinuities in the first derivative are commonly introduced at each data point by piecewise linear interpolation unless the data series is entirely linear.

CHAPTER 4

Kinematic and Reconstruction Modeling with Continuously Varying Parameters

4.1 Aside

There is a kind of frustration involved in interpolation of any numerical quantity, particularly if there are uncertainties involved in the input data. Uncertainties have been a passion for many plate tectonic workers; a number of important advances in the estimation of uncertainties in oceanic plate reconstructions and their composition into global reconstruction sets provide some level of confidence in the quality of the reconstructions. However, interpolations between reconstructions and corresponding, derivative kinematics have not been satisfactorily explored. There has been little progress in kinematic inference since the pioneering finite difference calculations of Pitman and Talwani (1972) and Weissel and Hayes (1972).

The idea of using spline interpolation instead of finite difference parameters occurred to me in the mid-1980s, inspired by a marvelous derivation of plate kinematics using analytic formulae by Smith (1981). The obvious limitation in Smith's application of his original derivation to plate kinematics was the use of simple polynomials. "Why not use cubic splines?" I asked myself. Aside from a simple abstract published in 1985, only in recent years have I had the opportunity to more fully pursue application of this technique.

4.2 Continuous Parameter Modeling

Smith (1981) introduced a derivation of instantaneous plate kinematic parameters from continuously varying finite reconstruction parameters that provides an alternative to finite difference interpolation. He showed that the instantaneous rotation pseudovector is related to a continuous, analytic function describing the total reconstruction parameters:

$$\omega_T = \lambda' \Omega_T + \sin\lambda u - (1+\cos\lambda)\Omega_T \times \Omega'_T \tag{4.1}$$

in which:

ω_t = instantaneous angular velocity pseudovector for time t; magnitude is equal to the instantaneous rotation rate in radians,

λ_t = total finite rotation angle for time t,

Ω_t =unit vector representation of the finite pole of rotation for time t,
× denotes the vector cross-product,

and the prime (′) denotes the first time derivative.

The derivation of Smith's (1981) equation is provided in Appendix 1; note that Smith used a non-standard form for the relation of the angular velocity pseudovector and the instantaneous velocity; the standard (right-handed) form is used here, resulting in a differences in sign of the second and third terms of the equation.

```
As a C routine:
// subroutine invpcoor3
// given ww[], vector equivalent of rotation pole, magnitude = average
rotation rate
// and first time derivative of the vector, pw[],
// compute instantaneous pseudovector, *xx-*yy-*zz,
// and spherical coordinate equivalent, *rlt,*rln,*rlm
void invpcoor3(double age,double *ww,double *pw,double *xx, double
 *yy, double *zz, double *rlt, double *rln, double *rlm)
 {
 extern double deg,rad;
// deg, rad: as defined above (Chapter 3)
 double ang,cang,cang1,pang,pxi,pxx,pyy,pzz,
  rlmi,rlmi2,rlmt,sang,wmagi,www,wx,wy,wz,pww,pwx,pwy,pwz,
 wxpywypx,wypzwzpy,wzpxwxpz;
 www=xnorm(ww[0],ww[1],ww[2])
 wmagi=divide(1.,www);
 wx=ww[0]*wmagi; wy=ww[1]*wmagi; wz=ww[2]*wmagi;
 sang=sin(ang=www*age);
 cang=cos(ang);
 cang1=cang-1;
 pww=wx*pw[0]+wy*pw[1]+wz*pw[2];
 pang=www+pww*age;
 pwx=(pw[0]-wx*pww)*wmagi;
 pwy=(pw[1]-wy*pww)*wmagi;
 pwz=(pw[2]-wz*pww)*wmagi;
 // vector cross product:
 wypzwzpy=wy*pwz-wz*pwy;
 wzpxwxpz=wz*pwx-wx*pwz;
 wxpywypx=wx*pwy-wy*pwx;
 *xx=pang*wx+sang*pwx+cang1*wypzwzpy;
 *yy=pang*wy+sang*pwy+cang1*wzpxwxpz;
 *zz=pang*wz+sang*pwz+cang1*wxpywypx;
 invcdtrn3(*xx,*yy,*zz,rlt,rln,rlm);
 }
```

In order to calculate the finite rotation vector and its time derivative, a continuous analytic function is required. Smith (1981) proposed a simple vector polynomial of degree n with coefficients a_{ij} derived by least-squares fitting to the vector-converted total rotation parameters (magnitude, $\Lambda(t)$, equal to the average rotation rate):

$$\Lambda(t)\Omega_j(t)=\sum_{i=1,n} a_{ij}t^i \tag{4.2}$$

For each age[i]:
- Compute x_x, y_y, z_z for reconstruction parameter set i, using cdtrnx, above.
- Calculate the magnitude of the average rotation vector, w_w=ang[i]/age[i].
- Compute $w_1[i]=x_x*w_w$, $w_2[i]=y_y*w_w$; $w3[i]=y_y*z_z$.
- Derive the polynomial parameters a_{ij}, j=1,n, by least-squares.

Then, to calculate the rotation parameters and the vector first derivative for any age:

```
// code fragment:
w1=w2=w3=0.0;
t=1;
for (i=0;i<n;i++)
 {
 w1+=a1[i]*t;
 w2+=a2[i]*t;
 w3+=a3[i]*t;
 t*=age;
 }
dw1=dw2=dw3=0.0;
t=1; m=n=1;
for (i=1;i<n;i++)
 {
 dw1+=a1[i]*t*m;
 dw2+=a2[i]*t*m;
 dw3+=a3[i]*t*m;
 t*=age;
 n++;
 m*=n;
 }
```

To calculate the spherical rotation parameters, use invcdtrn3, above, then multiply the resulting rotation angle by the age. To calculate the instantaneous rotation parameters, use invpcoor3, above.

The use of a simple polynomial has the limitation of either misfit to some or all of the input data, or, if the degree is high enough, thereby fitting the data exactly, the introduction of undesirable curvature in the function. An alternative to a high-order polynomial is the use of cubic splines.

Vector cubic splines for a set of reconstruction parameters, as applied here, consist of three sets of piecewise continuous and differentiable polynomials of degree three, corresponding with the x, y, and z dimensions. Each spline function is bounded by data points with ordinate of time, which join it to the previous and/or next function. The abscissa corresponds to the x, y, z coordinate for the appropriate vector component. The abscissa and first and second derivatives are continuous at each data point (or knot).

The procedure involves computing the vector equivalent of each finite parameter triple, creating three n-point arrays of time versus x-, y-, or z-component. Three successive calls to the spline routine (below) generate additional (to the array values) spline parameters, the first derivatives at time 0 and n, and the second derivatives and times 1 through n−1.

```
C routine:
// subroutine spline
// compute spline parameters for vector of data.
// modified after Press et al. (1992)
// calls utility routine vector, below
void spline(double *x, double *y, int n, double *yp1,double *ypn, double
 *y2)
 {
 int i,k;
 double h,hi,p,qn,sig,un,*u;
 u=vector (0,n);
 if (*yp1>0.99e30)
  y2[0]=u[0]=0.0;
  else
   {
  y2[0]=-0.5;
  u[0]=(3.0/(x[1]-x[0]))*((y[1]-y[0])/(x[1]-x[0])-*yp1);
  }
 for (i=1;i<=n-2;i++)
  {
  sig=(x[i]-x[i-1])/(x[i+1]-x[i-1]);
  p=sig*y2[i-1]+2.0;
  y2[i]=(sig-1.0)/p;
  u[i]=(y[i+1]-y[i])/(x[i+1]-x[i])-(y[i]-y[i-1])/(x[i]-x[i-1]);
  u[i]=(6.0*u[i]/(x[i+1]-x[i-1])-sig*u[i-1])/p;
  }
 if (*ypn>0.99e30)
  qn=un=0.0;
   else
   {
   qn=0.5;
   un=(3.0/(x[n-1]-x[n-2]))*
```

```
    (*ypn-(y[n-1]-y[n-2])/(x[n-1]-x[n-2]));
   }
  y2[n-1]=(un-qn*u[n-2])/(qn*y2[n-2]+1.0);
  for (k=n-2;k>=0;k–)
   y2[k]=y2[k]*y2[k+1]+u[k];
  h=x[1]-x[0];
  hi=divide(1.,h);
  *yp1=(y[1]-y[0])*hi-(y2[1]+y2[0]+y2[0])*h/6.0;
  h=x[n-1]-x[n-2];
  hi=divide(1.,h);
  *ypn=(y[n-1]-y[n-2])*hi+(y2[n-1]+y2[n-1]+y2[n-2])*h/6.0;
  free_vector(u,0);
  return;
  }
 // function vector
 // modified after Press et al. (1992)
 // calls utility routine nrerror
 double *vector(int nl,int nh)
  {
  double *v;
 // external: #define NR_END 1
  v=(double *)malloc((size_t) ((nh-nl+1+NR_END)*sizeof(double)));
  if (!v) nrerror("allocation failure in vector()");
  return v-nl+NR_END;
  }
 // routine nrerror
 // modified after Press et al. (1992)
 void nrerror(char error_text[])
  {
 // modified after Press et al. (1992)
 /*fprintf(stderr,"Numerical Recipes run-time error...\n");
  fprintf(stderr,"%s\n",error_text);
  fprintf(stderr,"... now exiting to system..."); */
  printf("Numerical Recipes run-time error...\n");
  printf("%s\n",error_text);
  printf("... now exiting to system...\n");
  exit(1);
  }
```

Typically, the time spacing of the knots is irregular. As it is desirable that they be computed at a regular interval, new arrays are calculated at a constant interval, using the splint routine, below.

```
// routine splint
// modified after Press et al. (1992)
// use spline parameters to interpolate at desired times x
// (call routine for each desired time)
// calls utility routine nrerror, above
void splint(double *xa,double *ya,double **y2a,int ind,int n,double
 x,double *y)
 {
 int klo,khi,k;
 double h,b,a;
 klo=0;
 khi=n-1;
 while (khi-klo>1)
  {
  k=(khi+klo) >>1;
  if (xa[k]>x) khi=k;
   else
    klo=k;
  }
 h=xa[khi]-xa[klo];
 if (h==0.0)
  nrerror("Bad xa input to routine splint");
 a=(xa[khi]-x)/h;
 b=(x-xa[klo])/h;
 *y=a*ya[klo]+b*ya[khi]+
  ((a*a*a-a)*y2a[ind][klo]+(b*b*b-b)*y2a[ind][khi])*(h*h)/6.0;
 return;
 }
```

Then, new additional spline parameters are recomputed at a regular interval (Warning: there is a chance of aliasing with this procedure if the sampling interval is coarser than the data time spacing).

```
// routine dtspline
// modified after Press et al. (1992)
void dtspline(double dx,int knd,int kflag,int knd0,int ind,double **y2)
// dx: time interval (my); knd: plate index;
// kflag: 0 – only one plate, 1 – two plates;
// kndo: second plate index, if kflag =1;
// ind: vector component – 0, 1, or 2.
// input spline master coefficients for multi-plate set
// stored in external vector pp
// output supplementary coefficients stored in array y2.
 {
 extern double *pp;
```

```
extern int npars,npars3,nparst;
int i,k,kk,n,n1,n2,nn;
double dti,p,qn,temp,un,*u;
if (kflag==0)
 kk=0;
  else
   kk=3;
kk+=ind;
if (knd>0) nn=nparst[knd-1];
 else
  nn=0;
nn+=npars[knd]*ind;
n=npars[knd]-2;
n1=n;
n2=n1+1;
u=vector (0,n2);
dti=divide(1.0,dx);
y2[kk][0]=-0.5;
temp=3.0*dti;
u[0]=temp*((pp[nn+1]-pp[nn])*dti-pp[nn+n1]);
for (i=1;i<=n-2;i++)
 {
 p=0.5*y2[kk][i-1]+2.0;
 y2[kk][i]=-0.5/p;
 u[i]=dti*(pp[nn+i+1]-pp[nn+i]-pp[nn+i]+pp[nn+i-1]);
 u[i]=(3.0*u[i]*dti-0.5*u[i-1])/p;
 }
qn=0.5;
un=3.0*dti*(pp[nn+n2]-(pp[nn+n-1]-pp[nn+n-2])*dti);
y2[kk][n-1]=(un-qn*u[n-2])/(qn*y2[kk][n-2]+1.0);
for (k=n-2;k>=0;k–)
 y2[kk][k]=y2[kk][k]*y2[kk][k+1]+u[k];
free_vector(u,0);
return;
}
```

Note that the additional parameters could be directly calculated within the interpolation routine by including terms for the first and second derivatives, skipping the recomputation step. However, the recomputation step is necessary for the implementation of the spline application described here.

Finally, using the regularly spaced spline parameters, the rotation parameters and their first derivatives can be calculated for any time within the time interval that the input data exist:

```
// routine dydtsplint
// modified after Press et al. (1992)
// compute interpolated vector and its first derivative
// for desired age by three successive calls (ind=0,1,2)
void dydtsplint(double dt,int ind,int knd,int kflag,int knd0,int *npars,int
  *nparst,
 double *pp,double **y2,double x,double *y,double *dy)
// dt: time interval (my), ind: vector component – 0, 1, or 2;
  knd: plate index;
// kflag: 0 – only one plate stored in y2, 1 – two plates in y2;
// npars: number of parameters in plate set;
// nparst: cumulative number of parameters in plate set;
// pp: master coefficent vector for multi-plate set
// y2: input supplementary coefficients
// x: input time (my)
// y: output vector component value
// dy: output vector component first derivative value
 {
 int khi,kk,klo,lhi,llo,nn,npar;
 double a,b,dt26,dti,xalo,xahi;
 if (kflag==0)
  kk=0;
   else
    kk=3;
 kk+=ind;
 npar=npars[knd]-2;
 dti=1./dt;
 klo=(int)(x*dti);
 khi=klo+1;
 if (khi>npar-1)
  {
  klo=npar-2;
  khi=npar-1;
  }
 if (klo<0)
  {
  klo=0;
  khi=1;
  x=0.;
  }
 llo=klo;
 lhi=khi;
 xalo=klo*dt;
 xahi=xalo+dt;
 nn=npars[knd]*ind;
 if (knd>0) nn+=nparst[knd-1];
```

```
klo+=nn;
khi+=nn;
a=(xahi-x)*dti;
b=(x-xalo)*dti;
dt26=dt*dt/6.0;
*y=a*pp[klo]+b*pp[khi]+
 ((a*a*a-a)*y2[kk][llo]+(b*b*b-b)*y2[kk][lhi])*dt*dt/6.0;
*dy=dti*(pp[khi]-pp[klo]+
 ((3.0*b*b-1.0)*y2[kk][lhi]-(3*a*a-1.0)*y2[kk][llo])*dt26);
return;
}
```

CHAPTER 5

An Inverse Model

5.1 Aside

When Bernard Minster and Tom Jordan, together with Peter Molnar and Ian Haines, first published their inverse plate kinematic model in 1974 (Minster et al. 1974), *the affect on the tectonics community was extraordinary. Jason Morgan had already published a global plate tectonics model parameterized in Cartesian vectors* (Morgan 1972a), *while Clement Chase had similarly produced an inverse model accompanied by a neat stereoscopic pseudovector portrayal of contemporary plate motions. But the Minster et al. model incorporated uncertainties. Interestingly, their application of non-linear least-squares to plate tectonics also stimulated similar applications in non-plate-tectonics applications. Subsequently, Chase introduced uncertainties in his updated formulation using, in my opinion, a somewhat more realistic formulation that included ridge-azimuth as part of the spreading rate datum, an approach also followed by DeMets et al. (1990) in the current contemporary plate kinematic standard model. While uncertainties are an obviously desirable product of such an analysis, the ability to construct a reasonable model that fits the data, even in a subjective analysis, is quite an accomplishment.*

I'm still not convinced that uncertainties in plate reconstruction parameters are adequately handled in any automated inferential model of plate tectonics. Rather, reconstruction parameters are derived first, and then uncertainties are estimated. Further, only paleoplate boundary data are used (magnetic isochrons and fracture zone offsets), while, in theory, at least, fracture zone orientations could be incorporated into a combined reconstruction and plate kinematics model, with a suitable kinematics and reconstruction parameterization. Perhaps the one presented in Chap. 4 would suffice.

5.2 A Grand Model

The sets of derivations and algorithms described in Chaps. 3 and 4 set the stage for the development of a local, regional, or even global plate model, combining instantaneous kinematic and finite reconstruction constraints. Given a set of plate reconstructions for a series of ages and instantaneous rotation parameters for the present, it is possible to apply the spline parameterization of Chap. 4 to interpolate reconstruction and instantaneous parameters for any age within the set of reconstruction parameters provided. However, there are a number of other kinematic and reconstruction constraints that could be used to refine the spline model, if the model is well-conditioned. That is, existing relative plate reconstructions are based on magnetic isochron and fracture zone crossings. However, the trend of the fracture zones could also be incorporated if plate kinematics were also to be solved for.

Following the familiar discrete inverse model formulation, the desired outcome is a model solution, D_i, which best fits the observations, d_i, in a least squares sense:

$$\Phi^2=\Sigma_{i=1,n}(d_i-D_i)^2/\sigma_i^2 \qquad (5.1)$$

Consider what these various data types are, the reference frames they are defined in, and how they are calculated for inclusion in the model.

5.3 Finite Plate Reconstruction Data – Magnetic Isochrons and Paleofracture Zones

Magnetic isochrons are typically interpreted at midpoints of or boundaries between normal and/or reverse polarity intervals. Visual inspection of interpreted isochrons suggests that great circle segments can approximate them. Typically, most fracture zones are interpreted as paleotransform faults, although, in some cases, propagating ridges (Hey et al. 1977) can truncate isochrons.

Paleotransform faults are ideally parallel to the direction of relative motion at the time they formed. However, changes in trends of fracture zones, combined with independent evidence for changes in plate motion, suggest that fracture zones accommodate significant changes in plate motion by increasing the dimensions of the fracture zone. Thus, especially if relative motions change continuously, the finite length of fracture zones suggests it is unrealistic to expect that the zones will precisely follow small circles. Therefore, fracture zones, too, can be interpreted as great circle segments, rather than small circles.

Hellinger (1981) and Pilger (1978a) both introduced plate reconstruction methods that utilized great circle segments. Pilger's approach, an extension of McKenzie et al.'s (1970) continental margin fitting method, used great circle segments connecting adjacent plate margin (isochron or fracture zone offsets) data points, and then fit individual rotated data points from the corresponding margin to the nearest great circle segment. In this way, individual isochron segments between fracture zones could be allowed to deviate from a single great circle.

Hellinger (1981assumed that a single great circle approximated an individual isochron segment between fracture zones (or a fracture zone between the offset corresponding isochron). Thus, in addition to the rotation parameters, Hellinger's method requires solution for the parameters of each great circle segment.

5.4 Instantaneous Proper Motion Indicators

There are a number of particular geological phenomena that may bear on lithosphere-mesosphere kinematics and the possible existence of a proper frame of reference that corresponds with the hotspot reference frame:

5.4.1 Intraplate Stress

Early suggestions of a correspondence between driving mechanisms and hotspot motion models (e.g., Forsythe and Uyeda 1975; Solomon and Sleep 1975) were strengthened by correlation of observed intraplate stress field orientations with plate-hotspot motions (Sykes 1978; Richardson et al. 1979; Zoback and Zoback 1980). Contemporary maximum horizontal principal compressive (MHPC) stresses are commonly oriented parallel with the direction of contemporary proper plate motion in the stable and moderately elevated portions of continental plates (e.g., Zoback et al. 1989) Observations from the interiors of oceanic plates are not numerous enough to infer or contradict a similar correlation. Unfortunately, little apparent effort has been directed towards assessing the geohistorical relationship between proper plate kinematics, intraplate paleostresses, and hotspot traces prior to the present epoch. Perhaps this state of affairs reflects inadequacy of plate-hotspot kinematic models and/or relative scarcity of critical data. Nevertheless, it would seem to be critical to determine whether inferred proper motions and MHPC stresses in the recent geological past also correspond, as they appear to do at present.

The state of stress within plates is presumably irrelevant to the assessment of plate-hotspot models if hotspots are of deep-mantle origin, unless the hotspots/plumes are a major driving mechanism for plate motion. Efforts to model relative plate motions and intraplate stresses from boundary forces have not seen the need to incorporate a plume/hotspot contribution. Forsythe and Uyeda (1975) and Solomon and Sleep (1975) argued that the principal driving forces have predictable kinematic consequences, dominated by ridge "push" (the force due to the gravitational potential of uplifted ridges and the dynamics of dike emplacement), trench "pull", and "drag" of plates against the underlying asthenosphere. Richardson et al. (1979) elaborated on this theme, going beyond interplate kinematics by also comparing intraplate stress indicators with stress fields predicted from various combinations of ridge push, trench pull, and drag of the plates against the asthenosphere. In these various modeling efforts, the role hotspots/plumes play in the driving mechanisms was not clearly demonstrated. Nevertheless, comparisons of contemporary intraplate stress indicators with instantaneous hotspot models show correspondence of the MHPC stress direction with the azimuth of plate motion predicted by many such models. This is especially apparent for the interior of North America and portions of other continental plates (e.g., Richardson et al. 1979; Nakamura and Uyeda 1980; Zoback and Zoback 1980; Zoback et al. 1989).

Hotspot traces provide the basis for the proper plate motion model, and the contemporary North American plate stress field corresponds with the predicted plate motions. Is there a closer, genetic connection between the hotspots and the stresses, or are they essentially independent? In order to begin to examine this question, not only the contemporary stress field on many more plates needs to be studied, but also the paleostress history of each plate, for comparison with hotspot-plate motion models.

5.4.2 Volcanic-Bathymetric Lineaments

Schouten et al. (1987) suggested that certain distinctive fine-scale volcanic-bathymetric lineaments observed on the flanks of mid-ocean ridges, while oblique to the direction of spreading, still show orientations consistent with the motion of the ridge crests (not the plates) in the hotspot frame of reference. One consequence of this suggestion is that uncoupling of lithosphere and asthenosphere probably occurs at relatively shallow depths, since in order for anomalous volcanism to record plate-mesosphere motion, the source of the volcanism must be within more slowly moving asthenosphere or even mesosphere, but shallow enough for diapiric rise and, eventually, melting to occur. Further, they proposed that if a ridge remains essentially fixed in relation to the mesosphere, prolonged diapirism to replace material that has become new oceanic lithosphere will result in tapping progressively deeper sources in the mantle. Schouten et al. (1987) cited evidence of geochemical anomalies, which imply deeper sources and which correspond with ridge segments, that have apparently moved very slowly in the hotspot frame of reference [using Morgan's (1983) model for the Atlantic and Indian Oceans, and Jurdy and Gordon's (1984) and Schult and Gordon's (1984) models for the Pacific Ocean].

The volcanic lineaments interpreted by Schouten et al. (1987) are latest-Cenozoic in age. They suggested that the geochemical trends represent stability of certain ridge segments for as much as 100 my. Are there bathymetric anomalies, that record similar processes in the earlier Cenozoic and Late Mesozoic?

5.4.3 Parallel Gravity Anomaly Lineations

Another kind of apparent lithospheric anomaly interpreted to reflect lithosphere-asthenosphere interaction has been described by Haxby and Weissel (1986). Parallel elongate gravity anomalies in the central Pacific Ocean, determined from Seasat altimeter measurements, trend oblique to paleotransform faults (which, of course, parallel the paleoseafloor relative spreading direction). They noted approximate parallelism of many of these anomalies with the inferred contemporary direction of Pacific plate motion in the hotspot frame. Haxby and Weissel (1986) further suggested that these anomalies may reflect longitudinal, horizontal cylindrical convection cells formed in the direction of plate motion, as predicted from convection modeling by Richter (1973). Winterer and Sandwell (1987) suggested that at least a part of the reason for these "cross-grain" lineaments is progressive tensional fracture of the plate, for which they presented independent documentation: post-seafloor spreading ridges associated with volcanic sills and flows, whose ages can be constrained in at least two different locations. Tensional stress orientations inferred by Winter and Sandwell correspond with either the longitudinal convection cell model of Richter or the intraplate drag mechanism postulated by Solomon and Sleep (1975).

While there is a gross parallelism of the gravity anomalies noted by Haxby and Weissel (1986) and the ridges identified by Winterer and Sandwell (1987) with present day motion of the Pacific plate relative to the hotspots, there are, nevertheless, some departures from this direction apparent in the mapped gravity field, as recognized by Kroenke et al. (1991). Is it possible that some of the anomalies represent paleoplate features, formed at the time that the Pacific seafloor formed by a mechanism equivalent to that proposed by Schouten et al. (1987), rather than features reflecting contemporary convection patterns? And, do these anomalies record proper plate motion?

5.4.4 Back-Arc Spreading

In the same volume in which Morgan (1972a) introduced his fully documented plume hypothesis, a related hypothesis was advanced in order to explain the origin of back-arc basins. Moberly (1972) suggested that subduction zones resist lateral displacement and that the zones themselves tend to sink vertically; as a consequence, if proper motion of the upper plate is inadequate to keep up with the resulting displaced trench axis, back-arc extension results. Morgan noted that his preliminary hotspot model produced motion estimates of several subduction zone upper plates, which correspond with Moberly's suggestion either in the existence, or nonexistence of back-arc extension. Several other workers have made similar subsequent observations, and Cross and Pilger (1978a) extended the hypothesis to the Basin Range Province of western North America. They suggested that the back-arc extension, which characterizes the region, could represent very slow motion of the North American plate, which inadequately paced the developing continental margin subduction zone. Cross and Pilger's suggestion was largely speculative, in the absence of adequate proper motion models. Do newer proper motion models support this hypothesis for the origin of back-arc extension in the Late Cenozoic of western North America?

CHAPTER 6

Hotspot Trace Patterns

6.1 Aside

Data and patterns–so much of what we do in science reduces to that x versus y plot. Sometimes, it's predicted and observed longitude and latitude (for example, reconstructed magnetic anomalies across a spreading center). When it comes to hotspot traces, it's age versus distance (but there's more to it than that). Despite what we may have learned in junior-year geostatistics, the central-limit theorem doesn't always apply. Yes, experimental dates (or model ages, to the purists) can demonstrate a Gaussian distribution, but there's more to the tectonic meaning of an age than what emerges from the laboratory. When the 43-Ma age for the Hawaiian–Emperor bend was first pronounced, I had misgivings about its tectonic interpretation, especially since there were no obvious manifestations of tectonic changes of that age elsewhere in the Pacific. A few others shared my doubts (published before and after the 1981 paper I wrote with David Handschumacher, and a 1982 paper), but most discussions of the hotspot traces of the Pacific continued to cite the 43-Ma age almost to the present day.

Averaging isotopic ages of volcanic rocks, weighted by analytical uncertainty, provides an estimate of the weighted mean age of those ages, but it doesn't tell us when the volcanic intrusion began; it provides only a minimum age. Neither does such an average give us a mean age estimate of when a tectonic event–a change in plate (or mantle) motion direction–occurred; again, it is a minimum age for that event.

There is no 43-Ma plate tectonic event recorded by the Hawaiian–Emperor bend; there may have been an earlier event whose surficially dated volcanic record is some millions of years later.

Unfortunately, volcanic rocks that isotopic chemists can date at the surface provide minimum age constraints on underlying tectonic events. We need to learn to deal with this constraint.

6.2 Traces

The literature on inferred hotspot traces has, naturally enough, emphasized progressive age trends inferred to record motion of plates over hotspots without consideration of the detailed history, such as it is, of magmatism recorded within traces. Consequently, there have been explicit or, more commonly, implicit assumptions about the relationship of isotopically dated surface or near-surface samples and the timing of initial encounter of a point on a plate and a hotspot. It is worthwhile reexamining age–distance relationships of hotspots and their inferred locus from kinematic modeling. As noted by Pilger and Handschumacher (1981) and Pilger (1982), some traces seem to demonstrate persistence of magmatism for several million years after inception. Add an uncertain amount of time from first encounter of a plate with a hotspot to magmatic inception and then cessation (which is most likely to be sampled) and the kinematic age is that much less certain. Finally, some Ar/Ar results from key hotspot traces have been recently reinterpreted with interesting consequences for the fixed hotspot hypothesis.

To reiterate, within the context of the hotspot-plume hypothesis, accurate and precise isotopic dates provide only a minimum age for a kinematic event. That is, impingement of a moving plate upon a plume occurs prior to emplacement of magma within and upon the plate. Further, we typically sample the shallowest and, therefore, the youngest lava from the trace. Similarly, the actual location of a hotspot in the mesosphere will be displaced an uncertain amount from the locus of current volcanism except for that case (if any) in which the plate is stationary over the hotspot.

The persistence of magmatic activity for a sustained period after magmatic inception provides challenging interpretive skills. Of course, the origin of the hotspot magmatism is the crux of the matter. If indeed a low-viscosity plume is the primary hotspot culprit, channeling of plume flow along the trace, as suggested for the Yellowstone trace (e.g., Suppe et al. 1975), is one possibility. Entrapment of magma within the plate (Morgan 1972a) is another. If the hotspot trace represents progressive fracturing of the moving plate, a minority view at present, the persistence of magmatism could be inferred to represent prolonged extension focused along the trace and normal to its trend.

One of the great uncertainties in the interpretation of hotspot traces is the duration of the volcanic episode for constructing an edifice. Early arguments for rapid construction of each Hawaiian island or seamount (e.g., Clague and Jarrard 1973) are commonly cited, but, as noted by Pilger and Handschumacher (1981) and Pilger (1982), persistence of volcanism along traces can be documented from a number of Pacific Ocean examples.

Each of the observed dated traces has distinctive properties that involve some degree of consistency with the fixed hotspot hypothesis, relative to either the Atlantic–Indian Ocean (Pangean) hotspot set or the Pacific Ocean set. The patterns observed provide critical constraints on the ultimate origin of hotspot traces. It is therefore worth reviewing available isotopic ages from each of the principal hotspot traces, including a few that are not typically interpreted as traces.

6.2.1 Pacific Hotspot Traces

The Pacific Ocean is the birthplace of the hotspot hypothesis, with recognition by Wilson (1963a, b, 1965a) that the inferred progressive increase in age of the

Hawaiian Islands to the west-northwest of Hawaii could be explained by motion of the Pacific crust relative to a hotspot in the mantle. Subsequent isotopic dating of the islands and seamounts of the Hawaiian–Emperor island-seamount chain has confirmed the inference that the islands and seamounts are progressively older with increasing distance from the active volcanoes of the big island.

Isotopic age dating of many of the inferred traces of the Pacific, Nazca, and Cocos plates indicate a progressive increase in age, although, in some cases, persistence of volcanic activity for some time after inception seems to be indicated. Figure 6.1 illustrates predicted hotspot loci, according to the model of Raymond et al. (2000), interpolated at 5-my increments using the spline method presented in Chap. 2, together with the location of isotopically data samples from the Pacific Ocean. In most cases, the location of the inferred hotspot is speculative or weakly constrained. Each proposed hotspot trace is examined in more detail in terms of available isotopic ages and its predicted locus.

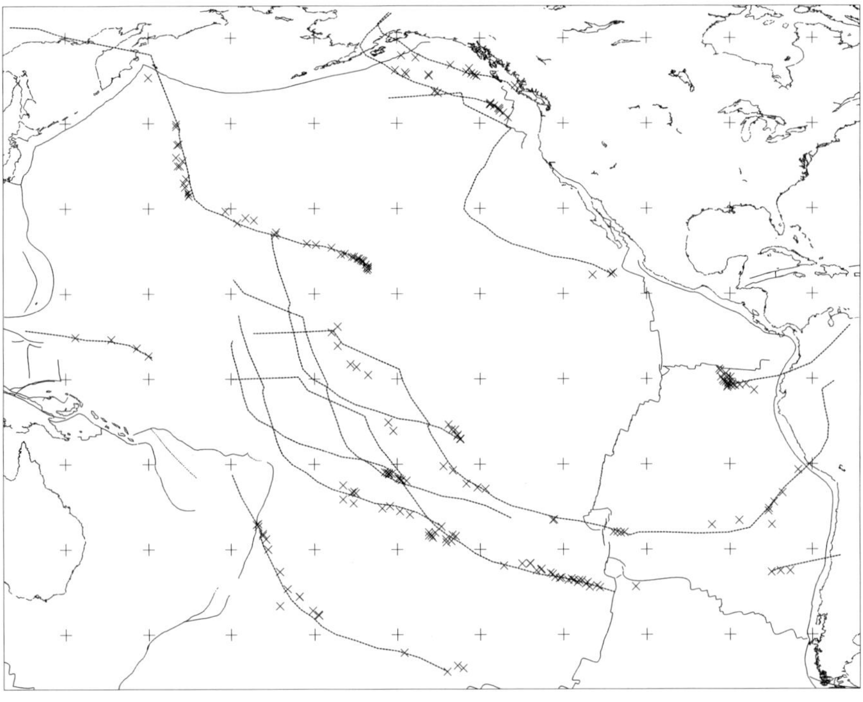

Fig. 6.1. Predicted loci of hotspots of the Pacific Ocean, together with the location of isotopically dated samples (or, in a few cases, magnetic isochron intersections with traces inferred to have formed at spreading centers). Loci are anchored by speculatively inferred locations of hotspots.

6.2.1.1
Hawaiian–Emperor Island-Seamount Chain (Hotspot: 19.6°N, 155.5°W)

In Fig. 6.2, the age–distance plot implies a slightly older predicted age for inception of volcanism than observed, particularly for distances greater than 30° from Kilauea and observed ages greater than 43 Ma. However, such an apparent discrepancy is not unreasonable if allowance is made for a time lag between kinematic encounter of a point on a plate with an underlying hotspot, followed by melting and rise into the plate. The apparent mismatch could be improved somewhat by relocation of the inferred hotspot further to the northwest of Kilauea, as well as implied by comparison of the locus with sample locations in map view (Fig. 6.3). Comparison of the age–distance pattern of the Hawaiian trace with that of the Louisville trace is particularly pertinent (see below).

Note that in some locations isotopic ages imply persistence of volcanism for as much as 5 my after inception; in some cases dates 10 my years younger than the oldest age are observed. It is not clear whether these age ranges accurately represent the duration of volcanism or can be attributed to argon loss (most of the available ages are K/Ar.) It has been argued that the bulk of individual island/seamount edifices was emplaced in a short period of time, but, nevertheless, the latest volcanism could have occurred several million years after inception of activity (e.g., Clague and Dalrymple 1987, 1989). It is the latest volcanism that is sampled from most seamounts and islands.

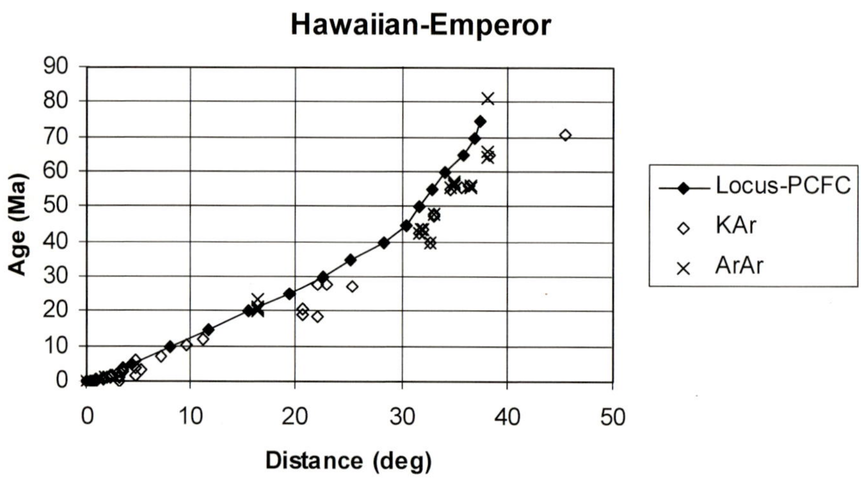

Fig. 6.2. Hawaiian–Emperor: plots of isotopic age from proposed hotspot traces and predicted loci ages versus distance from the inferred hotspot. The locus is based on the model of Raymond et al. (2000) (extended to the Nazca and Cocos plates) interpolated at 5-my increments using the spline method presented in Chap. 2.

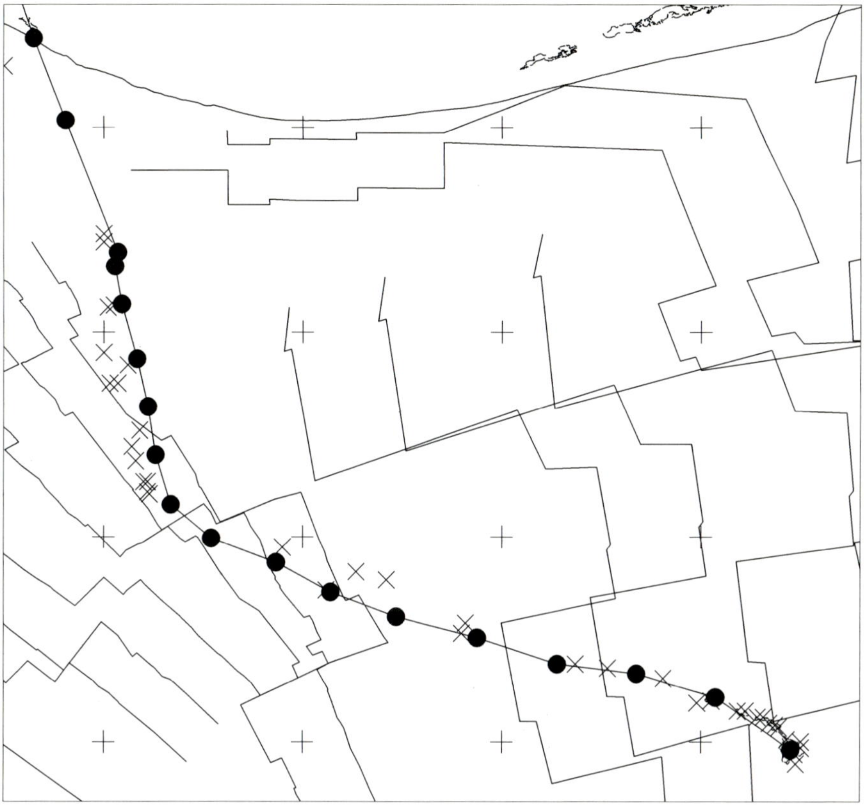

Fig. 6.3. Hawaiian–Emperor trace: map of predicted locus and dated samples. Circles locus, crosses sample locations.

6.2.1.2
Louisville Ridge (Hotspot: 51.5°S, 141.0°W)

The age–distance pattern for Louisville Ridge (Figs. 6.4, 6.5) is broadly consistent with that of the Hawaiian–Emperor ridge; several different kinematic models have been proposed to fit the two chains and their age dates. The model of Raymond et al. (2000) is used here, but with slightly different apparent hotspot locations than they proposed.

The age–distance plot (Fig. 6.4) could be adjusted a bit by choosing a different hotspot location (several different locations have been proposed). There is one Ar/Ar date that appears older than predicted and off the trend of the other Ar/Ar dates, as implicitly recognized by Raymond et al. (2000; examine their Plate 1b), the data point at 39.95°S, 171.19°N, dated at 53.4 Ma. Dates older than predicted are harder to explain than younger dates. Further, this date is approximately 9 my older than the date at 36.95°S, 169.83°N: 44.6 Ma, which is almost 120 km farther from the inferred hotspot. If these dates are accurate, this implies 10 my duration of magmatic activity along this stretch of the Louisville trace.

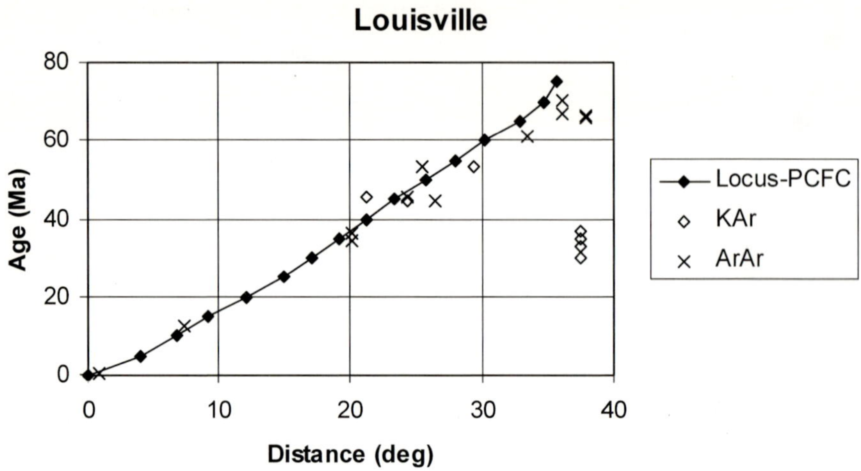

Fig. 6.4. Louisville: distance from inferred Louisville hotspot versus age of dated samples from the Louisville Seamounts and predicted locus as in Fig. 6.2.

At 27.28°S, 174.21°W, there is an age range of 3.4 Ma (between 70.1 and 66.7), although the uncertainty in age could explain the range. A smaller age range of 2.2 Ma is observed at 41.61°S, 164.2°W, again possibly attributable to age uncertainty.

6.2.1.3
Easter–Tuamotu (Hotspot: 27°N, 109°W)

Morgan (1972a) first proposed that the Tuamotu Ridge and Line Islands on the Pacific plate, along with the Sala y Gomez island-seamount chain and Nazca Ridge on the Nazca plate, were formed by a hotspot beneath Easter Island. Subsequent isotopic age dating from the Line Islands (Schlanger et al. 1984) indicated that the Line Islands could not easily be explained in terms of a single hotspot. However, Pilger and Handschumacher (1981) showed that the Nazca and Tuamotu ridges could be interpreted as having formed as mirror-images of each other with a common origin of a hotspot centered on the Pacific–Nazca spreading center. The intersection of identified magnetic anomalies with the ridges provides an estimate of the age of the formation of the ridge at each intersection. Age–distance plots of the few available isotopic ages and magnetic anomaly intersections from the inferred trace on the Pacific plate indicate fair correspondence with the locus predicted from the model of Raymond et al. (2000) as modified and interpolated in this volume (Figs. 6.6, 6.7). While not included in the plots, paleontologic data from the Tuamotu Ridge (as summarized by Jarrard and Clague 1977) are consistent with a hotspot origin, including the model presented here. Data points from the southern Line Islands do not correspond with the inferred locus (Figs. 6.6, 6.7), however, implying (in the context

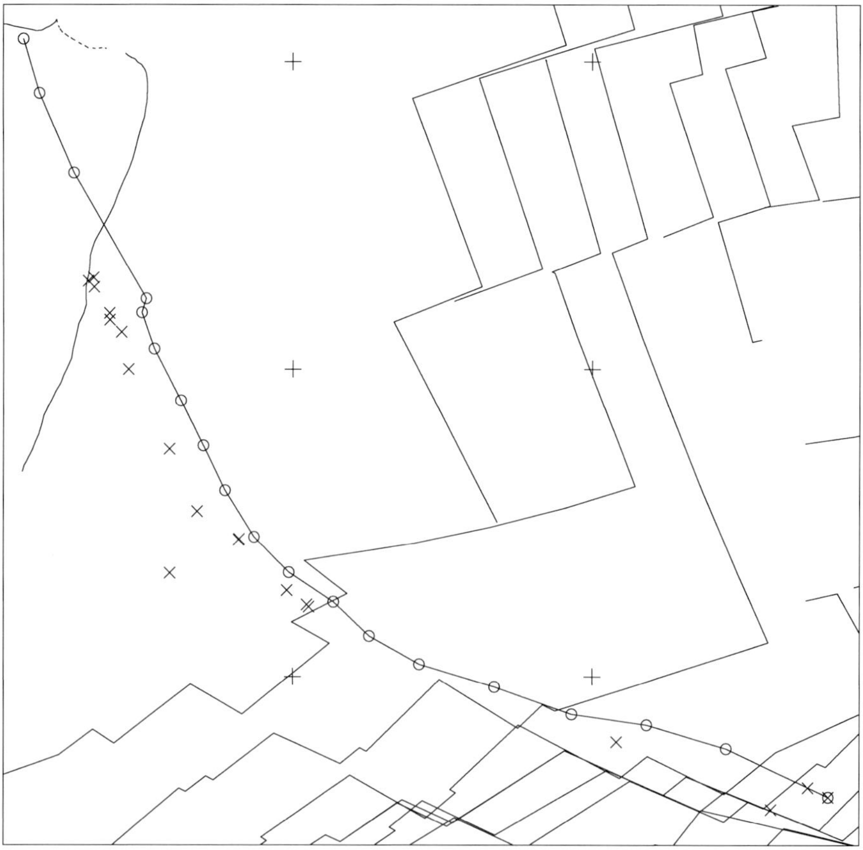

Fig. 6.5. Louisville: map of predicted locus and dated samples as in Fig. 6.3.

of the hotspot hypothesis) an error in the hotspot-plate kinematic model, relative movement of the Easter hotspot, or origin of the southern Line Islands from another hotspot.

6.2.1.4
Easter–Sala-y-Gomez Chain–Nazca Ridge (Hotspot: 27°N, 109°W)

As noted in the section above, the Sala y Gomez island-seamount chain and the Nazca Ridge are inferred to have formed from the Easter hotspot. Available isotopic dates and magnetic anomaly-ridge intersections show a fair correspondence with the predicted locus (Figs. 6.8, 6.9). A better fit of the locus to the magnetic anomaly-ridge intersections could be achieved by relocating the hotspot farther east, but at the expense of degrading the fit of the locus to the Tuamotu Ridge on the Pacific plate (Figs. 6.6, 6.7). An alternative would involve modifications of the magnetic anomaly time scale or improved magnetic anomaly identifications adjacent to the Tuamotu and/or Nazca ridges.

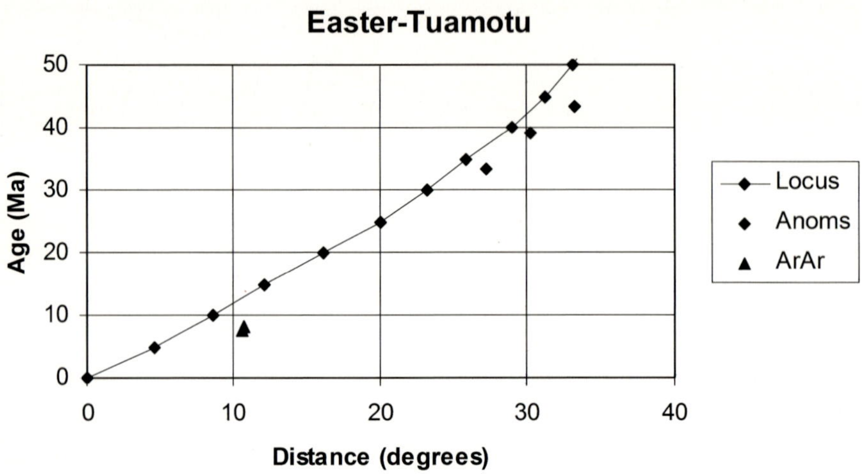

Fig. 6.6. Easter–Tuamotu: distance from inferred Easter hotspot versus age of dated samples from the Tuamotu Ridge and Seamounts and predicted locus as in Fig. 6.2.

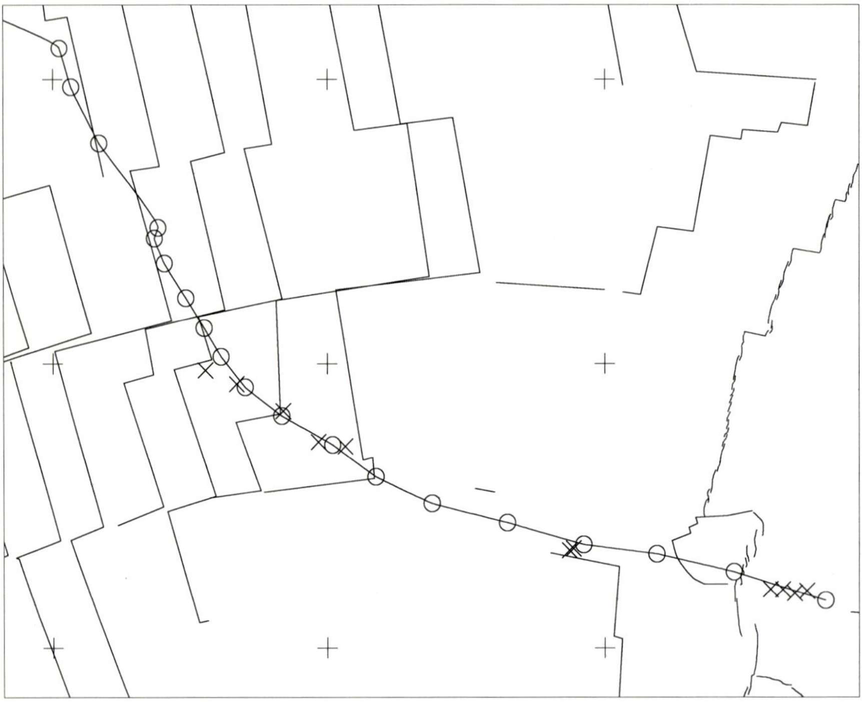

Fig. 6.7. Easter–Tuamotu: map of predicted locus and dated samples as in Fig. 6.3.

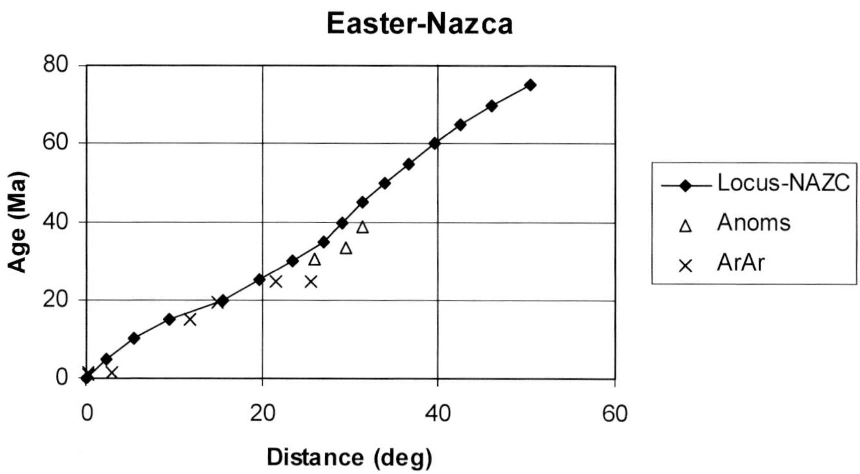

Fig. 6.8. Easter–Sala y Gomez–Nazca: distance from inferred Easter hotspot versus age of dated samples from the Nazca Ridge and Sala y Gomez islands and seamounts and predicted locus as in Fig. 6.7.

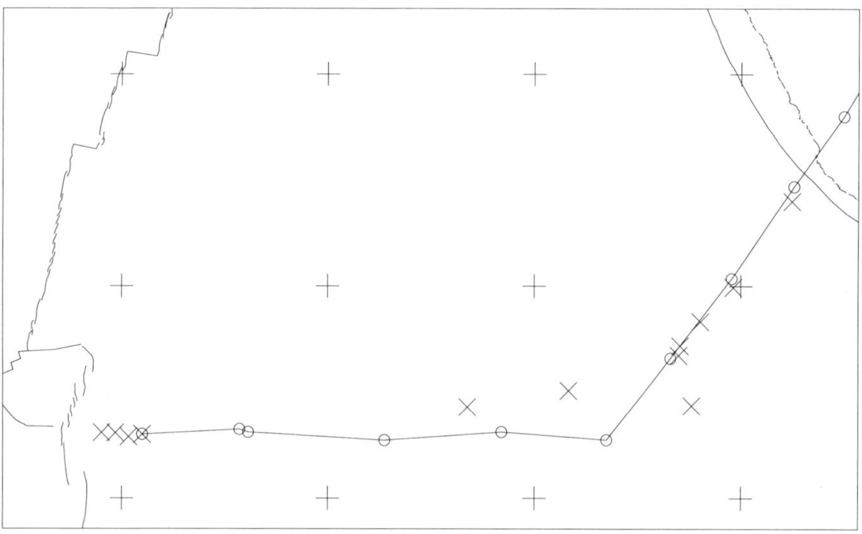

Fig. 6.9. Easter–Sala y Gomez–Nazca: map of predicted locus and dated samples as in Fig. 6.3.

Plate reconstructions of the Pacific and Nazca are poorly constrained between magnetic anomalies 3 and 7 due to limited anomaly identifications in that age range (4–25 Ma) in the east-central Pacific Ocean. In all probability, jumping ridges and transient microplates characterized this time interval, just as the Easter microplate is an important contemporary tectonic element. As a consequence, the calculated locus of Nazca plate motion relative to the Easter hotspot is quite uncertain between 25 and 4 Ma.

Few models of the motion of the Pacific plate relative to hotspots have taken into account the role of the Nazca plate, and especially Nazca and Tuamotu ridges complementarity. As the tectonic complexities of the east-central Pacific are slowly resolved, the relation of the Sala y Gomez chain to the Easter hotspot should become clearer.

6.2.1.5
Line Islands (Hotspots: Presently Unknown Locations)

As recognized by Schlanger et al. (1984), the Line Islands cannot be simply explained in terms of one hotspot. Plots of available isotopic ages and loci of postulated hotspots (Figs. 6.10, 6.11) indicate that multiple existing hotspots cannot easily explain the Line Islands, either. Nevertheless, two apparent age trends observed in the Line Islands are parallel with those inferred from hotspot models. Note that sample Ar/Ar dates between the Equator and 20 N include two age groups, between 65 and 95 Ma and between 35 and 60 Ma. Both groups of dates can be interpreted in terms of two migrating loci. The older group is parallel to calculated loci of the inferred Pitcairn and Marquesas hotspots for the same age range. The younger age group, corresponding with the Line–Cross island-seamount chain, is only subparallel to the loci for its age-range, however.

If the older group of dates originated from a hotspot, the present location of the hotspot would be roughly 19°S, 123°W, for which there is no contemporary evidence. This location was determined by back-tracking each data point by interpolation of the hotspot parameters using its analytically determined age by the techniques presented in this volume; the most southeasterly location was then chosen, which corresponds with: (1) the envelope of the locus around the oldest point(s) at any location along the locus, and (2) the locus that corresponds best with the trend of the Line Islands.

The few dates from the southern Line Islands, south of the Equator, could have been produced by another hotspot. Back-tracking provides a location of approximately 30°S, 114°W, for which there is no present-day indication.

The Line–Cross data points produce a diffuse cluster of predicted hotspot loci centered on 5°S, 135°W. Two old ages (~125 Ma) predict a hotspot at 17°S, 108°W. Again, there is little evidence for contemporary hotspots at either location.

6.2.1.6
Pitcairn–Gambier (Hotspot: 24.5°2, 129.5°W)

The age pattern of dates from the Pitcairn–Gambier island-seamount chain (Figs. 6.12, 6.13) records a progression of inception and cessation of volcanism

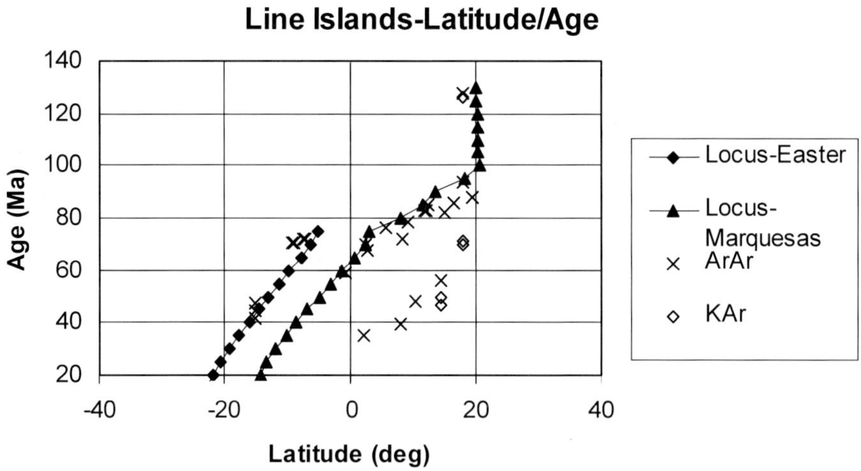

Fig. 6.10. Line Islands: distance from inferred Easter hotspot versus age of dated samples from the Tuamotu island-seamount chain and Line Islands and predicted locus. Line Islands could be attributed to one or more additional hotspots.

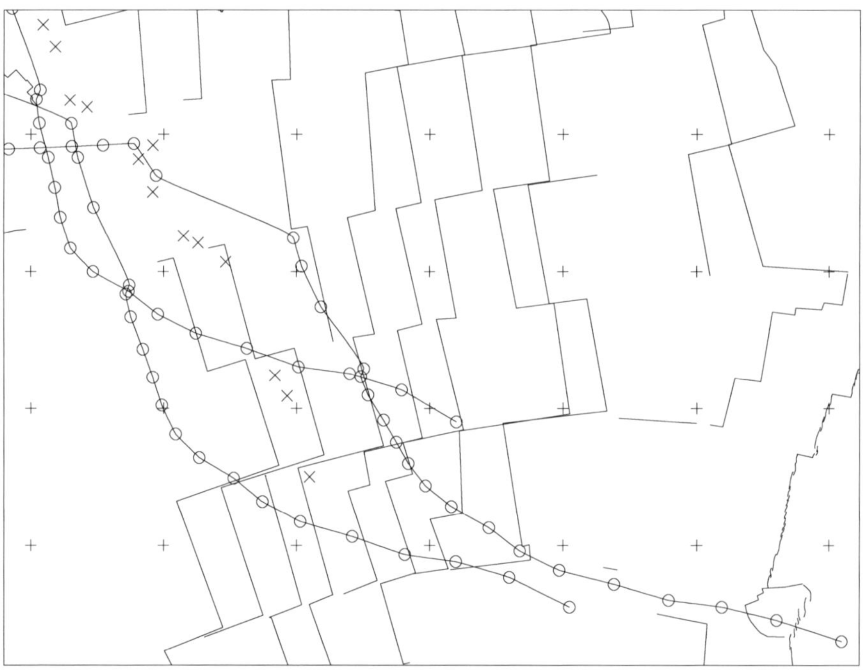

Fig. 6.11. Line Islands: map of predicted locus and dated samples as in Fig. 6.3.

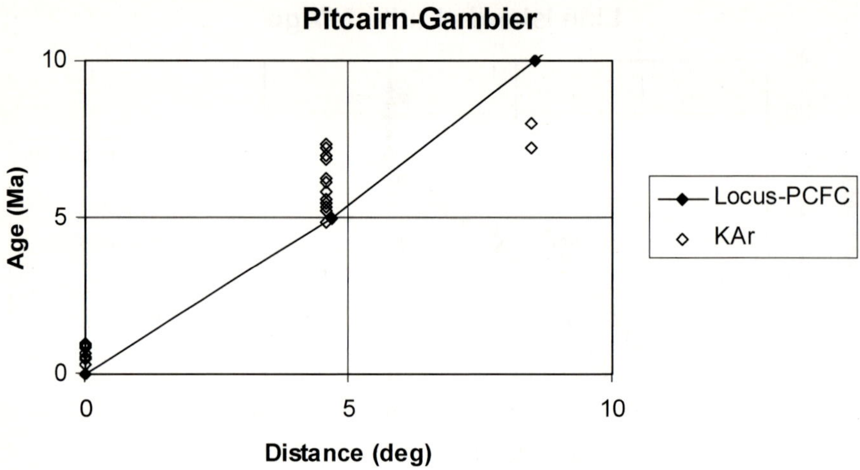

Fig. 6.12. Pitcairn–Gambier: distance from inferred Pitcairn hotspot versus age of dated samples from the Gambier island-seamount chain and predicted locus.

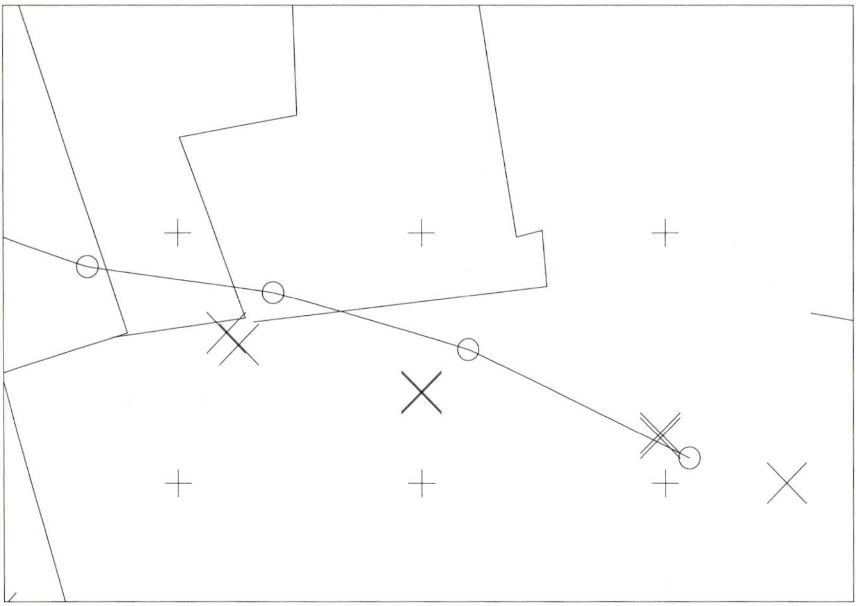

Fig. 6.13. Pitcairn–Gambier: map of predicted locus and dated samples as in Fig. 6.3.

from west to east. At Fangataufa and Mururao Islands, the range of ages, between 7 and 13.6 Ma, implies duration of volcanism for more than 6 my, assuming the dates are crystallization ages. There are no known older extensions of the chain that have been dated.

6.2.1.7
Marquesas Islands (Hotspot: 11.0°S, 138.0°W)

The age patterns of the Marquesas Islands indicate a steeper rate of inception of volcanism than predicted by most hotspot models (Fig. 16.14). Further, they are one of the most divergent hotspot chains, with a more northwesterly trend than predicted by the models (Fig. 16.15). The inferred location of the hotspot could be farther to the southeast than shown in either figure. The oldest dates from the chain imply a range of volcanism on the order of 4 my, assuming they are crystallization ages.

Interestingly, when the physiography of the Marquesas is considered, the chain can be interpreted in terms of three or four en echelon subchains. The subchains are oriented parallel with the calculated locus of Pacific plate motion in the hotspot frame.

6.2.1.8
Austral–Cook Chain (Hotspot: 26.0°S, 142,0°W)

The Austral–Cook island-seamount chain records nearly 30 my of magmatism. Progressive inception of magmatism to the southwest is implied from about 20 Ma to the present, with intermittent activity continuing to the present over much of the 2,000-km length of the chain (Fig. 16.16). An earlier period of vol-

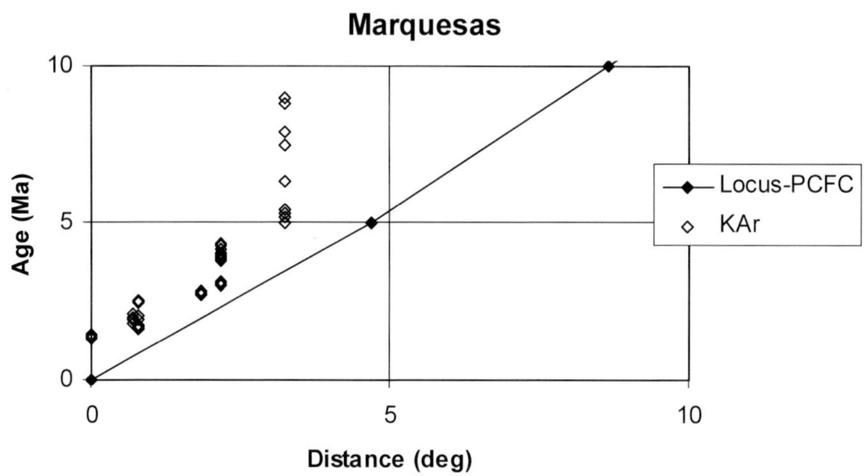

Fig. 6.14. Marquesas Islands: distance from inferred Marquesas hotspot versus age of dated samples from the Marquesas Islands and predicted locus.

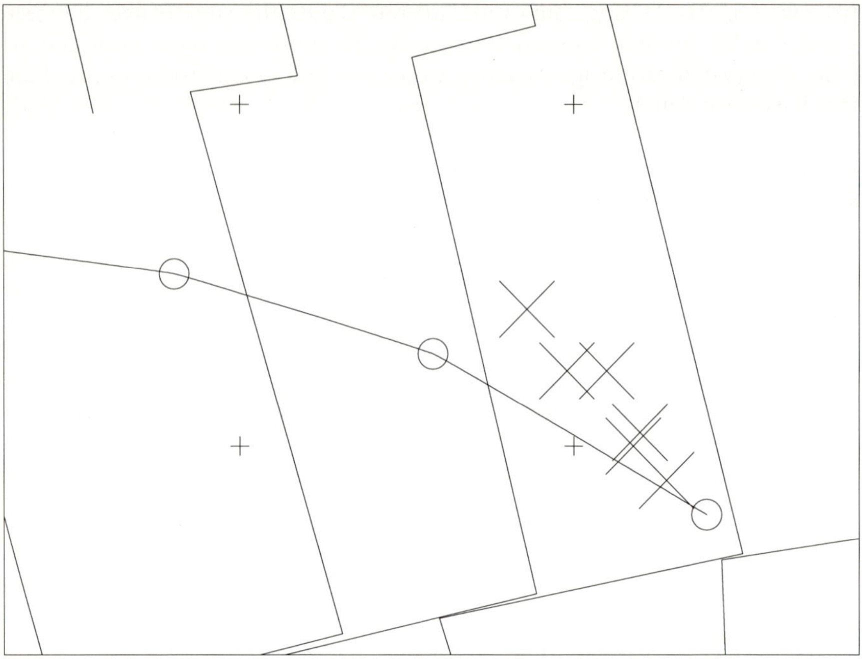

Fig. 6.15. Marquesas Islands: map of predicted locus and dated samples as in Fig. 6.3.

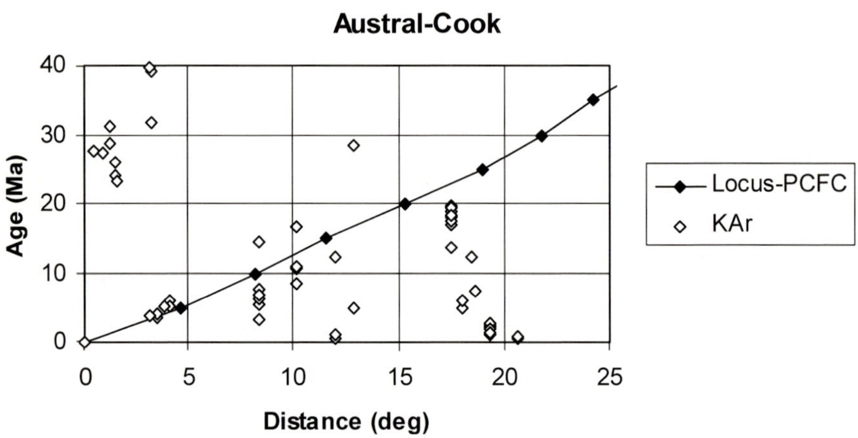

Fig. 6.16. Austral–Cook chain: distance from inferred Macdonald hotspot versus age of dated samples from the Austral–Cook island-seamount chain and predicted locus.

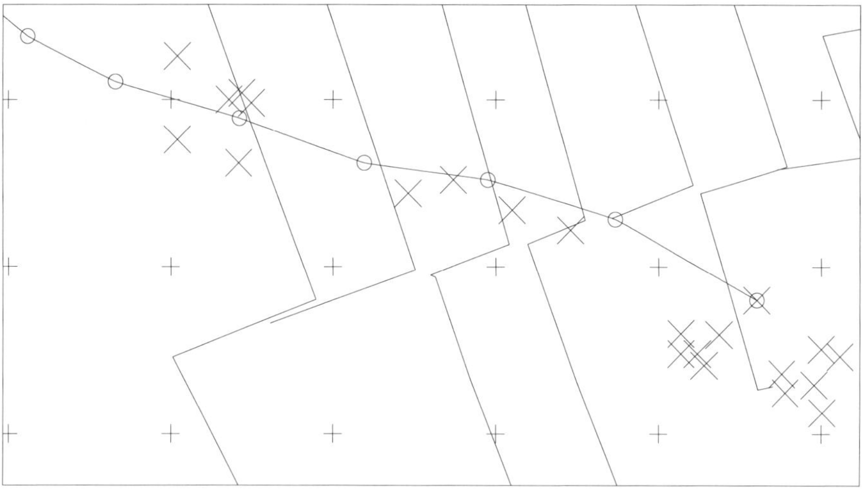

Fig. 6.17. Austral–Cook chain: map of predicted locus as and dated samples as in Fig. 6.3.

canic activity between 24 and 30 Ma is apparent near the inferred present-day location of the hotspot; an isolated date in this range is also present some 1,200 km to the west along the chain. The latter (western) date (K/Ar) could represent excess argon, but the other (eastern) dates are Ar/Ar and are presumably reliable. The eastern dates are from the Foundation trace.

There is a slight divergence of the predicted locus and the concentration of dated islands and seamounts (Fig. 6.17). In the context of the hotspot model, the hotspot could be located farther to the northeast than inferred.

6.2.1.9
Foundation Seamounts (Hotspot: 36.6°S, 110.8°W)

The Foundation Seamount chain demonstrates a southeasterly progression of volcanism (Fig. 6.18). There is a slight divergence between the calculated and observed seamounts (Fig. 6.19), even if a different hotspot location is chosen.

6.2.1.10
Society Islands (Hotspot: 17.9°S, 148.3°W)

The Society Islands, including Tahiti, imply a southeasterly progression of volcanic inception (Figs. 6.20, 6.21), except for a wide range of dates from Tahiti–Iki, between 24 and 1 Ma. Conceivably, the range of older dates represents argon retention as they are K/Ar. Ignoring the possibly spurious dates, the rate of inception is close to that predicted by the hotspot model, although the inferred location of the hotspot could be as much as 200 km to the southeast. Further, the duration of volcanism could be as long as 4–5 my.

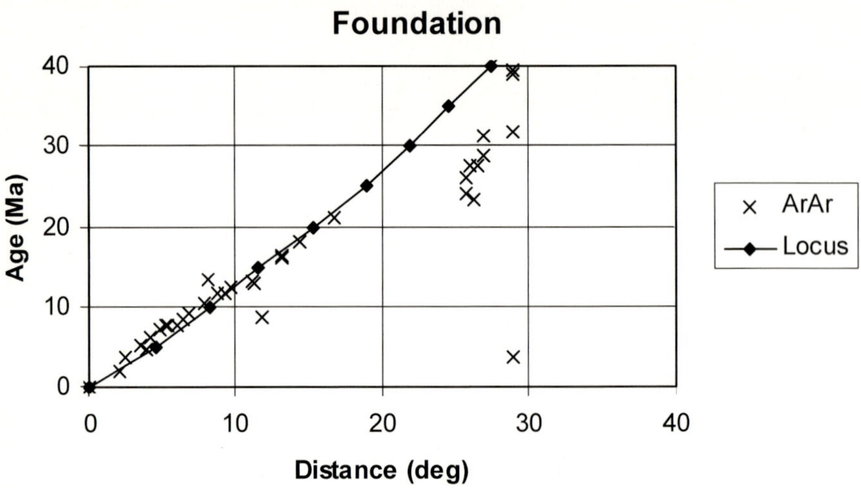

Fig. 6.18. Foundation Seamounts: distance from inferred Foundation hotspot versus age of dated samples from the Foundation seamount chain and predicted locus.

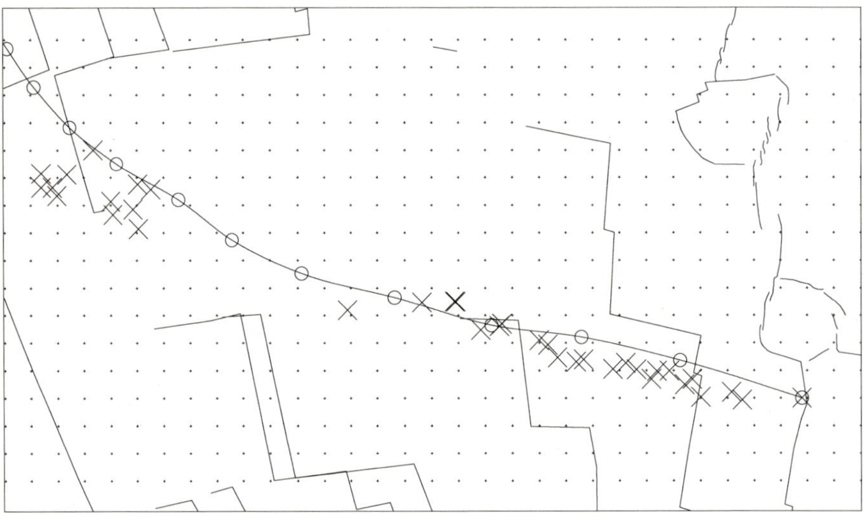

Fig. 6.19. Foundation Seamounts: Map of predicted locus and dated samples as in Fig. 6.3. Hotspot location is from Harada and Hamano (2000).

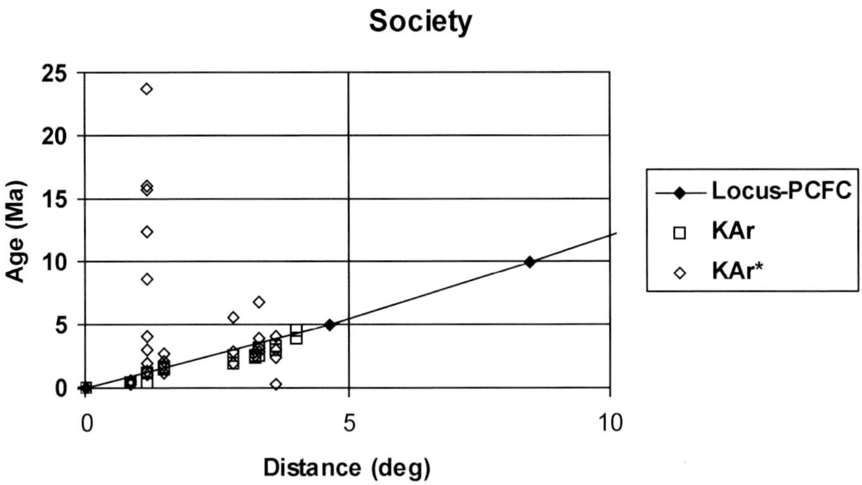

Fig. 6.20. Society Islands: distance from inferred Tahiti hotspot versus age of dated samples from the Society Islands and predicted locus.

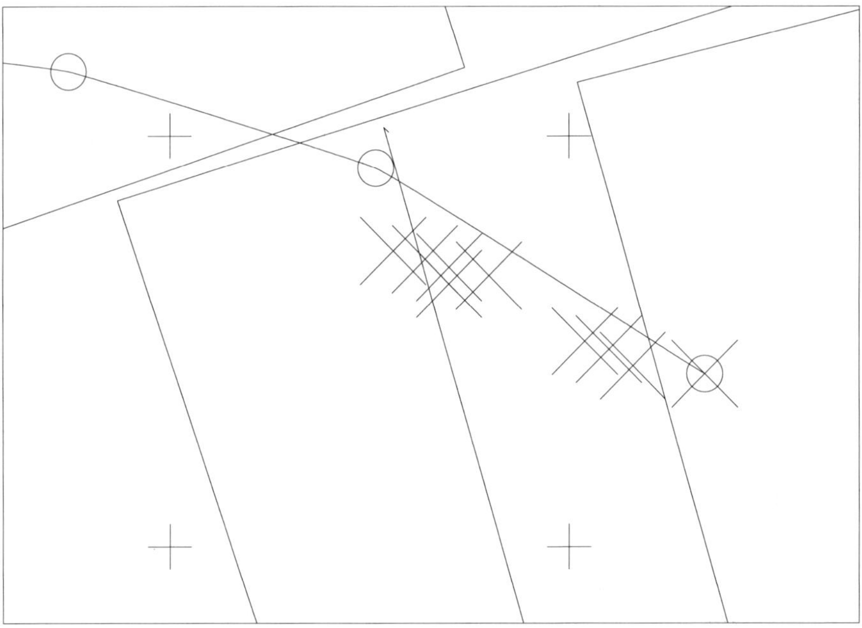

Fig. 6.21. Society Islands: map of predicted locus and dated samples as in Fig. 6.3.

6.2.1.11
Pratt-Welker (Hotspot: 50°N, 128°W)

The Pratt-Welker seamount chain in the Gulf of Alaska shows an apparent progressive inception of magmatism from northwest to southeast (Figs. 6.22, 6.23). However, there is evidence of younger activity some 7–12 my after inception near both ends of the chain. The inferred present location of the hotspot is based on trial-and-error fitting of the locus to the older dates of the chain.

6.2.1.12
Cobb Seamounts (Hotspot: 46°N, 130°W)

A southwest progression in inception and cessation of magmatism is observed in the Cobb seamount chain (Figs. 6.24, 6.25). The chain could be interpreted as a composite of two chains; samples from a set of seamounts to the north of and parallel with the main chain suggest, with one exception (the southeasternmost seamount), that the northern set might be younger than the main trend. The present location of the hotspot could be a few hundred kilometers to the southeast of the inferred location, based on the oldest ages determined along the chain. Duration of volcanism could be as much as 8 my at one point along the chain.

6.2.1.13
Caroline Islands (Hotspot: 3.1°N, 164°W)

Isolated from the principal traces of the Pacific plate to the east, the Caroline Islands show progressive inception and cessation of volcanic activity from west

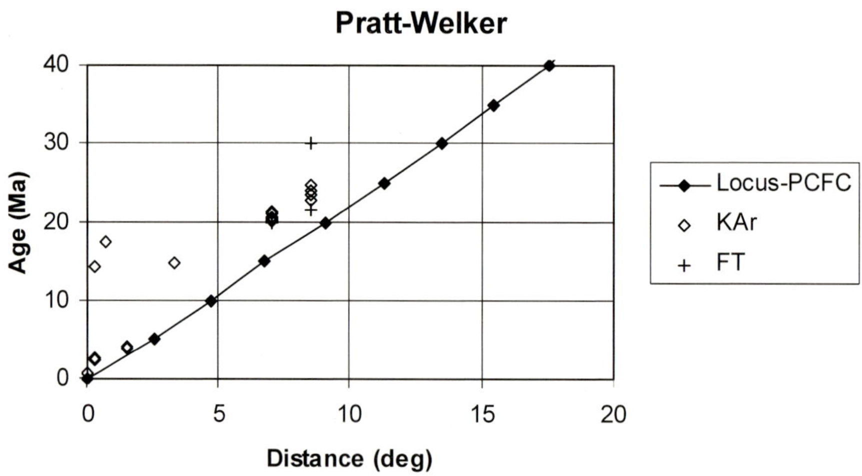

Fig. 6.22. Pratt-Welker: distance from inferred Pratt-Welker hotspot versus age of dated samples from the Pratt-Welker seamount chain and predicted locus.

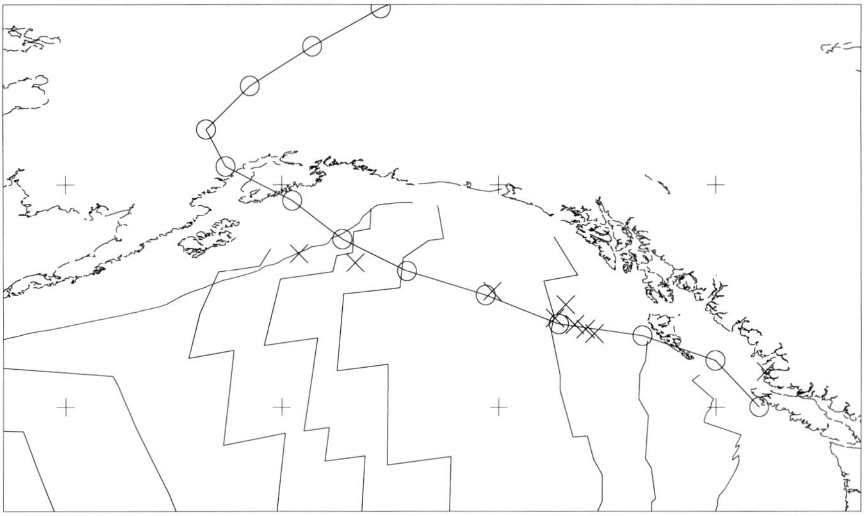

Fig. 6.23. Pratt-Welker: map of predicted locus and dated samples as in Fig. 6.3.

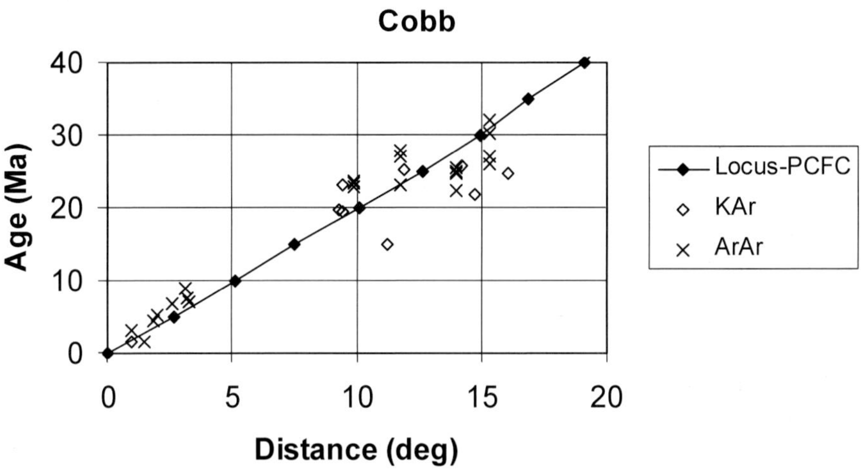

Fig. 6.24. Cobb Seamounts: distance from inferred Cobb hotspot versus age of dated samples from the Cobb Seamount chain and predicted locus.

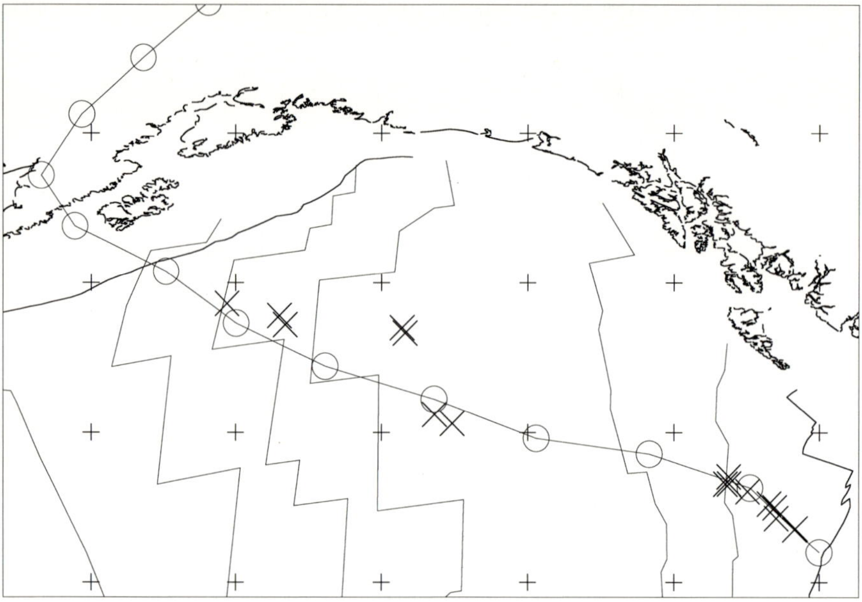

Fig. 6.25. Cobb seamounts: map of predicted locus and dated samples as in Fig. 6.3.

to east (Figs. 6.26, 27). However, the few dated samples do not indicate whether volcanism was persistent between inception and cessation of activity. Nevertheless, the few dates of inception are close to that predicted by the hotspot model of Raymond et al. (2000) as interpolated using the methods of this volume.

6.2.1.14
Juan Fernandez (Hotspot: 34°S, 82.5°W)

The Juan Fernandez Islands and Ridge, offshore central Chile, produce a few sampled dates (Figs. 6.28, 6.29) that indicate west to east age progession as predicted by the hotspot model of Gripp and Jordan (1991) and Raymond et al. (2000), extended to the Nazca plate in this volume, although there is a difference in observed and calculated rates. There is empirical evidence for continuation of the ridge (or seamount chain) within the subducted Nazca plate, beneath the South American Andes, in that substantially lower-angle subduction occurs beneath central Chile (Barazangi and Isacks 1977; Pilger 1981b, 1984). The location of the hotspot responsible for the ridge is conjectural.

6.2.1.15
Galapagos Islands–Carnegie Ridge (Hotspot: 1.5°S, 90.5°W)

The diverse islands of the Galapagos are inferred to be underlain by a mantle hotspot (e.g. Morgan 1972a) that has also affected seafloor spreading (produc-

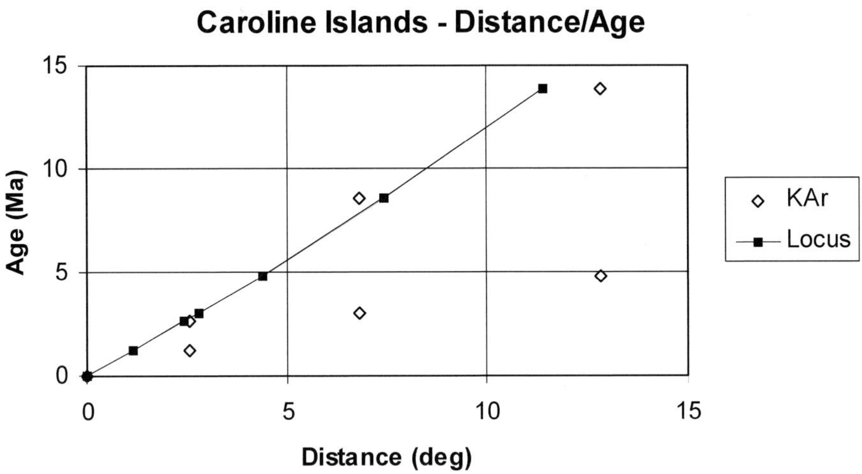

Fig. 6.26. Caroline Islands: distance from inferred Caroline hotspot versus age of dated samples from the Caroline Islands, and predicted locus.

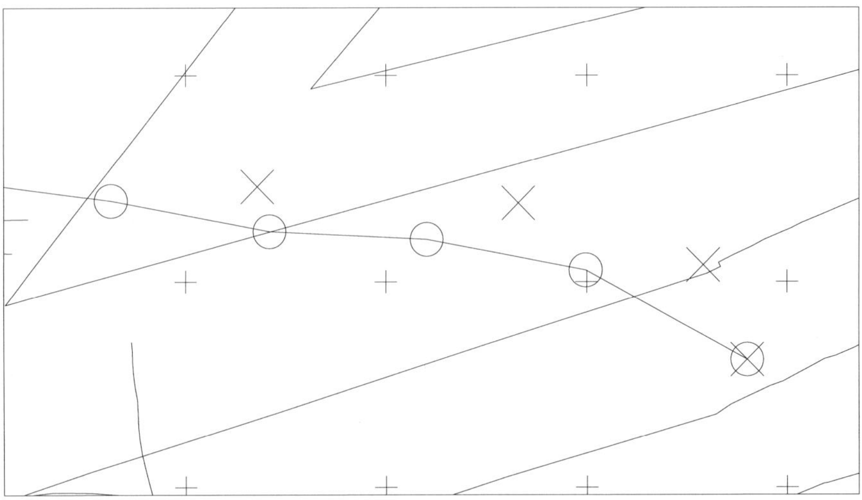

Fig. 6.27. Caroline Islands: map of predicted locus and dated samples as in Fig. 6.3.

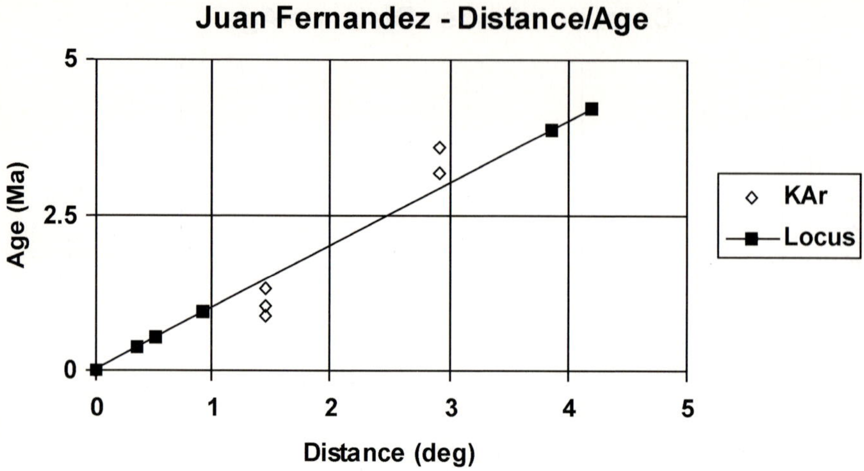

Fig. 6.28. Juan Fernandez Ridge: distance from inferred Juan Fernandez hotspot versus age of dated samples from the Juan Fernandez Ridge, Nazca Plate, and predicted locus.

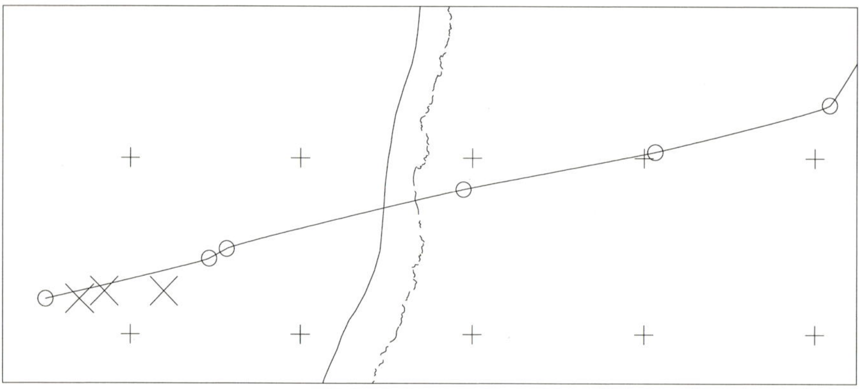

Fig. 6.29. Juan Fernandez Ridge: map of predicted locus and dated samples as in Fig. 6.3.

ing significant asymmetry and ridge jumps) at the nearby Cocos–Nazca ridge. In addition, the Carnegie and Cocos Ridges on the respective Nazca and Cocos plates are inferred to represent a conjugate (if not mirror-image) set produced by the hotspot when if was directly beneath the spreading center. It is difficult to recognize simple age progressions in the Galapagos (Figs. 6.30, 6.31), although a progression is observed to the east, along the Carnegie Ridge. One of the side effects of the reconstruction interpolation method presented here is a suggestion of reversal of plate motion between 10 and 8 Ma (Figs. 6.28, 6.29). Unfortunately, there are not enough reconstructions over the past 15–20 my to more pre-

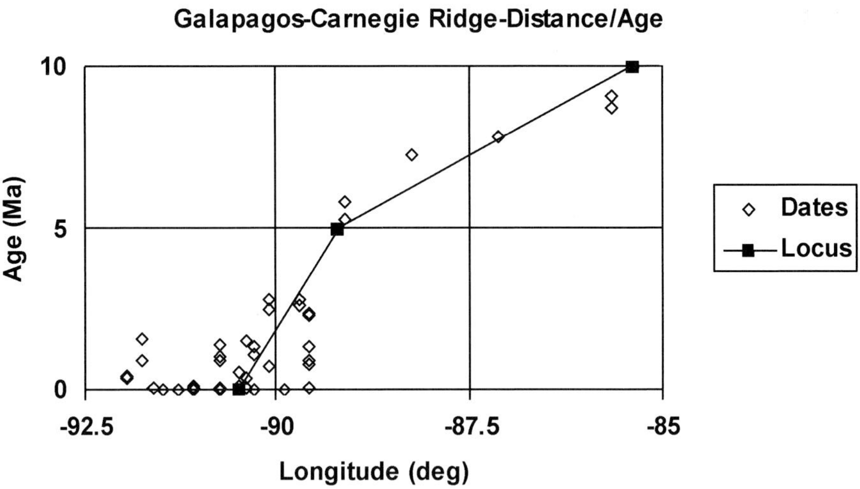

Fig. 6.30. Galapagos Islands–Carnegie Ridge: distance from inferred Galapagos hotspot versus age of dated samples from the Galapagos Islands and Carnegie Ridge, Nazca Plate, and predicted locus.

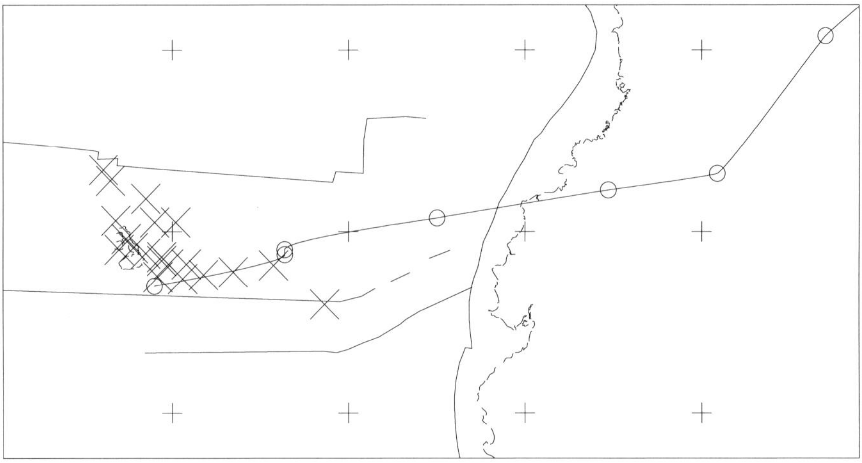

Fig. 6.31. Galapagos Islands–Carnegie Ridge: map of predicted locus and dated samples as in Fig. 6.3.

cisely constrain the plate kinematics. The reversal is interpreted as an artifact of the reconstruction interpolation technique.

6.2.2
Indian Ocean

6.2.2.1
Reunion–Chagos–Laccadive (Hotspot: 21°S, 55.5°E)

The Reunion hotspot is inferred to have produced the Chagos Ridge as well as the Mauritus Islands on the central African plate and the Laccadive Ridge and Deccan Traps on the Indian plate (e.g., Morgan 1972a). Sampled ages from the conjugate traces (Figs. 6.32, 6.33) provide important constraints on hotspot reconstruction models (e.g., Müller et al. 1993). Baksi (1999) reinterpreted the analytical data from some of the samples and produced either revised ages (or age ranges) or eliminated some dates. However, the model of Müller et al. (1993) is still consistent with even a reduced set of data points (Fig. 6.30).

6.2.2.2
Kerguelen–Broken Ridge–Ninetyeast Ridge (Hotspot: 49°S, 63°E)

Along with the Reunion hotspot and its inferred traces (Sect. 6.2.2.1) and the Tristan da Cunha hotspot and its inferred traces (Sect. 6.2.3.1), the Kerguelen hospot and its derivative traces, Broken and Ninetyeast Ridge on the Antarctic

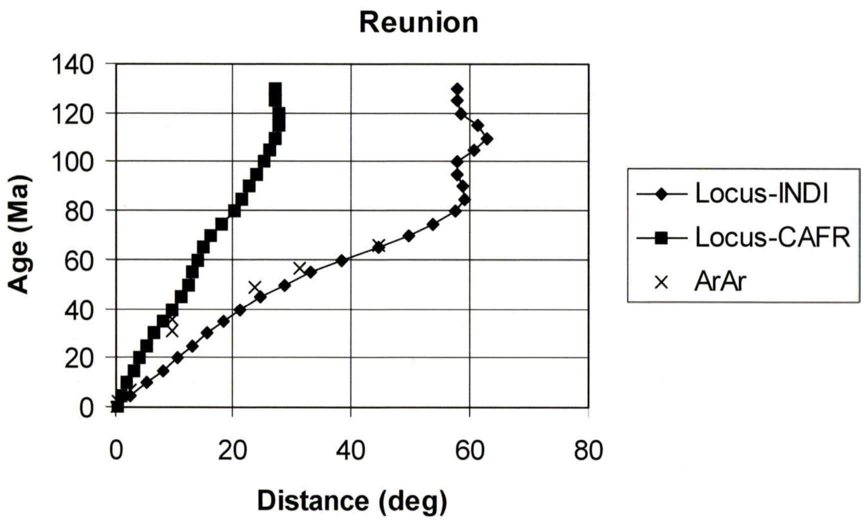

Fig. 6.32. Reunion Island–Chagos–Laccadive Ridges: distance from inferred Renuion hotspot versus age of dated samples from the Chagos and Laccadive Ridges, Central African and Indian Plates, and predicted loci from model of Müller et al. (1993) interpolated by methods described in this volume.

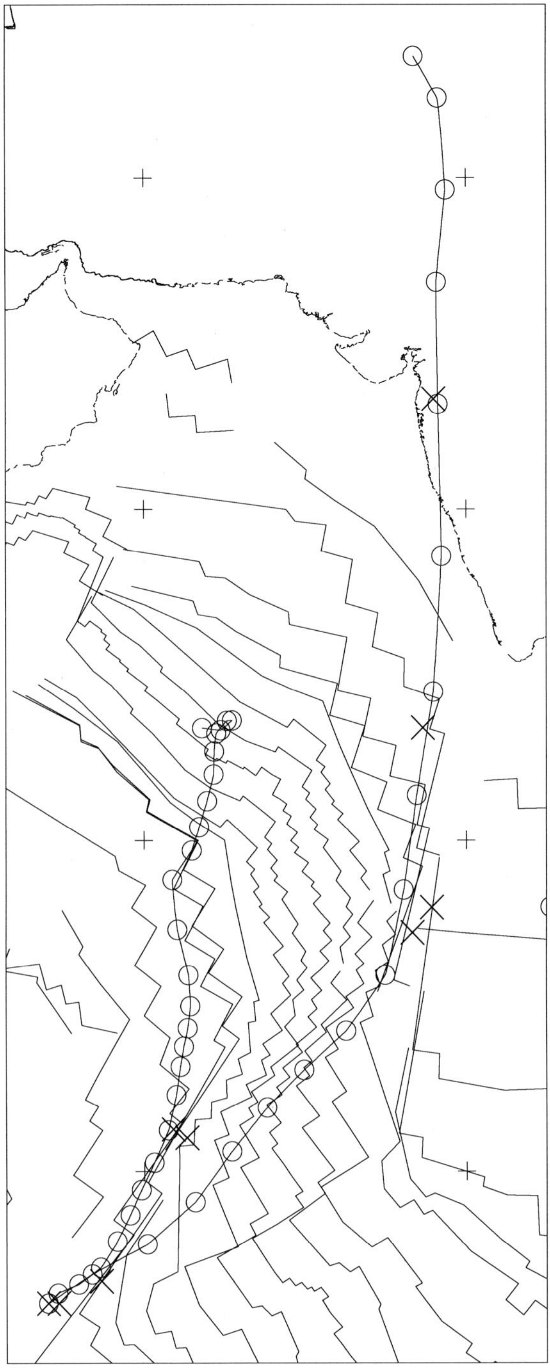

Fig. 6.33. Reunion Island–Chagos–Laccadive Ridges: map of predicted loci and dated samples as in Fig. 6.3.

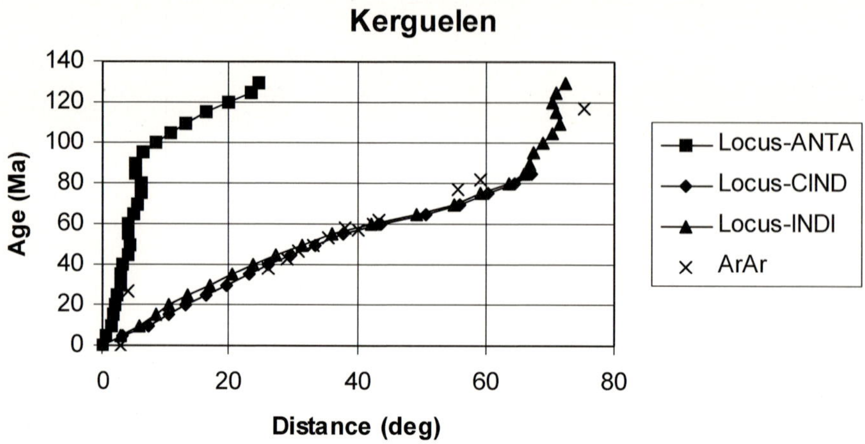

Fig. 6.34. Kerguelen Archjipelego–Ninetyeast Ridge: distance from inferred Kerguelen hotspot versus age of dated samples from the Kerguelen Archipelego and Ninetyeast Ridge, Antarctic, Central Indian, and Indian Plates, and predicted loci from model of Müller et al. (1993) for Africa extended via methods described in this volume.

and Indian (and central-Indian) plates, provide critical constraints on hotspot reconstruction models (e.g., Müller et al. 1993). Available age data (Figs. 6.34, 6.35) constrain the model, although there are some apparent discrepancies between predicted and observed data points. There is evidence for jumping ridges along the Ninetyeast ridge, which might explain some of the discrepancies. Baksi's (1999) reinterpretation of some of the analytical data reduces the number of data points but does not introduce any further discrepancies.

6.2.2.3
Comores (Hotspot: 11.7°S, 43.3°E)

The few islands of the Comores trace, on the Central African plate, provide date samples that imply progressive inception of volcanic activity from east to west (Figs. 6.36, 6.37). However, the predicted locus and the observed trend are divergent by more than 20°. Similarly, the observed and calculated rates are not that similar.

6.2.3
South Atlantic Ocean

6.2.3.1
Tristan da Cunha–Walvis Ridge (Hotspot: 38°S, 13°W)

The active volcanic island of Tristan da Cunha and its conjugate traces – the older portion of the Walvis Ridge on the central African plate and the Rio Grande Rise on the South American plate – are among the archetypal hotspot-

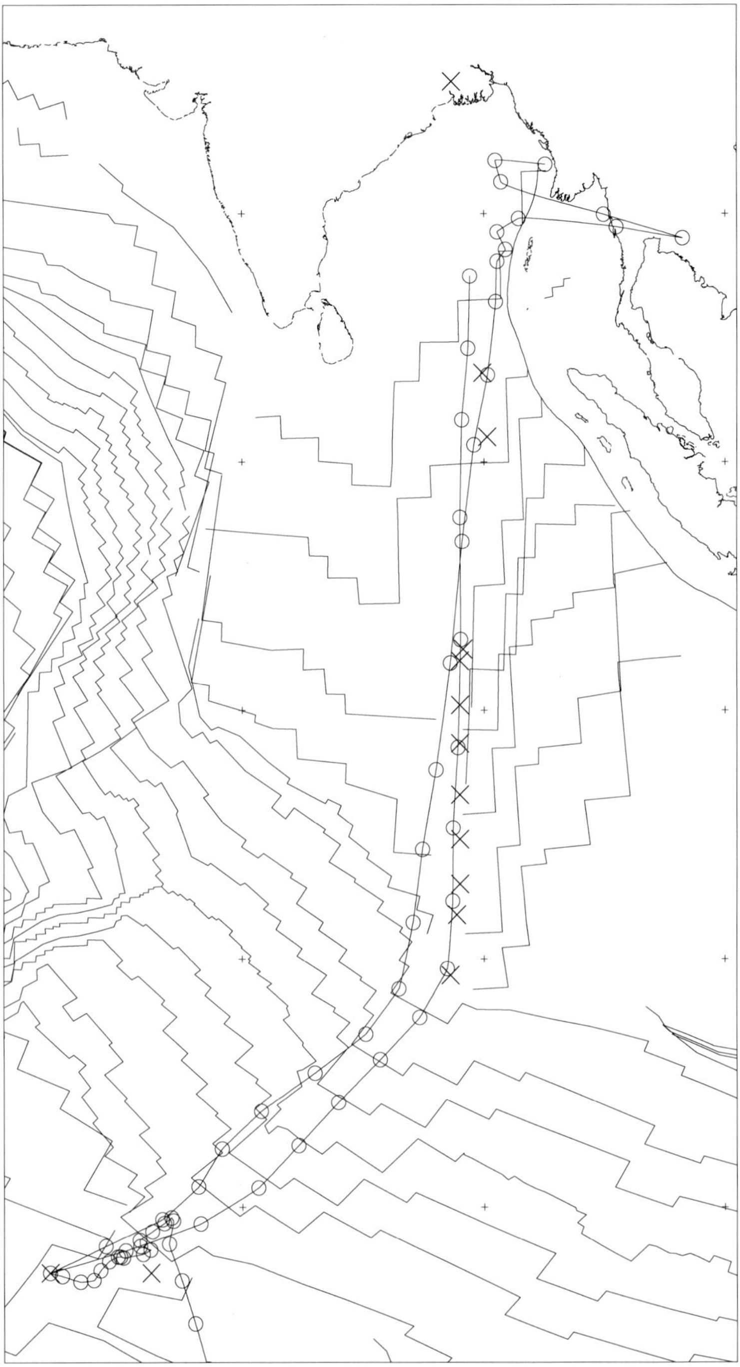

Fig. 6.35. Kerguelen Archjipelego–Ninetyeast Ridge: map of predicted loci and dated samples as in Fig. 6.3. Locus relative to central Indian plate ends at 80 Ma and fits dated samples better.

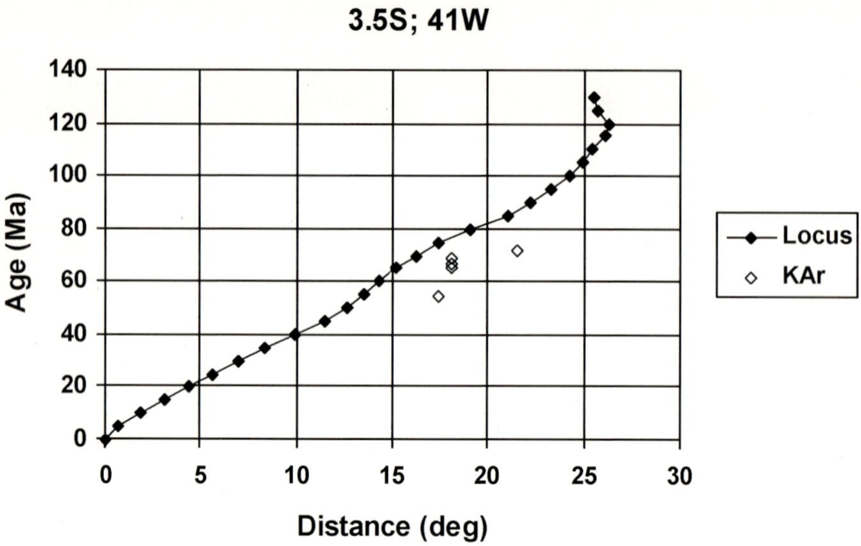

Fig. 6.36. Comores: distance from inferred Comores hotspot versus age of dated samples from the Comores Islands, Central African (or Nubian) Plate, and predicted locus from model of Müller et al. (1993) interpolated via methods described in this volume.

traces within the hotspot hypothesis (e.g., Morgan 1972a). Plots of dated samples from the central African trace, including the Walvis ridge, imply progressive inception of magmatism as predicted by the hotspot model (Figs. 6.38, 6.39). There is an apparent divergence between the predicted and observed trace and in distance versus age. However, the nature of the younger portions of the diffuse trace makes determination of the present location of the hotspot uncertain. A more southewesterly location would produce a better fit of the predicted locus to available dates. Baksi's (1999) revisions reduce the number of data points and reinforce the idea that the present-day location of the hotspot is farther to the southwest, as noted above.

6.2.3.2
Tristan da Cunha–Rio Grande Rise (Hotspot: 38°S, 13°W)

As noted above (Sect. 6.2.3.1), the Rio Grande Rise in the South Atlantic and on the South American plate is inferred to be conjugate to the Walvis ridge on the Central African plate, with both features having originated from the Tristan da Cunha hotspot (e.g., Morgan 1972a). Only a few isotopic ages are available from the Rio Grande Rise, supplemented by a few identified magnetic isochrons on top of the Rise (Figs. 6.39–6.41). The dimensions of the Rio Grande Rise and the inferred complexity of the evolution of the adjacent oceanic crust, with jumping ridges and asymmetric seafloor spreading, limit the ability to test the hotspot hypothesis. If there have been no ridge jumps from the west of the few samples

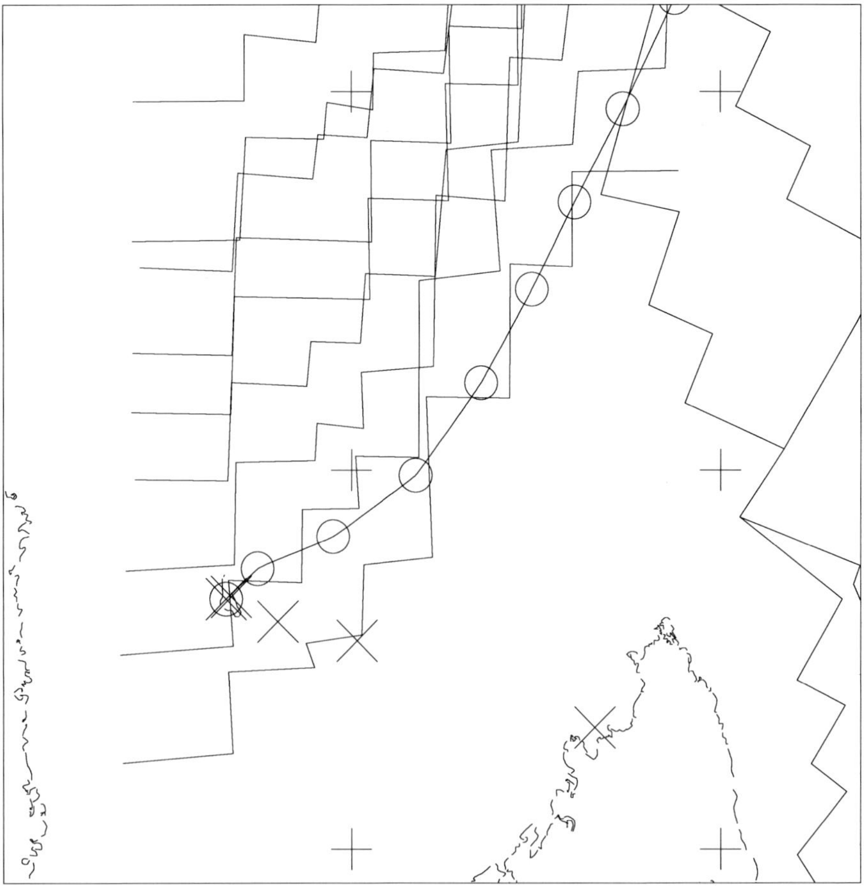

Fig. 6.37. Comores: map of predicted locus and dated samples as in Fig. 6.3.

from the Rio Grande Ridge to the east subsequent to the formation of the samples, then comparison of the age–distance plots with the predicted locus shows a fair correspondence; however, in map view, it is apparent that the predicted locus is southwest of two data points (one isotopic, one magnetic anomaly) but it corresponds moderately well with two other offshore data points plus a few continental samples at the older end of the locus. There are additional onshore ages to the north of the inferred trace that correspond in part with the age–distance plot. Farther to the north are additional data samples of the Ponte Grossa region, between the inferred onshore continuations of the Tristan da Cunha and Martin Vaz–Trinidade hotspots (Sect. 6.2.3.3).

A more southwesterly present-day location of the hotspot, as suggested for the Walvis Ridge (Sect. 6.2.3.1) would produce a somewhat poorer fit to the few age dates from the Rio Grande Rise, although the data points would remain close to the inferred locus (depending on which hotspot location is chosen). However,

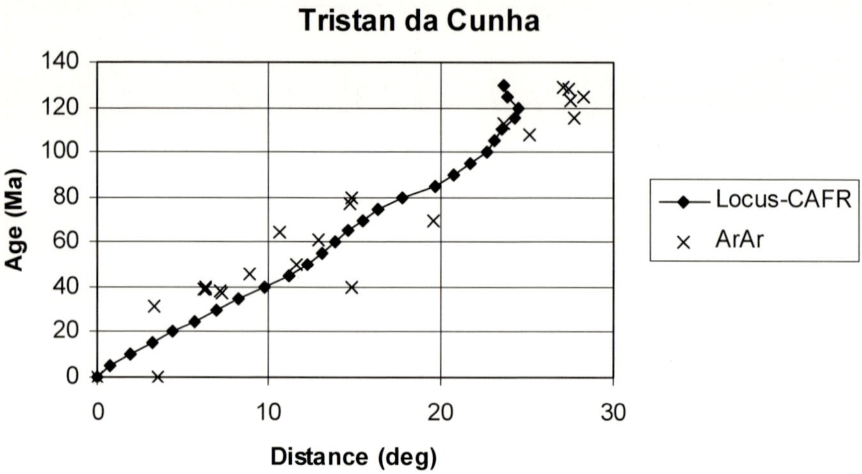

Fig 6.38. Tristan da Cunha–Walvis Ridge: distance from inferred Tristan hotspot versus age of dated samples from the Walvis Ridge, Central African plate, and predicted loci from model of Müller et al. (1993) interpolated and extended via methods described in this volume.

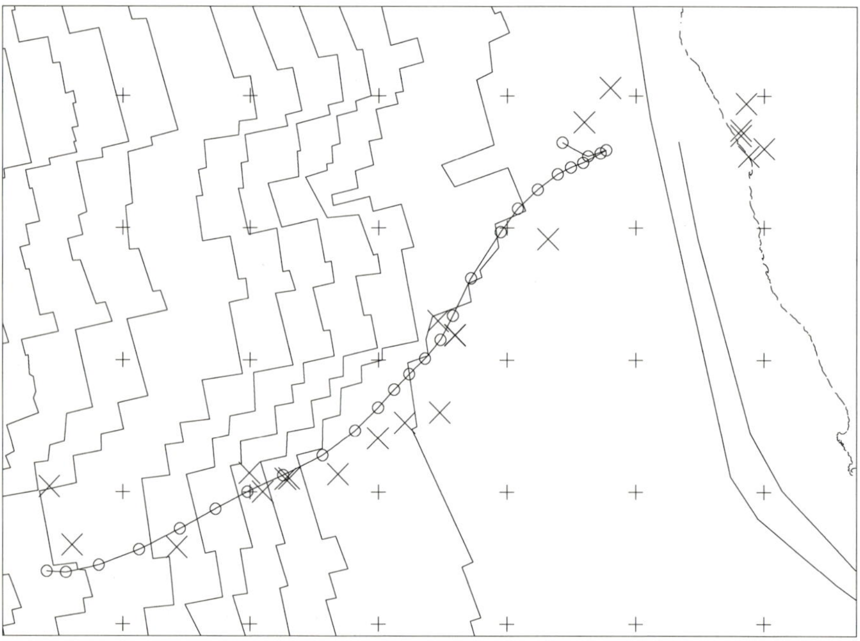

Fig. 6.39. Tristan da Cunha–Walvis Ridge: map of predicted loci and dated samples as in Fig. 6.3.

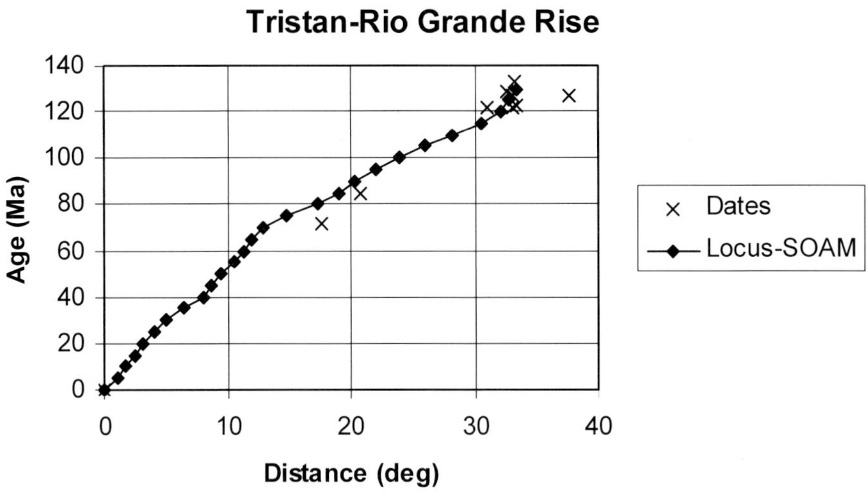

Fig. 6.40. Tristan da Cunha–Rio Grande Rise: distance from inferred Tristan hotspot versus age of dated samples from the Rio Grande Rise, South American plate, and predicted loci from model of Müller et al. (1993) interpolated and extended via methods described in this volume.

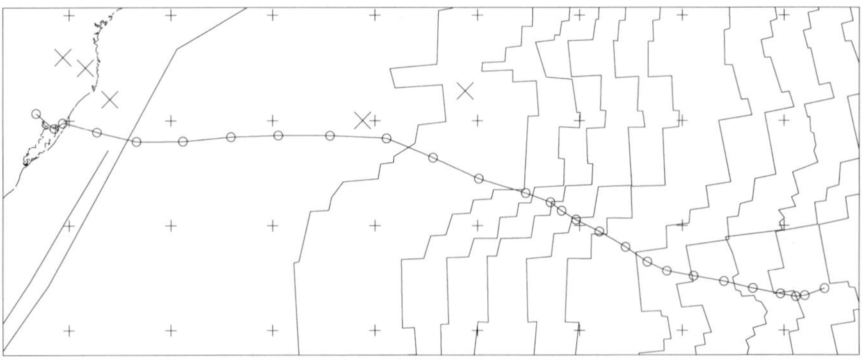

Fig. 6.41. Tristan da Cunha–Rio Grande Rise: map of predicted loci and dated samples as in Fig. 6.3.

there is evidence for ridge jumps in several locations, which would have the effect of producing a misfit (e.g., O'Connor and Duncan 1990). Further, Baksi's (1999) reinterpretation of the age data from the Rio Grande Rise would exclude the published data from consideration.

A few data points from Uruguay are included in both the map and age–distance plot (Figs. 6.40, 6.41). Despite their distance from the inferred map locus, they fall close to the predicted locus on the age–distance plot, like the data points to the north.

6.2.3.3
Martin Vaz–Trinidade (Hotspot: 21°S, 28.5°W)

The Martin Vaz–Trinidade trace (e.g., Morgan 1981) has only a few data points that come close to corresponding with the inferred locus (Figs. 6.42, 6.43). To the south of the trace there are a number of other data points in the age range of ~40 to ~90 Ma that could be interpreted in terms of another hotspot for which there is no recent evidence.

6.2.3.4
Bouvet (Hotspot: 49°S, 8°E)

The Bouvet hotspot trace in the South Atlantic Ocean (Figs. 6.44, 6.45) is poorly defined over much of its inferred oceanic extent. A number of age samples from South Africa can be interpreted as the landward continuation of the trace, although the spatial and temporal correspondence is incomplete. A more northerly location of the inferred hotspot would better fit the available age data.

6.2.3.5
Unnamed (17°S, 10°W)

A young seamount chain anchored by an inferred hotspot at 17°S, 10°W, shows an age progression with the same sense of younging as the calculated locus (Figs. 6.46, 6.47). Dates from a few older locations do not correspond with the older portion of the reconstructed locus, however. Lack of evidence for an intervening portion of the trace makes genetic association between the seamount chain and the older ages speculative at best.

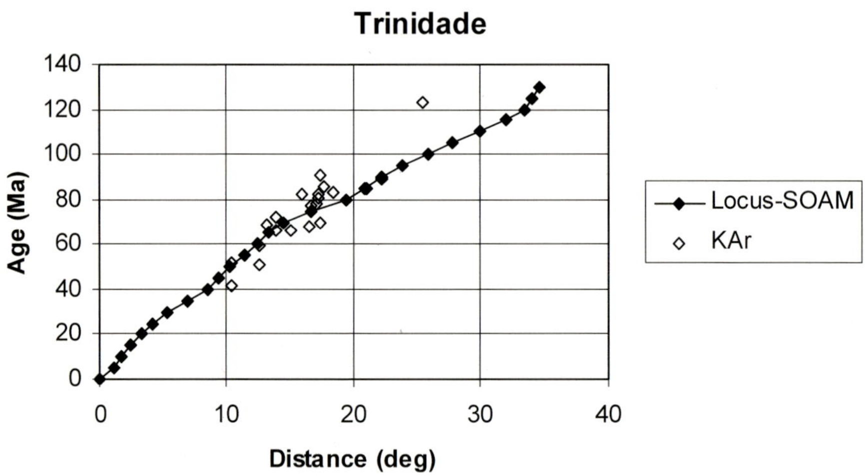

Fig. 6.42. Trinidade–Eastern Brazil: distance from inferred Trinidade hotspot versus age of dated samples from Eastern Brazil, and predicted locus from model of Müller et al. (1993) interpolated and extended via methods described in this volume.

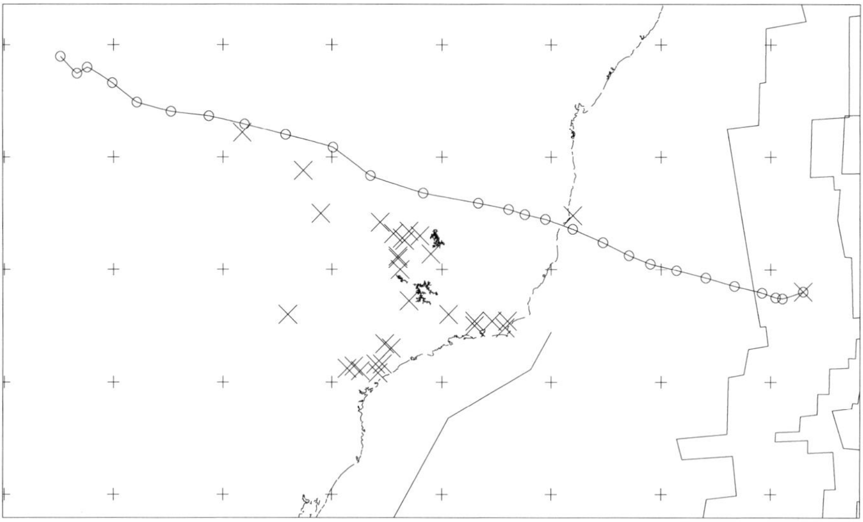

Fig. 6.43. Tristan Trinidade–Eastern Brazil: map of predicted loci and dated samples as in Fig. 6.3.

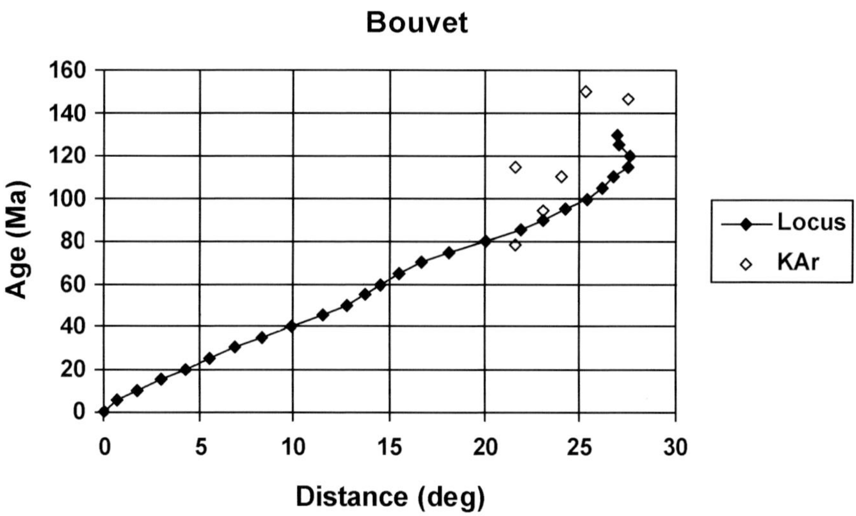

Fig. 6.44. Bouvet–South Africa: distance from inferred Bouvet hotspot versus age of dated samples from South Africa, and predicted locus from model of Müller et al. (1993) interpolated via methods described in this volume.

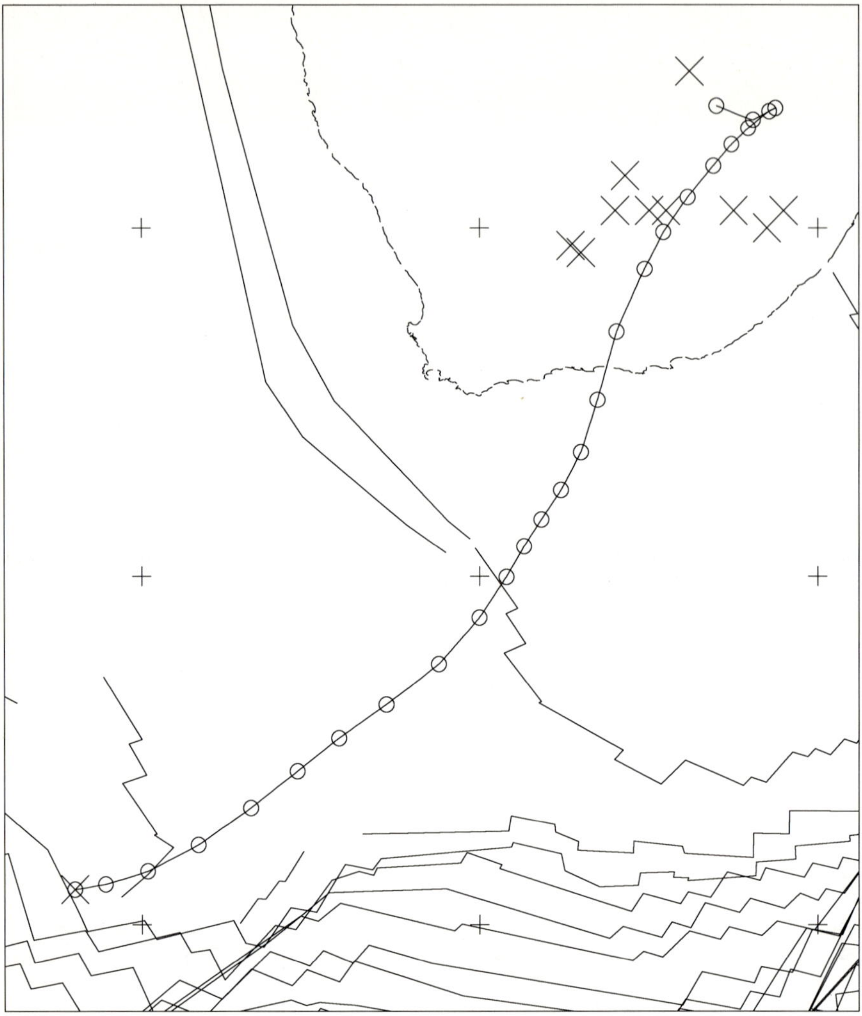

Fig. 6.45. Bouvet–South Africa: map of predicted loci and dated samples as in Fig. 6.3.

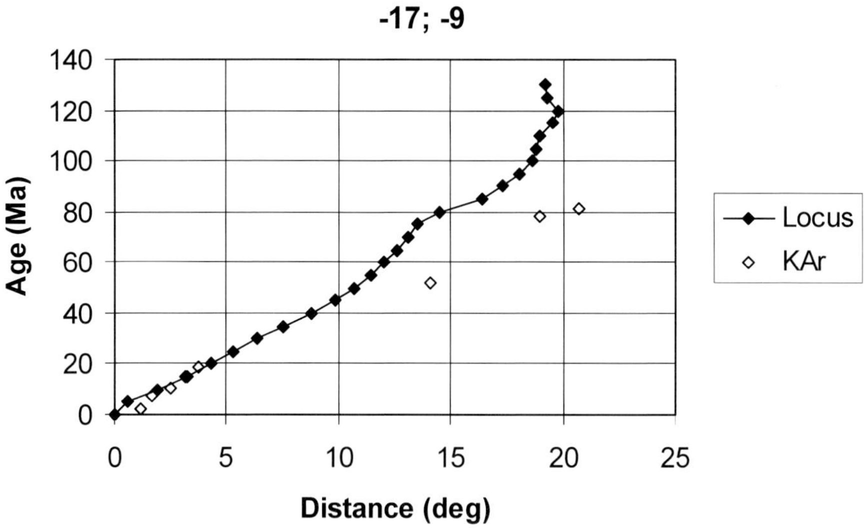

Fig. 6.46. Unnamed (17S, 10 W)–Africa: distance from unnnamed hotspot versus age of dated samples from South Atlantic and Central Africa, and predicted locus from model of Müller et al. (1993) interpolated via methods described in this volume.

6.2.3.6
Unnamed (41°S, 3.5°E)

A few aligned dates from Southwest Africa provide a suggestion of southwesterly younging (Figs. 6.48, 6.49). A predicted locus anchored on a young seamount a 41°S, 0.5°E shows poor correspondence with the few samples. A more easterly present location of a hotspot responsible for the intrusives would be required to fit the hotspot hypothesis.

6.2.4
Central North Atlantic Ocean

6.2.4.1
Great Meteor–New England Seamounts–White Mountains (Hotspot: 30°N, 28.5°W)

An important constraint on recent hotspot reconstruction models (Duncan and Richards 1991; Müller et al. 1993) is the New England seamount chain and White Mountains of New England. Müller et al. (1993) inferred a hotspot location beneath Great Meteor Seamount as the origin of the trace. Available age dates support the hotspot model, both in distance versus age and in map view (Figs. 6.50, 6.51). Baksi (1999) reinterpreted ages from the New England seamount chain and discarded most of the dates. However, the few dates he retained are still consistent with the hotspot hypothesis.

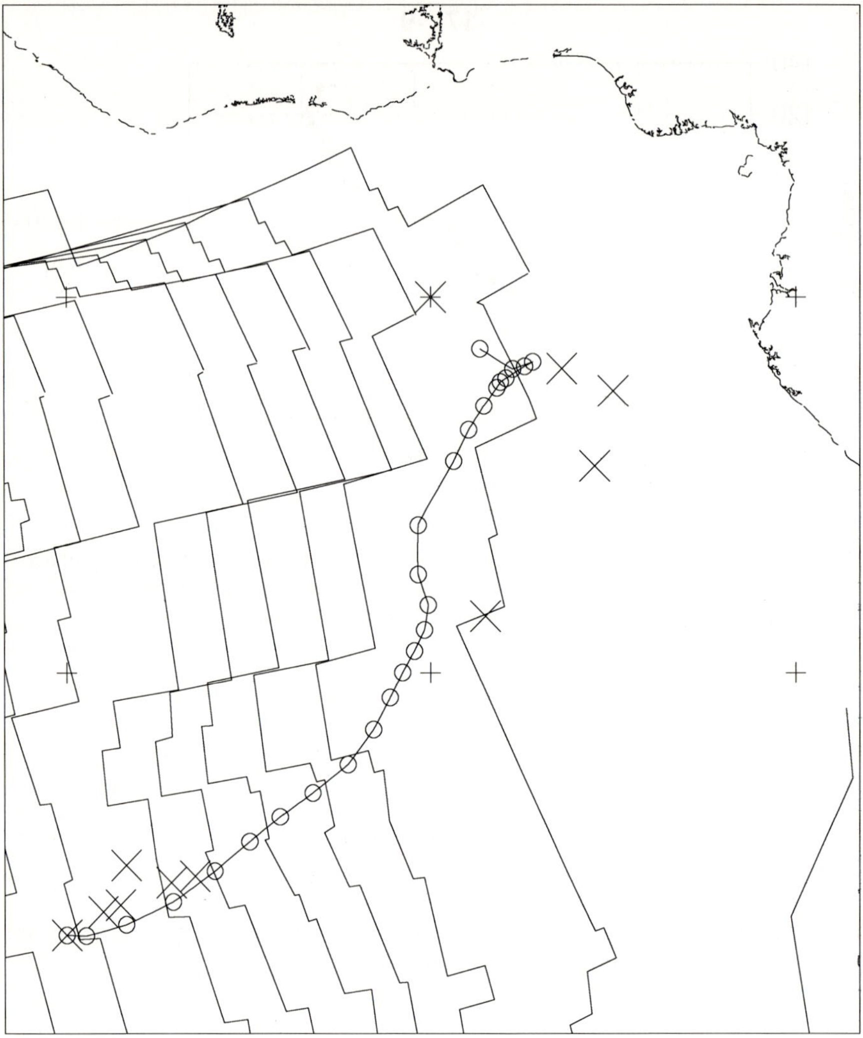

Fig. 6.47. Unnamed (17°S, 10°W)–Africa: map of predicted loci and dated samples as in Fig. 6.3.

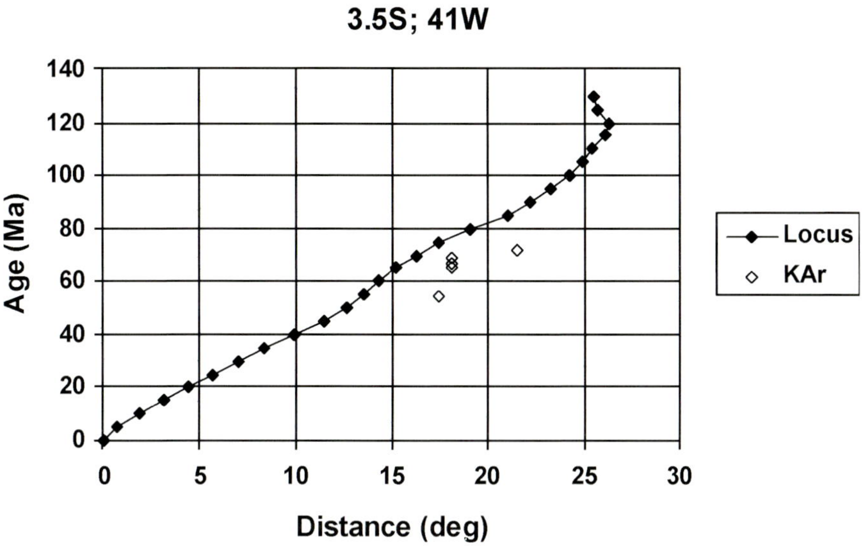

Fig. 6.48. Unnamed (0.5°S, 41°W)–South America: distance from unnnamed seamount versus age of dated samples from Brazil, and predicted locus from model of Müller et al. (1993) interpolated and extended via methods described in this volume.

6.2.4.2
Canary Islands (Hotspot: 32.6°N, 17.3°E)

The available data from the Canary Islands are poorly fit by predicted hotspot loci (Figs. 6.52, 6.53). The arcuate shape of the islands and associated seamounts is similar in to the shape of the locus (a horizontally flipped "C"). Some of the inferred data might represent a different origin than from the Canary hotspot in the Atlantic–Indian Ocean hotspot reference frame, however, as discussed further below (Chap. 11).

6.2.5
Central-Western North America

6.2.5.1
Yellowstone–Snake River Plain (Hotspot: 45°N, 110°W)

Morgan (1972a) included active volcanism in Yellowstone, western North America, as an active hotspot in his global hotspot model, with the Snake River Plain volcanics inferred to represent its trace (see also Suppe et al. 1975). Figures 6.54 and 6.55 provide support for the hotspot interpretation, although it is not clear as to whether the older continuation of the trace better fits the Hawaiian or Atlantic–Indian Ocean hotspot model (when restored through a global plate circuit, following the methods presented in this volume). The inferred locus in

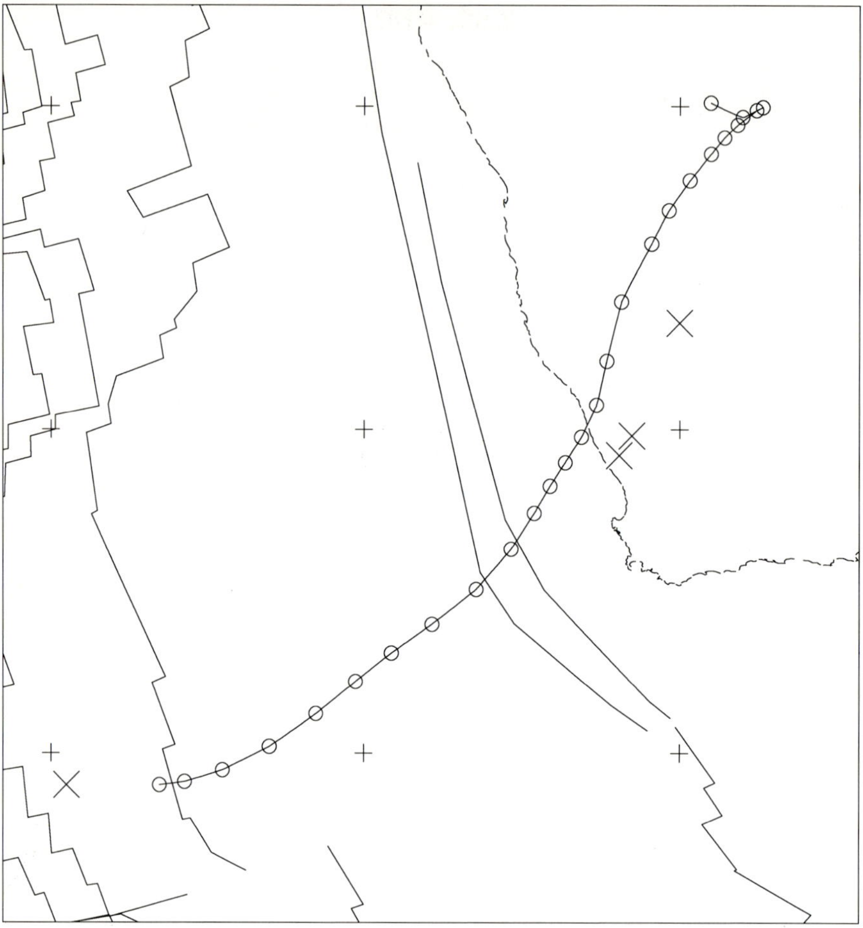

Fig. 6.49. Unnamed (41°S, 3.5°E)–South America: map of predicted loci and dated samples as in Fig. 6.3. Hotspot is displaced to east of seamount used in Fig. 6.48.

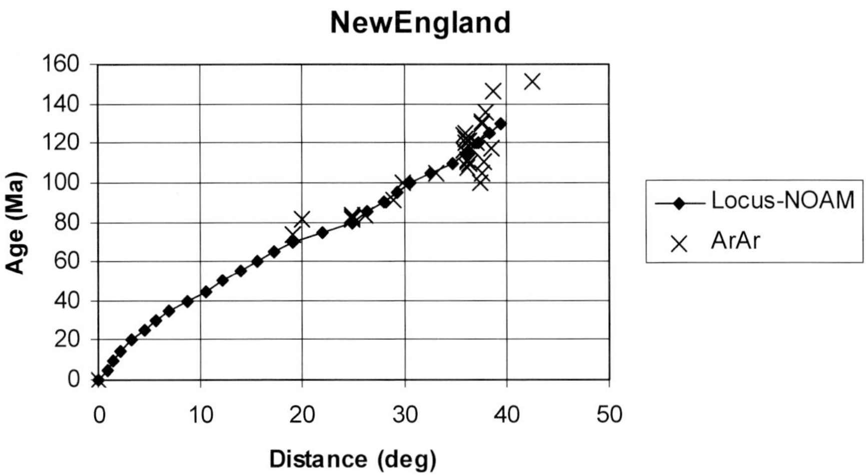

Fig. 6.50. Great Meteor -New England Seamounts–White Mountains: distance from Great Meteor hotspot versus age of dated samples from New England Seamounts and White Mountains, North African and North American plates, and predicted locus from model of Müller et al. (1993) interpolated and extended via methods described in this volume.

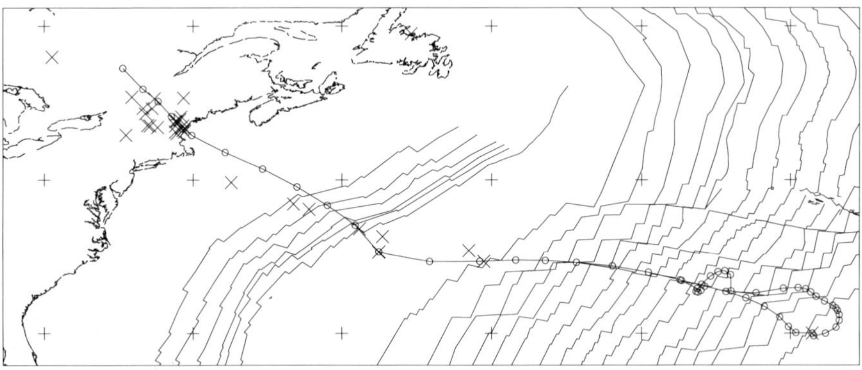

Fig. 6.51. Great Meteor–New England Seamounts–White Mountains: map of predicted locus (North American Plate) and dated samples as in Fig. 6.3.

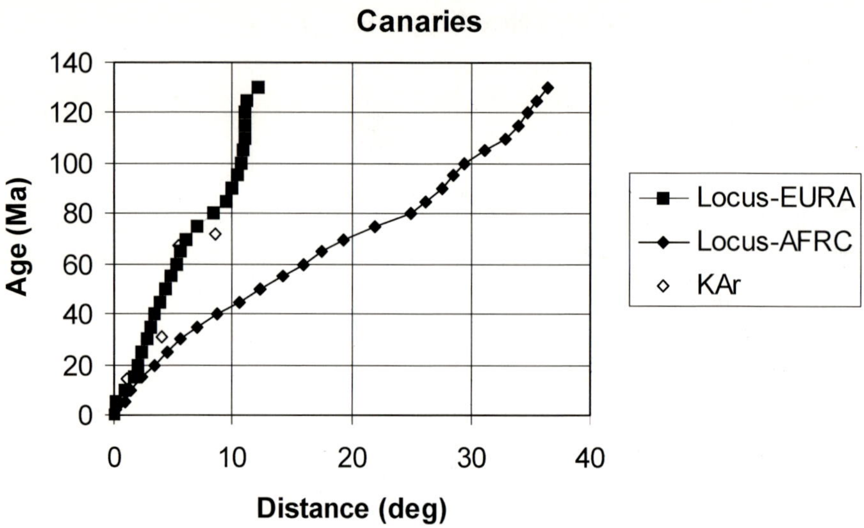

Fig. 6.52. Canary Islands: distance from Canary Islands hotspot versus age of dated samples from Canary Islands and nearby islands, seamounts, and southern Spain, North African and Eurasian plates, and predicted loci from model of Müller et al. (1993) interpolated and extended via methods described in this volume.

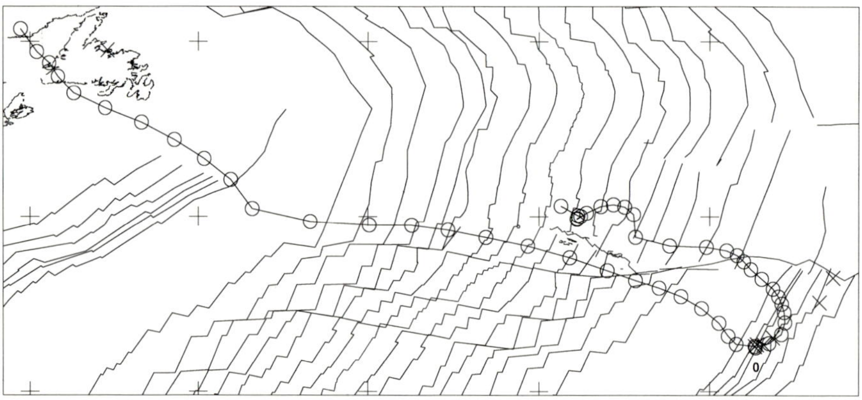

Fig. 6.53. Canary Islands: map of predicted loci (North African and Eurasian Plates) and dated samples as in Fig. 6.3.

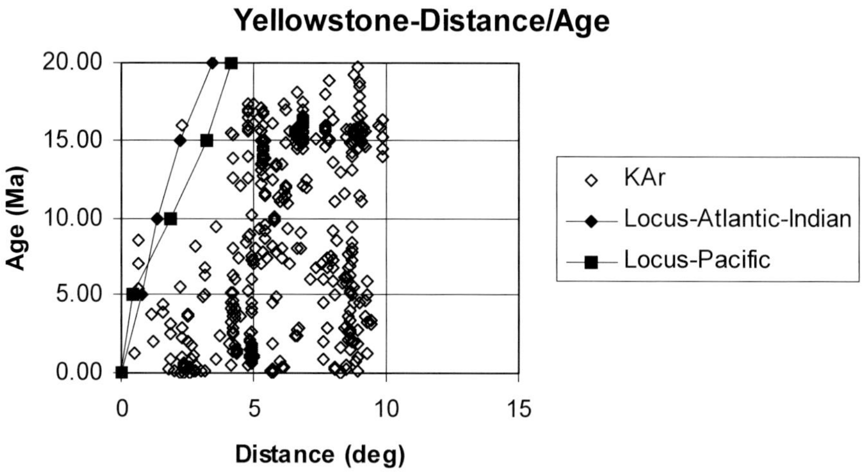

Fig. 6.54. Yellowstone–Snake River Plain: distance from Yellowstone hotspot versus age of dated samples Yellowstone and Snake River Plain, and predicted loci from models of Müller et al. (1993; Atlantic–Indian) and Raymond et al. (2000; Pacific) interpolated and extended via methods described in this volume.

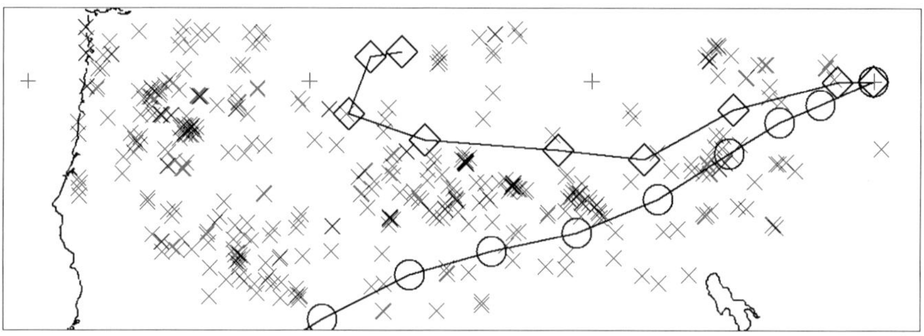

Fig. 6.55. Yellowstone–Snake River Plain: map of dated samples from Yellowstone and Snake River Plain, and predicted loci from models of Müller et al. (1993; Atlantic–Indian) and Raymond et al. (2000; Pacific) as in Fig. 6.3. Diamonds-Pacific (Hawaiian) locus, circles Atlantic-Indian (Tristan).

the Hawaiian hotspot reference frame appears to fit the trace over the past 20 my or so. The older extension of the trace is uncertain, however. Conceivably, the Yellowstone hotspot came into existence with the onset of Columbian River basalt implacement. Alternatively, it may have had an early history beneath the now subducted Farallon or Kula plates, with remnants preserved in the Oregon and Washington Coast Ranges, as many authors have suggested.

6.2.5.2
Raton (Hotspot: 37.5°N, 103°W)

Like Yellowstone, the Raton hotspot is inferred to underlie northern New Mexico. Available isotopic ages from northwestern New Mexico and northeastern Arizona suggest that progressive inception of igneous activity is recorded along the inferred trace (Figs. 6.56, 6.57), with continuation almost to the present. Other sources of igneous activity, especially along the Rio Grande rift, obscure the trend of the trace in the age–distance plot, but when identified, the trend becomes clearer (Fig. 6.54). The predicted locus based on the Atlantic–Indian Ocean hotspot model better fits the Raton trace than does the Pacific-hotspot model.

6.2.6
Northernmost Atlantic and Western Europe

6.2.6.1
Iceland (Hotspot: 64°N, 20°W)

The prominence of Iceland, accompanied by the submarine Iceland–Faeroes ridge and early Tertiary igneous activity in East Greenland and the northern British Isles, prompted its interpretation as a hotspot produced by a mantle plume (e.g., Morgan 1972a). Only a few age constraints are available for the inferred Iceland trace (Figs. 6.58, 6.59). However, as initially recognized by Molnar and Atwater (1973) with a different formulation, predicted loci based on Pacific hotspots do not correspond with the inferred Iceland traces on the Eurasian and Greenland–North American plates, nor do loci based on the Atlantic–Indian Ocean hotspots and their inferred traces.

Because of the strong internal consistency of the Pacific and the Atlantic–Indian hotspot systems, it is tempting to infer that the Iceland trace originated within a third reference frame. Unfortunately, there are few other apparent hotspots with traces that demonstrate progressive volcanic inception beneath Eurasia and northeastern North America to provide adequate constraints for a distinct kinematic and reconstruction model.

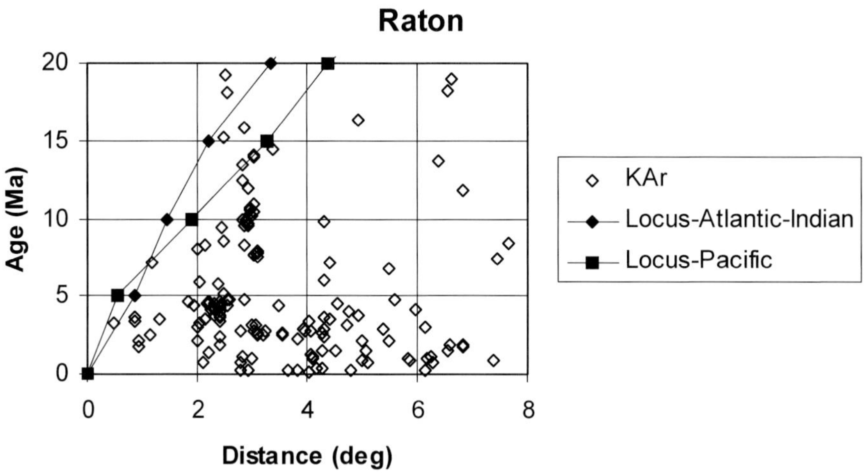

Fig. 6.56. Raton: distance from Raton hotspot versus age of dated samples from northern New Mexico and Arizona and predicted loci from models of Müller et al. (1993, Atlantic–Indian) and Raymond et al. (2000, Pacific) interpolated and extended via methods described in this volume.

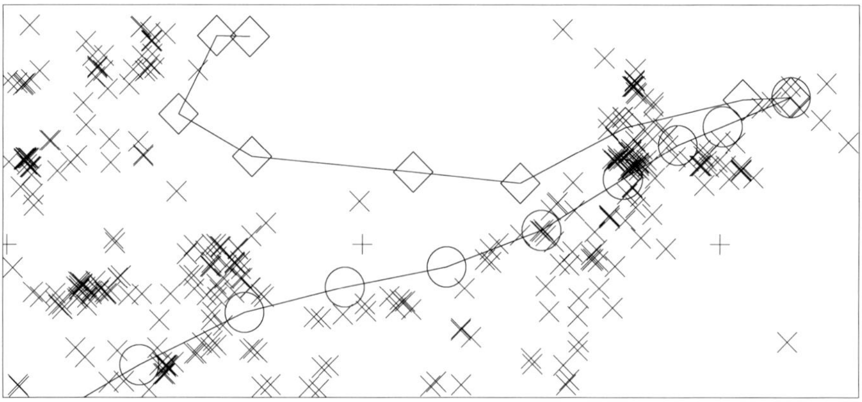

Fig. 6.57. Raton: map of dated samples from northern New Mexico and Arizona and predicted loci from models of Müller et al. (1993; Atlantic–Indian) and Raymond et al. (2000; Pacific) as in Fig. 6.3. Diamonds-Pacific (Hawaiian) locus, circles Atlantic-Indian (Tristan).

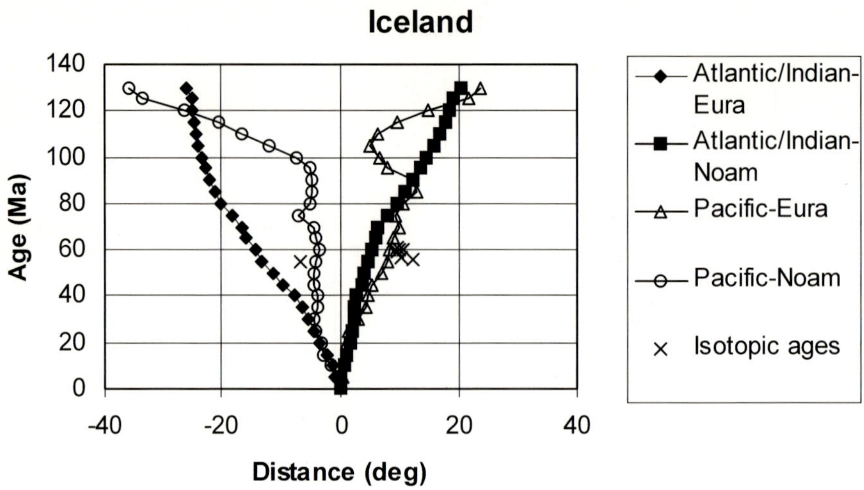

Fig. 6.58. Iceland: distance from center of Iceland to data samples from Greenland and British Isles, together with loci of Greenland and Eurasia relative to hotspots from models of Müller et al. (1993; Atlantic–Indian) and Raymond et al. (2000; Pacific).

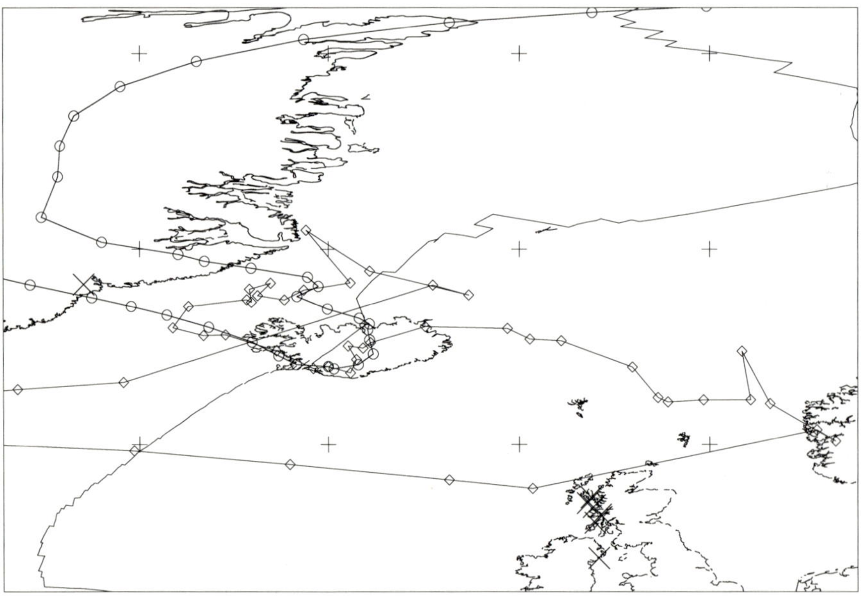

Fig. 6.59. Iceland: map of Iceland, dated sample locations and loci of Greenland and Eurasia relative to hotspots from models of Müller et al. (1993; circles, Atlantic–Indian) and Raymond et al. (2000; diamonds, Pacific).

6.2.7
Eastern Australia–Tasman Sea

6.2.7.1
East Australian Highlands

The volcanism of the East Australia Highlands provides an intriguing pattern, apparently unique among intraplate volcanic provinces (Figs. 6.60, 6.61). Basaltic volcanism in the province began in the Late Cretaceous, but there is no obvious indication of progressive inception of activity. Volcanism continued intermittently throughout much of the length of the province, until about 30 Ma, when a more alkaline "central volcano" magmatism (Wellman and McDougall 1974) began at the north end of the province and progressively migrated to the south. The temporary onset of central volcanism was accompanied by termination of all volcanism, except for resumption of activity in northern Queensland in the late Cenozoic, some 25 my after cessation.

Pilger (1982) interpreted the age pattern as a manifestation of stress-controlled processes. Sutherland (1981) proposed that the central volcanics represented a hot line that originated beneath the Early Cenozoic Coral Sea spreading center, although he did not attempt to explain the earlier volcanism and its cessation. Both noted that the central volcanics pattern cannot be explained in terms of a single hotspot; rather ,several aligned central volcanic centers seem to exist.

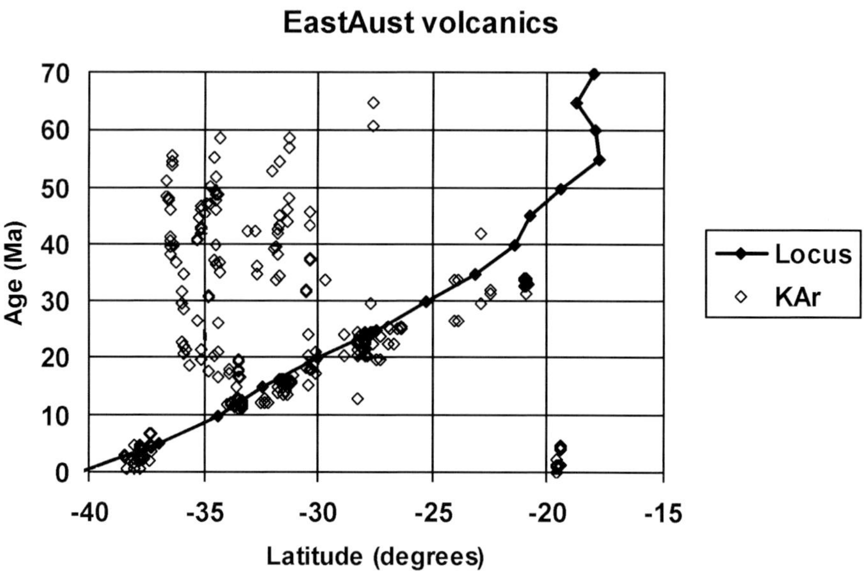

Fig. 6.60. Eastern Australia: Latitude of dated samples versus K/Ar age and predicted locus from model of Müller et al. (1993) interpolated and extended via methods described in this volume.

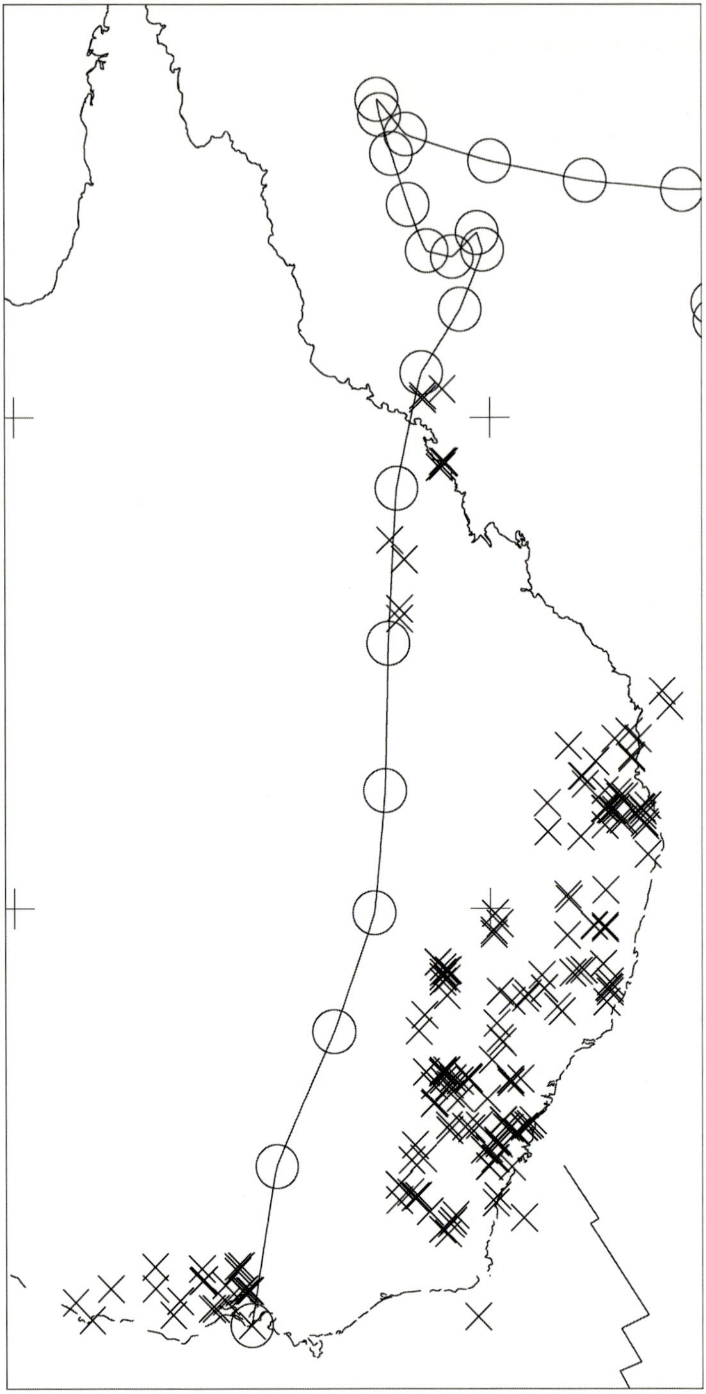

Fig. 6.61. Eastern Australia: map of predicted locus and dated samples as in Fig. 6.3.

6.2.7.2
Tasman Seamounts (40.4°S, 155°E)

The most prominent seamount chain of the Tasman Sea provides important constraints on the Late Cenozoic part of hotspot-reconstruction models for the Atlantic and Indian Ocean (e.g., Müller et al. 1993). Age–distance plots correspond with predicted loci, as expected (Figs. 6.62, 6.63). However, the locus based on Pacific hotspots is not very different. While Baksi (1999) was reluctant to accept the analytical results, the model does correspond moderately well with at least the trend of the seamounts.

Predicted loci based on either the Pacific or Atlantic–Indian Ocean hotspot reference frame show a fair correspondence with the observed patterns (Figs. 6.62, 6.63). The Atlantic–Indian Ocean locus appears to show a slightly better fit.

6.2.8
Africa

6.2.8.1
East African Rift

In most hotspot models, the magmatism associated with the East African Rift has been largely ignored. Turcotte and Oxburgh (1978) compiled available isotopic ages and noted that there is an apparent progressive inception of mag-

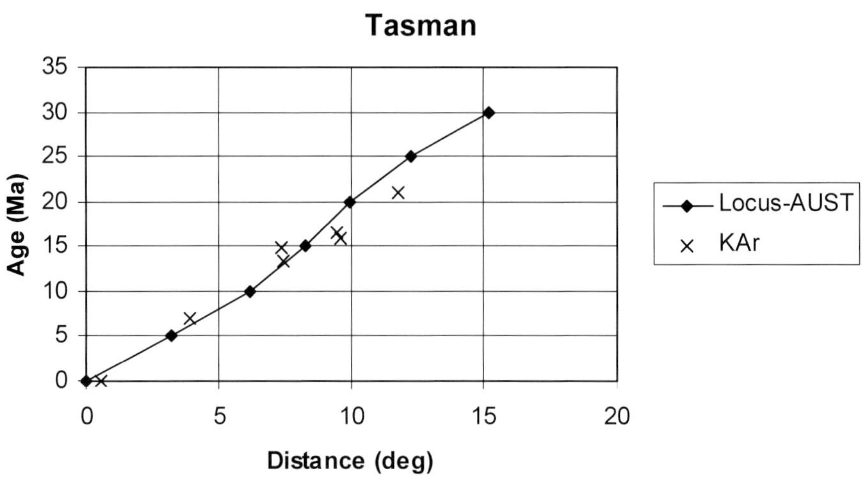

Fig. 6.62. Tasman Seamounts: distance from inferred Tasman hotspot versus age of dated samples from Tasman Seamounts, and predicted loci from model of Müller et al. (1993) interpolated and extended via methods described in this volume.

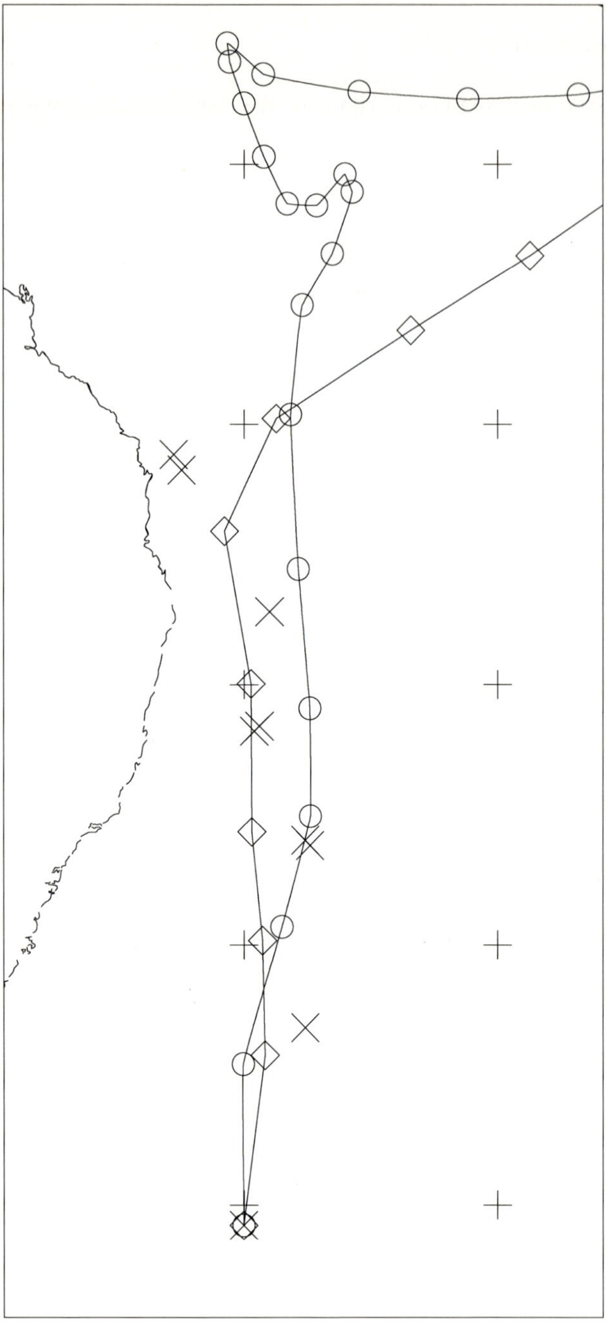

Fig. 6.63. Tasman Seamounts: map of predicted loci and dated samples as in Fig. 6.3. Diamonds Pacific (Hawaiian) locus, circles Atlantic-Indian (Tristan).

matic activity from north to south. Adding to the data they compiled, and augmented by predicted hotspot loci, plots of latitude and longitude versus age together with a map of samples and locus (Figs. 6.64, 6.66) strongly imply that the motion of Africa in the hotspot frame has controlled southward progression of activity since 60 Ma, with an apparent increase in intensity of magmatism occurring about 30 Ma.

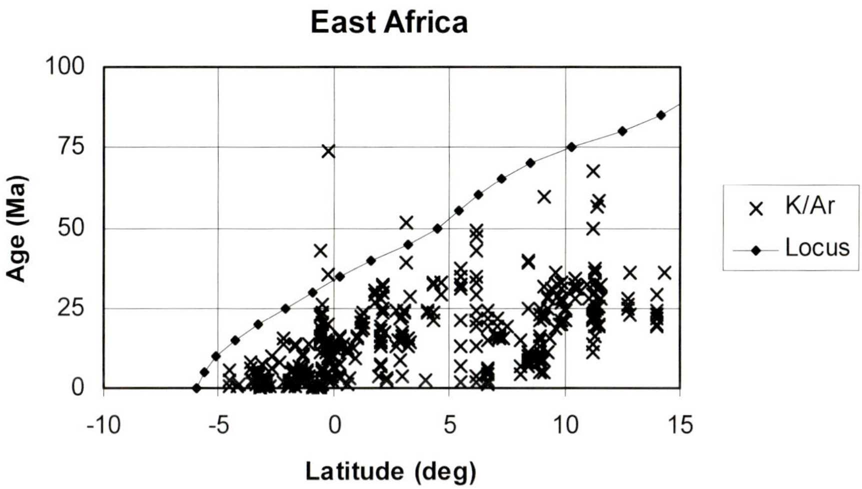

Fig. 6.64. East African Rift: latitude versus age of isotopically dated samples and locus in the Atlantic–Indian Ocean hotspot reference frame.

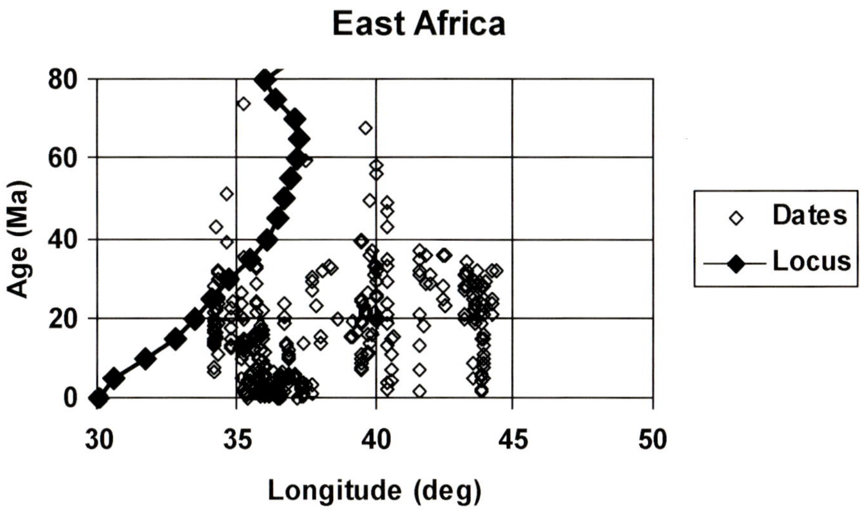

Fig. 6.65. East African Rift: longitude versus age of isotopically dated samples and locus in the Atlantic–Indian Ocean hotspot reference frame.

CHAPTER 7

Contemporary Stress Fields

7.1 Aside

Check out the first one or two chapters of a good structural geology or seismology textbook and you'll find the derivation of the elastic stress tensor. Read further and, while you might find some application of the total tensor; more typically the tensor is quickly reduced to principal stresses. Then, the principal stresses get reduced even more to their components referenced to local earth coordinates.

It's easy to infer that the reduction of the stress tensor is largely a result of the desire for simplicity. It's much easier to visualize two principal horizontal components and one vertical component as a vector than the nine components of the tensor, even if three of the tensor components are redundant. However, the simplicity of analysis is mirrored in observation, especially in the shallow crust of the Earth. One of the principal stresses tends to orient itself normal to the Earth's surface, with the other two in the locally horizontal (surface-tangent) plane, an observation that theory also would support. Beginning in the 1970s, however, it became clear that the horizontal principal components also tend to be oriented in regionally consistent directions. The simplicity of theory finds expression in contemporary natural observation, where even less complexity is found.

7.2 Compilations

The continuously growing database of contemporary stress field measurements within the continents (Fig. 7.1) demonstrates a remarkable consistency in orientation within large regions of the continents (e.g., Zoback et al. 1989). The consistency is particularly distinctive in continental regions of moderate elevation and which lack significant tectonism. Of course, tectonism is itself indicative of variations in lithospheric stress; what has been surprising is the persistence of constant orientation stress fields within "quiescent" regions of the continents.

7.3 North America: Contemporary Stresses

In North America (Fig. 7.2), the continent with the most measurements at present, a northeast-southwest directed, maximum horizontal principal component

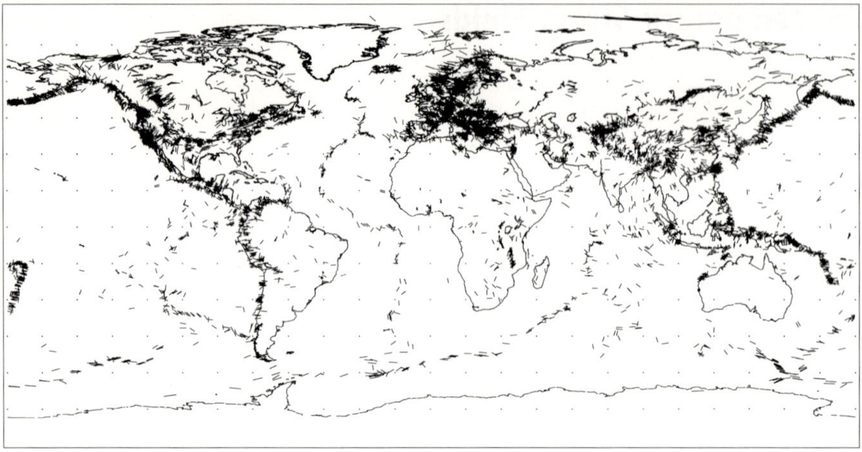

Fig. 7.1. Global set of contemporary maximum horizontal principal compressive (MHPC) stress orientations (http://www-wsm.physik.uni-karlsruhe.de/pub/introduction/introduction.html). Long axes of bars are parallel with the MHPC stress.

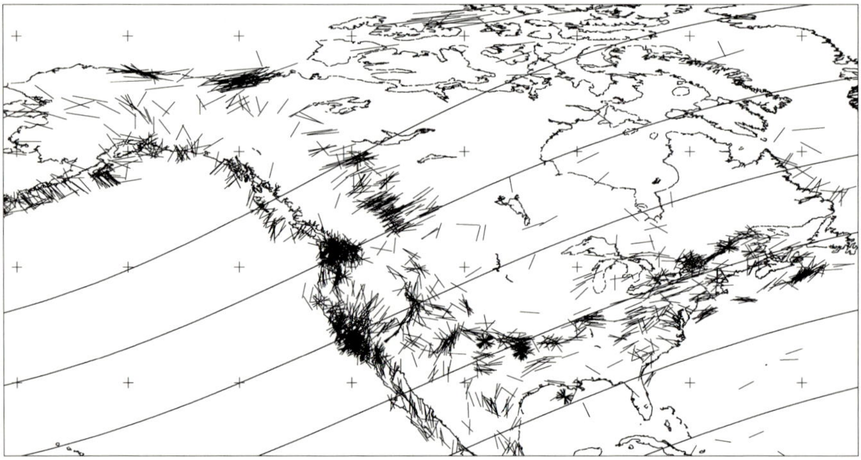

Fig. 7.2. Contemporary MHPC stress orientations from North America. See Fig. 7.1 for explanation.

(MHPC or σ) of stress characterizes the northeastern United States and the mid-continent from the upper Gulf Coastal Plain through the Midwest into the Canadian Great Plains. Only on the margins of the still tectonically active Cordillera and areas affected by deglaciation of the Canadian Shield are distinctly different stress fields observed. The most significant data type in North America is borehole breakouts; holes drilled for oil exploration and production typically demonstrate an elliptical shape when logged, with the short axis of the ellipse inferred to correspond with the MHPC stress orientation. Focal mechanisms of intraplate earthquakes, particularly along the mid-continent seismic zone, generally demonstrate the same MHPC stress orientation as borehole breakouts. Similarly, quarry breakouts, particularly in the northeastern United States, also show a northeast-southwest MHPC stress azimuth. The orientation of the stress field corresponds with the direction of motion of North America in the global hotspot reference frame (Fig. 7.2), as numerous workers have observed (e.g., Nakamura 1977; Zoback and Zoback 1980).

7.4 Western Europe: Contemporary Stresses

Western Europe (Fig. 7.3) possesses a dominant MHPC stress orientation of north-northwest–south-southeast to the north of the Alps and away from the glacially rebounding Fenno-Scandian Shield. This orientation corresponds with Eurasian plate motion in one global hotspot model (Minster et al. 1974), but subsequent models (Chase 1978; Minster and Jordan 1978; Gripp and Gordon 1991) are not consistent with the MHPC stress trend in the region. The slow rate of motion of Eurasia in all of the hotspot models introduces a large uncertainty in the

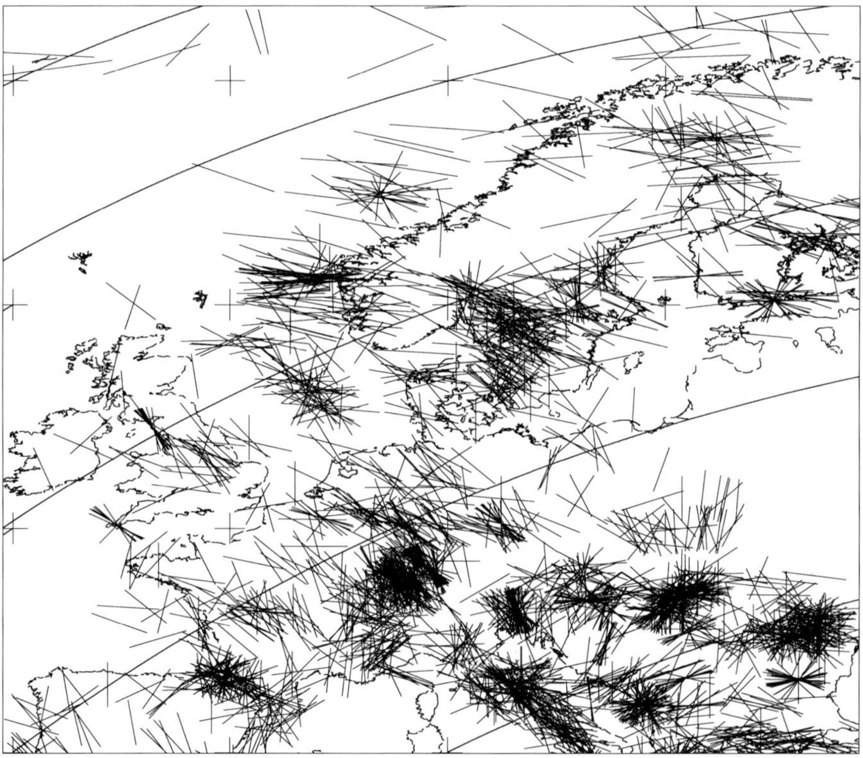

Fig. 7.3. Contemporary MHPC stress orientations from Western Europe. See Fig. 7.1 for explanation.

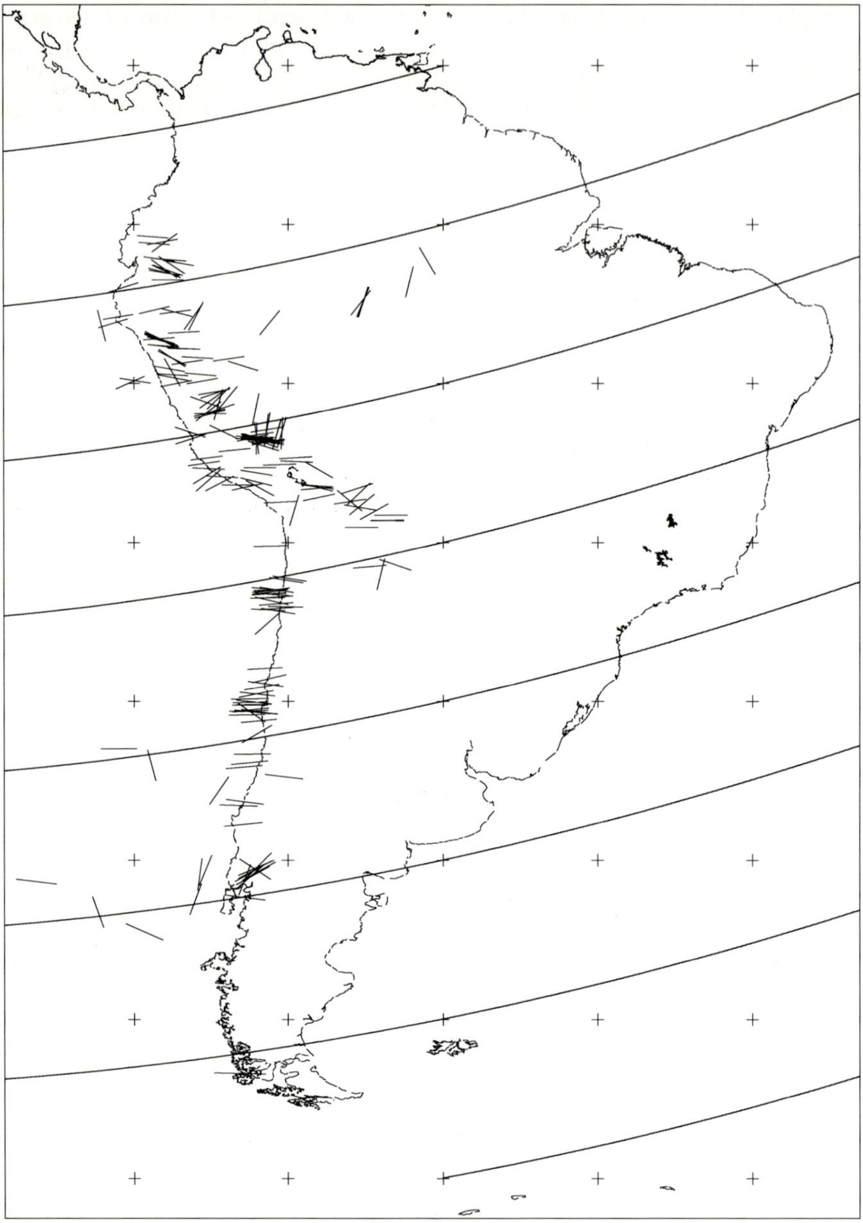

Fig. 7.4. Contemporary MHPC stress orientations from South America. See Fig. 7.1 for explanation.

direction of plate motion and also doubts in the expectation of any correlation at all.

7.5 South America: Contemporary Stresses

South America (Fig. 7.4) shows a greater degree of correspondence of contemporary stresses with the hotspot model, except within the paleo-Amazon Rift and the northern Andes. Almost all of the stress indicators are based on earthquake focal mechanisms.

7.6 Africa: Contemporary Stresses

Outside of North America and Western Europe, most available intracontinental stress measurements have been obtained from seismic focal mechanisms. There are limited correspondences of stress orientations with plate motions in the hotspot frame in Africa (Fig. 7.5), with substantial dispersion. Virtually all of the stress indicators are from earthquake focal mechanisms. Most intracontinental indicators are associated with the East African Rift and can be interpreted in terms of a developing plate boundary (e.g., Chu and Gordon 1999).

7.7 Australia: Contemporary Stresses

Australia (Fig. 7.6) has a limited set of observations, principally concentrated along the northwestern continental margin. Very little correspondence is observed, but the region of expected correspondence – the continental interior – is bereft of observations.

7.8 Other Areas

The remaining continental areas do not show an obvious correspondence with the hotspot model (Fig. 7.1). Eastern Asia involves complex intraplate deformation associated with the collision of India with Asia (e.g. Molnar and Tapponier 1977). There are few measurements available from within stable India or Antarctica (Fig. 7.1).

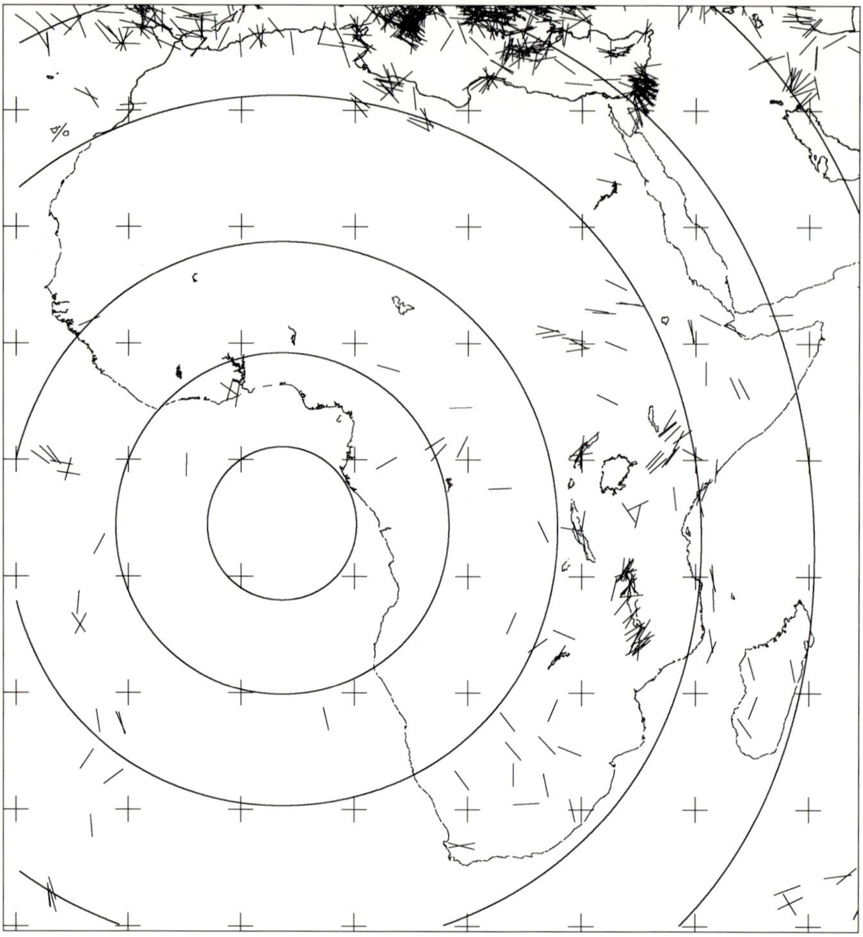

Fig. 7.5. Contemporary MHPC stress orientations from Africa. See Fig. 7.1 for explanation.

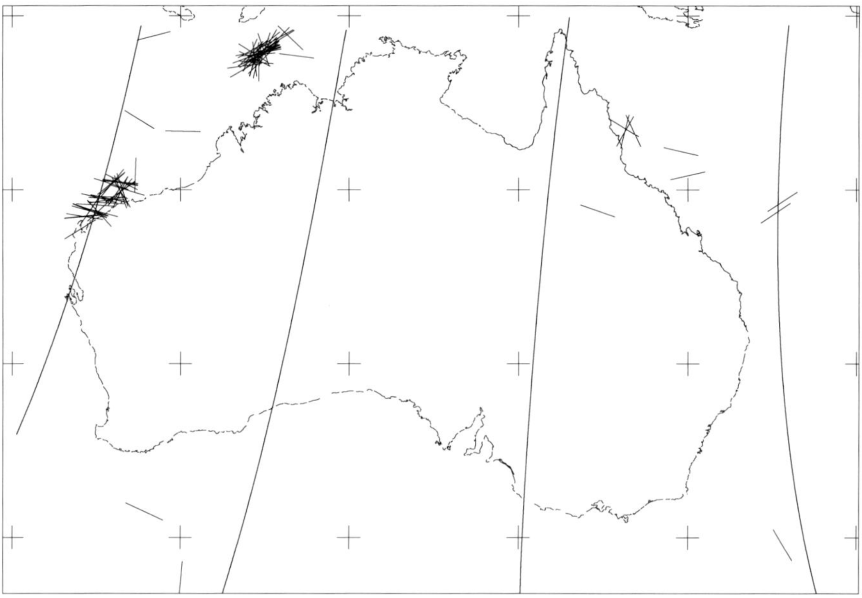

Fig. 7.6. Contemporary MHPC stress orientations from Australia. See Fig. 7.1 caption for explanation.

CHAPTER 8

Observed Paleostress Orientations and Predicted Plate-Hotspot Motions

8.1 Aside

As Plate Tectonics matured, a number of workers began to consider philosophical implications of the science. Allan Cox (1973), in particular, found Thomas Kuhn's "Structure of Scientific Revolutions", published in the same year as Harry Hess's "History of ocean basins – an essay in geopoetry", to be directly applicable to the experience of earth scientists in the 1960s. In a sense, Kuhn's paradigm hypothesis found a kind of confirmation in this new theory of the Earth. Sometimes earth scientists experience an inferiority complex in relation to the more "basic" sciences – physics and chemistry. Now, geology and geophysics had a revolution of their own to parallel Copernican heliocentricity, Newtonian physics, Darwinian natural selection, Einsteinian relativity, quantum mechanics, and the double helix.

Of course, the origin of geology as a predictive science had its own primal revolution with James Hutton's Principle of Uniformitarianism. In its modern sense, the basic assumption of historical geology is that the laws of physics and chemistry operable at present were also operable in the past. Further application of this principle to the interpretation of sedimentary rock sequences by Charles Lyell led quickly to the realization of the immense age of the earth, paving the way for Lyell's friend, Charles Darwin.

Ironically, at least in the context of those who initially opposed Plate Tectonics, the new theory of the Earth turns out to be eminently uniformitarian. Contemporary plate tectonics can be extended into the geologic past, at least into the Early Mesozoic, and probably into the Precambrian. Geohistorical observations of oceanic magnetic lineations, inactive fractures zones, belts of igneous rocks, and sedimentary basins form a coherent connection with present day seismicity, volcanism, physiography, and heat flow.

To the uniform coherence of plate tectonics from present into the past, can we add the coherence of contemporary intraplate stresses when we examine paleostresses? Do paleostresses correspond with plate motions in the hotspot reference frame, or is the present-day correlation a contemporary accident?

8.2 Empiricism

8.2.1 Rationale

The empirical correspondence between contemporary maximum horizontal principal component (MHPC or σ) stress measurements from the continents and the direction of plate motion in the hotspot (plate-mesosphere) frame of reference (Nakamura 1977; Zoback and Zoback 1980; Zoback et al. 1989), if extrapolated backwards in time, should result in parallelism of paleostress indicators (dated dikes, faults, aligned volcanoes) with predicted loci of plate motion from the hotspot model within the stable, moderate elevation portion of the continents. From compilations of such paleostress indicators (Appendix 3), regional maps of the inferred principal maximum horizontal stress directions have been constructed, together with synthetic proper motion loci, so as to compare likely directions of plate motion over the past 130 my with the stress history of each plate. The number of measured stress indicators, the complex stress histories of the plates, and frequent apparent changes in plate motion require that portrayal of the stress orientations and predicted plate-hotspot motions are limited to short time windows; 10-my windows are used here). Plate loci in the hotspot reference frame are small circles around instantaneous poles of motion calculated using the methods presented in Chap. 3.

8.2.2 Information Sources

The principal source of information concerning the paleostress field within a plate is the orientation of isotopically dated igneous dikes. In some regions, particularly Western Europe, considerable effort has been devoted to the study of minor faults and vertical stylolites as stress indicators, with age constraints determined from stratigraphic and cross-cutting relationships and biostratigraphy. In the Basin Range province and Rocky Mountains of North America, significant constraints on deformational age are also available for Late Mesozoic and younger events. Over the past two and one-half decades, abundant study and isotopic dating of igneous rocks, particularly in North America, has been accomplished. Fortunately, in many cases, the structural setting of dated dikes (including their orientation) has been published together with the isotopic dates, providing a valuable data base for this type of study (unfortunately, a larger number of isotopic dates on dikes has been published without a description of the orientation of the dikes). For North America, a few regional studies and compilations of Cenozoic and/or Mesozoic dike orientations and other paleostress indicators have been published for selected areas, including Arizona and southwestern New Mexico (Rehrig and Heidrick 1976; Heidrick and Titley 1982), New Mexico (Aldrich et al. 1986), the Late Cenozoic of the western United States (Zoback et al. 1981; Best 1988), Trans-Pecos Texas (Price and Henry 1984), New England and the Appalachians (McHone 1978), and Quebec and New England (Eby 1985).

For Europe, Letouzey (1986) and Bergerat (1987) have made extensive studies of the paleostress history, largely relying on detailed fault and vertical stylolite analyses in Cretaceous and younger sedimentary rocks. Pilger (1982) compiled published Cenozoic paleostress indicators for eastern Australia.

To add to these studies, an extensive compilation of isotopically dated and oriented dikes was made for North America, western Europe, South America, Africa, India, and Australia, supplemented by examination of secondary references when dike orientations were not described. In the compilation, dikes from radially symmetric swarms were explicitly excluded, as were measurements on those dikes that had not been dated. While compiled dike orientations included those from within areas of synchronous or subsequent deformation, interpretations of the paleostress orientations are focused on those indicators that appear to penetrate relatively undeformed cratonic crust. Similarly, data adjacent to plate boundaries active at the time the stress indicator was formed are discounted as indicative of intraplate stress.

In this chapter, Figs. 8.1–8.55 are organized by continent; both contemporary stresses and paleostresses (with line segments centered on the observation point and oriented in the direction of the maximum horizontal principal compressive stress) are illustrated, as are predicted proper motions (as small circle arcs) in 5-my increments (e.g., 5 Ma≤age≥10 Ma). The paleostresses are listed in Appendix V.

8.3 North America: Paleostresses

For North America, the first illustration includes contemporary intraplate stresses (from the Worldwide Stress Project) along with young paleostresses (less than or equal to 5 Ma). The remaining illustrations are in 5-my increments to 130 Ma.

The collections of paleostress orientations presented in Figs. 8.1–8.23 are interpreted as indicating remarkable uniformity in orientation of the interior North American MPHC stress azimuth within most of the selected time periods. (The stress orientations which do not correspond well with the kinematic model largely come from what is inferred to be non-stable North America, such as west of the Cretaceous Sevier belt or within the Cenozoic Basin Range province; that is, these measurements are not within the stable North American plate interior.) For convenience, even larger increments of time can be considered, in which the general orientation of the paleostresses varies only about ±20°. From 120 to 80 Ma, the predominant orientation is inferred to be west northwest. From 80 to 65 Ma, the orientation is approximately westerly. From 65 to 35 Ma, the predominant orientation is southwesterly. An essentially westerly orientation is inferred for 35 to 10 Ma; and from 10 Ma until the present, the orientation of MPHC stress azimuth has been southwesterly.

8.3.1 North America: 0–5 Ma

MHPC stress (σ_1) indicators correspond well with calculated flowlines in the hotspot frame, calculated using the methods outlined in this volume (Fig. 8.1). The high curvature flowline (in this map projection) represents 5 Ma; the other is the contemporary flowline. Deviations from the calculated flowlines occur on the margins of that portion of the Canadian Shield last affected by the glacial ice cap and in the tectonically active Basin Range Province and Rio Grande Rift.

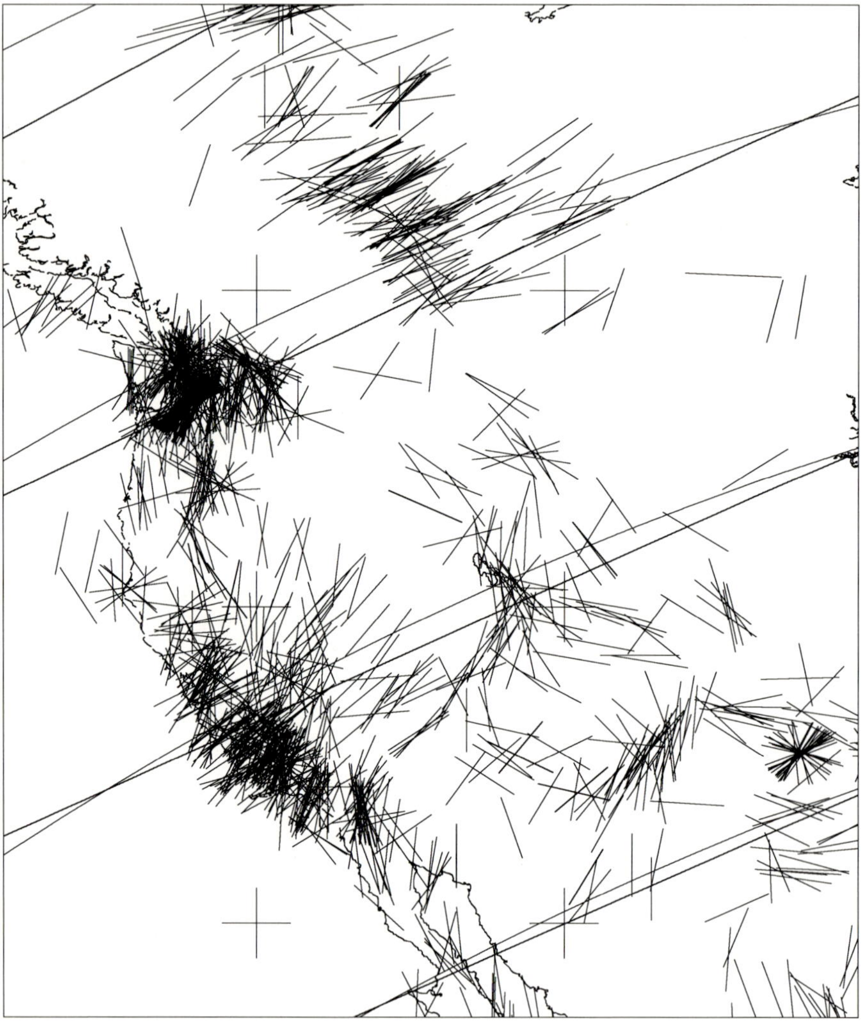

Fig. 8.1. North America: paleostresses and calculated flowlines in hotspot reference frame, 0–5 Ma. Bold flowline 0 Ma.

8.3.2
North America: 5–10 Ma

Few observations of σ_1 are apparent outside the Basin Range Province and the Rio Grande Rift for the interval from 5 to 10 Ma (Fig. 8.2). Since most observations are from dikes, some areas on the margins of stable western North America that exhibit volcanism of this age range have not been deeply eroded enough to expose feeder dikes.

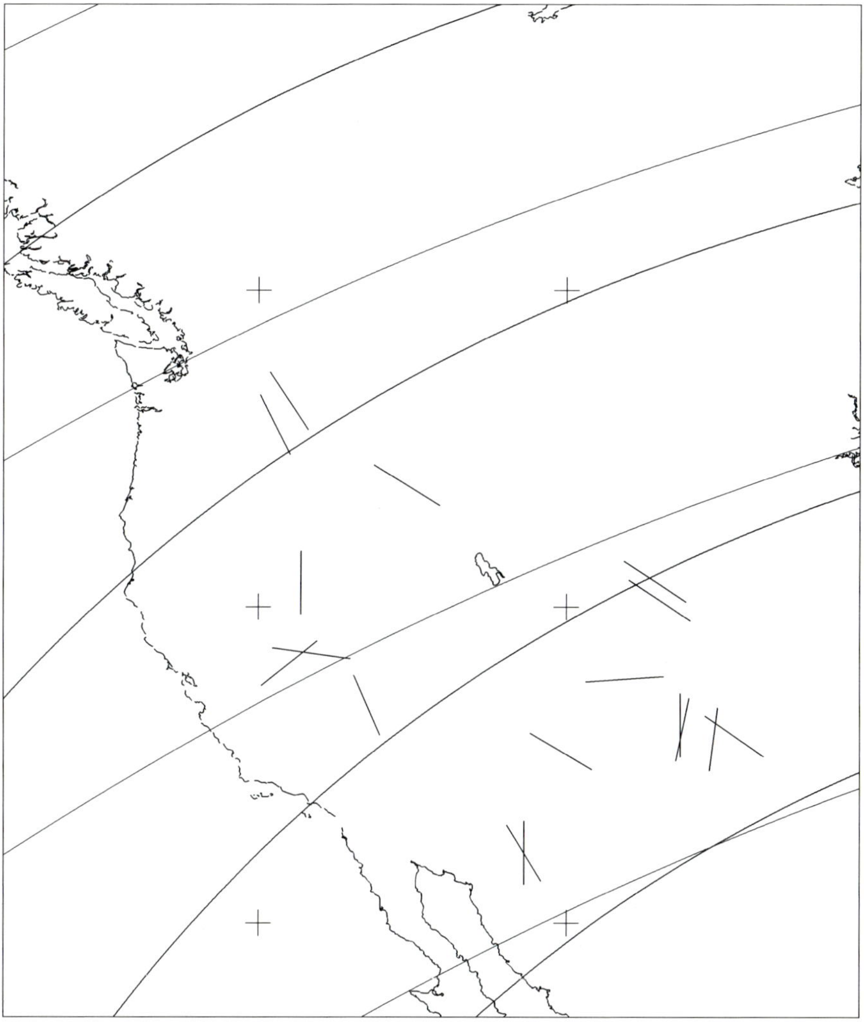

Fig. 8.2. North America: paleostresses and calculated flowlines in hotspot reference frame, 5–10 Ma. Bold flowline 10 Ma.

8.3.3 North America: 10–15 Ma

As with the 5–10 Ma period, few σ_1 measurements are available for 10–15 Ma from within stable North America (Fig. 8.3).

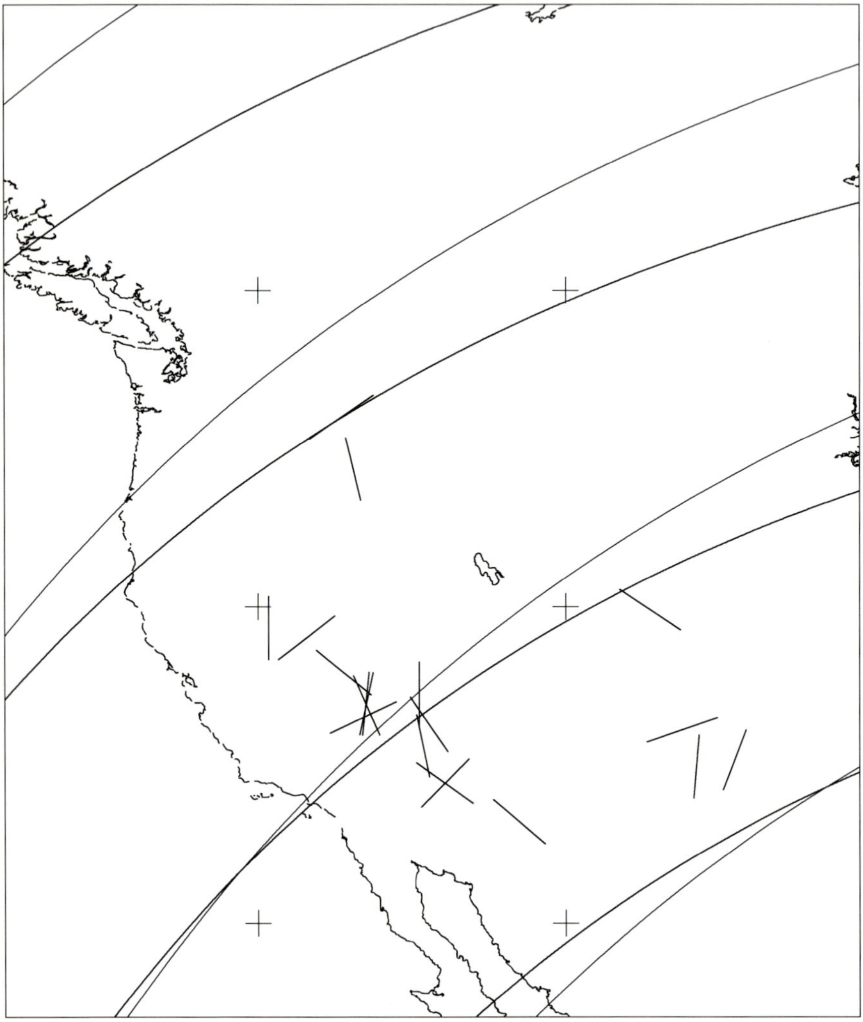

Fig. 8.3. North America: paleostresses and calculated flowlines in hotspot reference frame, 10–15 Ma. Bold flowline 10 Ma.

8.3.4
North America: 15–20 Ma

All paleostress observations between 15 and 20 Ma are within the active Cordillera (Fig. 8.4). A few east-west measurements of σ_1 within the Cordillera correspond with calculated flowlines.

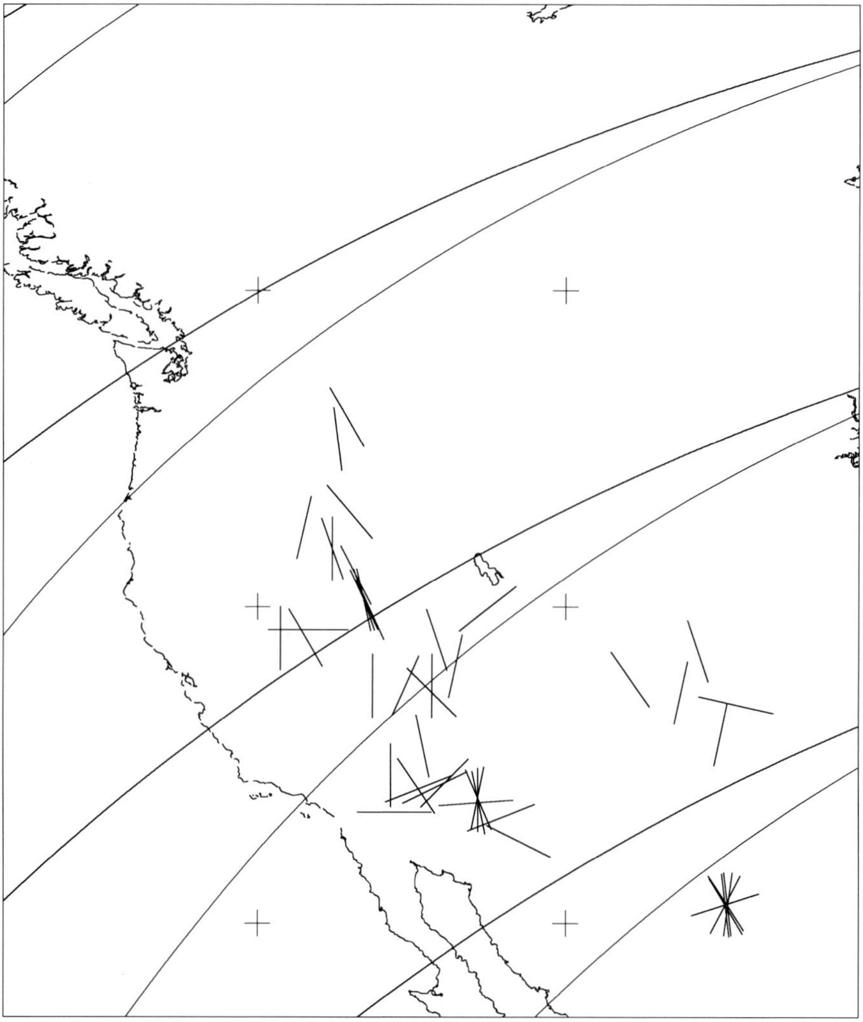

Fig. 8.4. North America: paleostresses and calculated flowlines in hotspot reference frame, 15–20 Ma. Bold flowline 20 Ma.

8.3.5 North America: 20–25 Ma

Most observations for 20–25 Ma are within the Cordillera (Fig. 8.5). A number indicate east-west extension (north-south σ_1), but there are several of dated dikes or dike swarms from the eastern part of the Corillera with an east-west orientation. The calculated flowline for 25 Ma is close to east-west in orientation. The dike swarms of Spanish Peaks in southern Colorado are the most familiar of these.

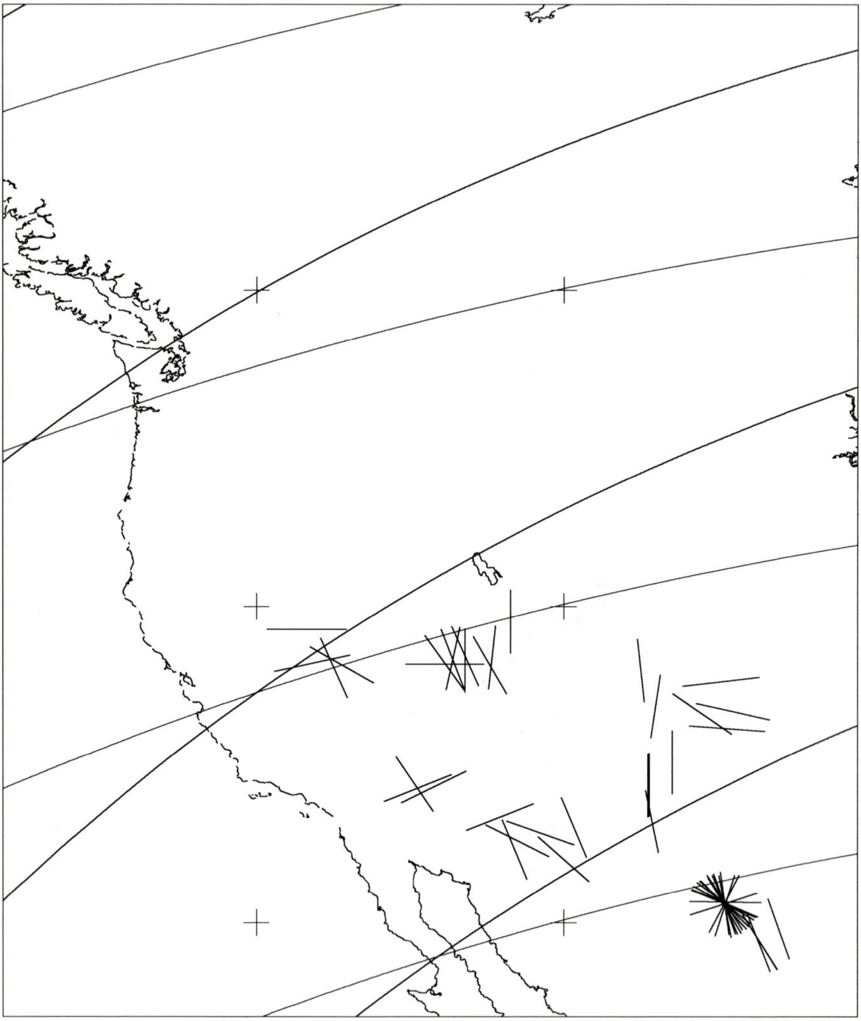

Fig. 8.5. North America: paleostresses and calculated flowlines in hotspot reference frame, 20–25 Ma. Bold flowline 20 Ma.

8.3.6
North America: 25–30 Ma

Dominant east-west to east-northeast-west-southwest dike (and therefore, σ_1) orientations are observed in the eastern U.S. Cordillera for the 25–30 Ma interval (Fig. 8.6). A secondary overlay of east-west extension (north-south σ_1) is apparent from dikes and age-constrained faults. Calculated flowlines correspond well with the predominant dike orientations. Similar dike orientations are observed from within the interior of the Cordillera as well.

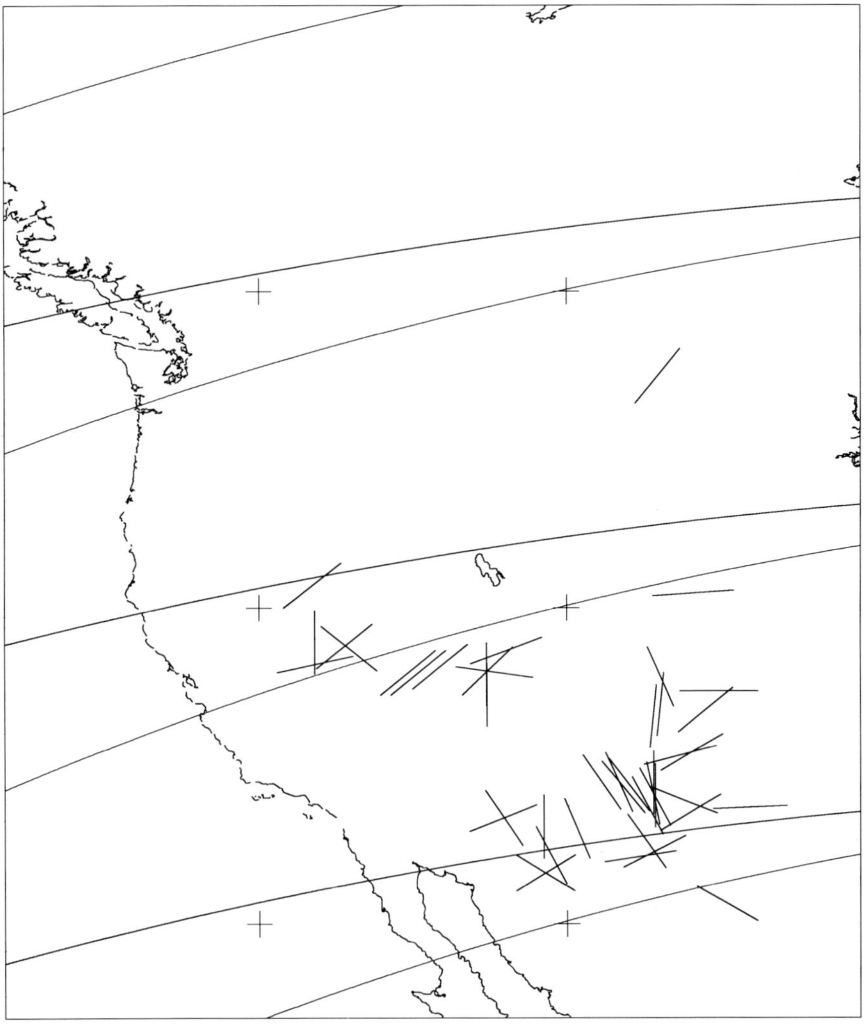

Fig. 8.6. North America: paleostresses and calculated flowlines in hotspot reference frame, 25–30 Ma. Bold flowline 30 Ma.

8.3.7
North America: 30–35 Ma

East-west to east-northeast–west-southwest σ_1 orientations are observed in the eastern Cordillera (Fig. 8.7), with a few southwest-northeast observed, along with a few, more diverse orientations. Calculated flowlines are consistent with the dominant orientations.

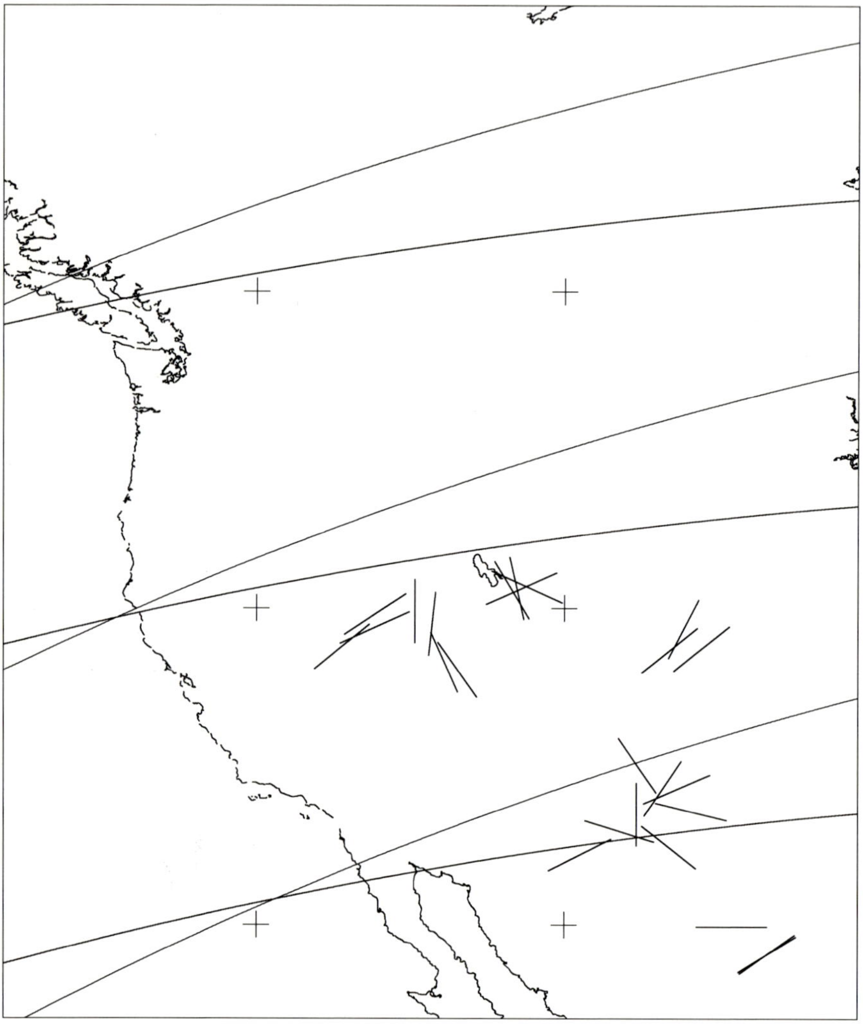

Fig. 8.7. North America: paleostresses and calculated flowlines in hotspot reference frame, 30–35 Ma. Bold flowline 30 Ma.

8.3.8
North America: 35–40 Ma

East-northeast–west-southwest σ_1 orientations are observed along the eastern edge of the Cordillera (Fig. 8.7), with more diverse orientations to the west. Calculate flowlines correspond fairly well with the observed orientations.

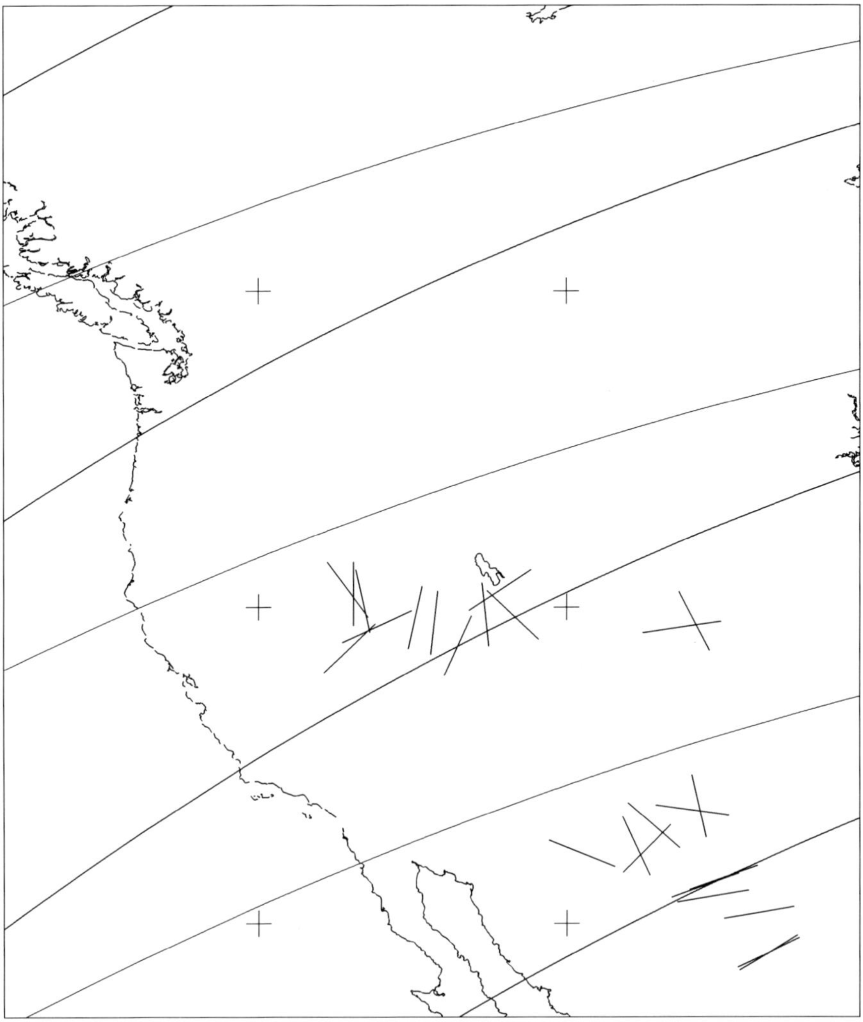

Fig. 8.8. North America: paleostresses and calculated flowlines in hotspot reference frame, 35–40 Ma. Bold flowline 40 Ma.

8.3.9
North America: 40–45 Ma

Southwest-northeast oriented σ_1 observations are present along the east edge of the Cordillera (Fig. 8.9). These observations are consistent with the more south-westerly orientation of the 45 Ma flowline, compared with subsequent intervals.

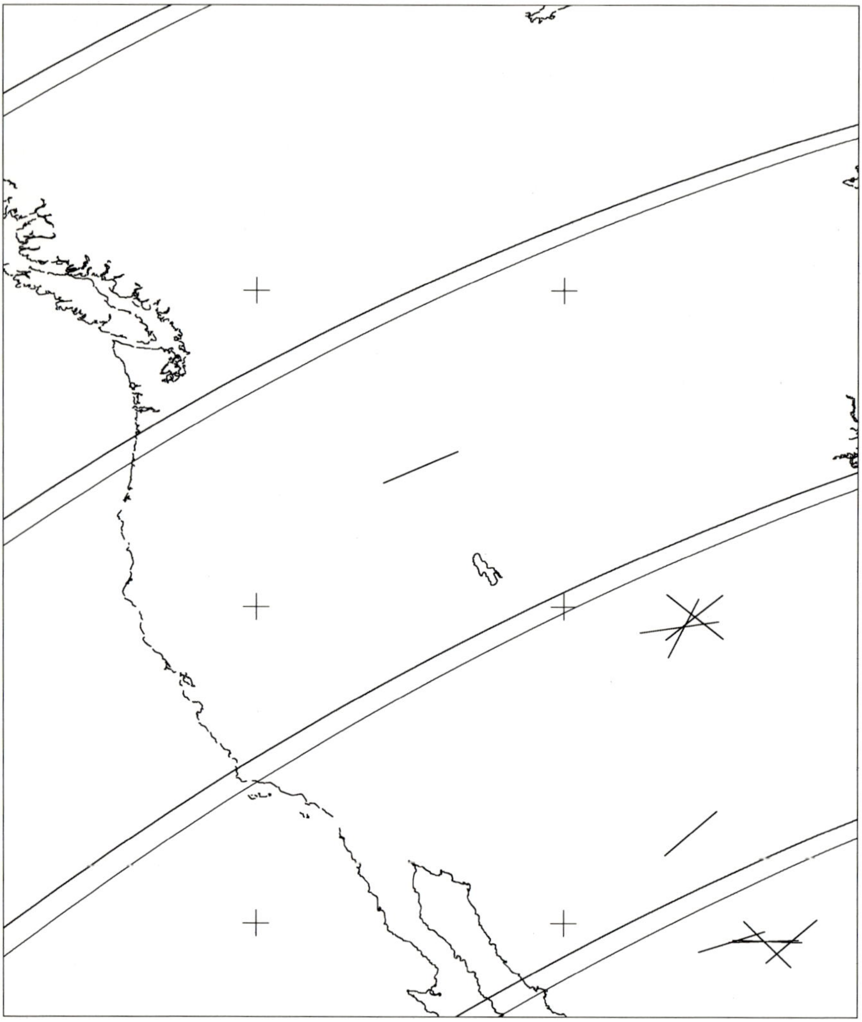

Fig. 8.9. North America: paleostresses and calculated flowlines in hotspot reference frame, 40–45 Ma. Bold flowline 40 Ma.

8.3.10
North America: 45–50 Ma

As with the 40–45 Ma interval, southwest-northeast σ_1 orientations are observed along the east flank of the Cordillera (Fig. 8.10). Calculated flowlines correspond well with the paleostress observations.

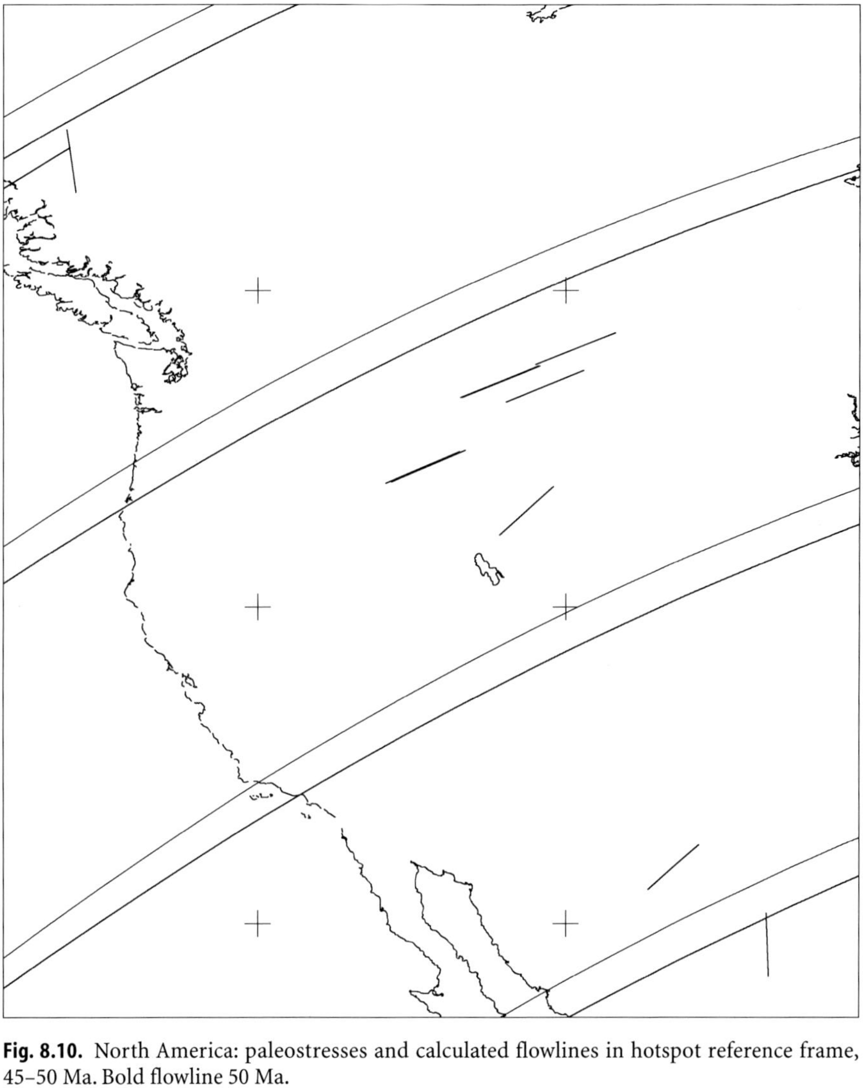

Fig. 8.10. North America: paleostresses and calculated flowlines in hotspot reference frame, 45–50 Ma. Bold flowline 50 Ma.

8.3.11 North America: 50–55 Ma

The few available σ_1 orientations between 50 and 55 Ma trend southwest-northeast (Fig. 8.11), from Montana, Wyoming, and Arizona. Similarly, the calculated flowlines correspond with the σ_1 measurements.

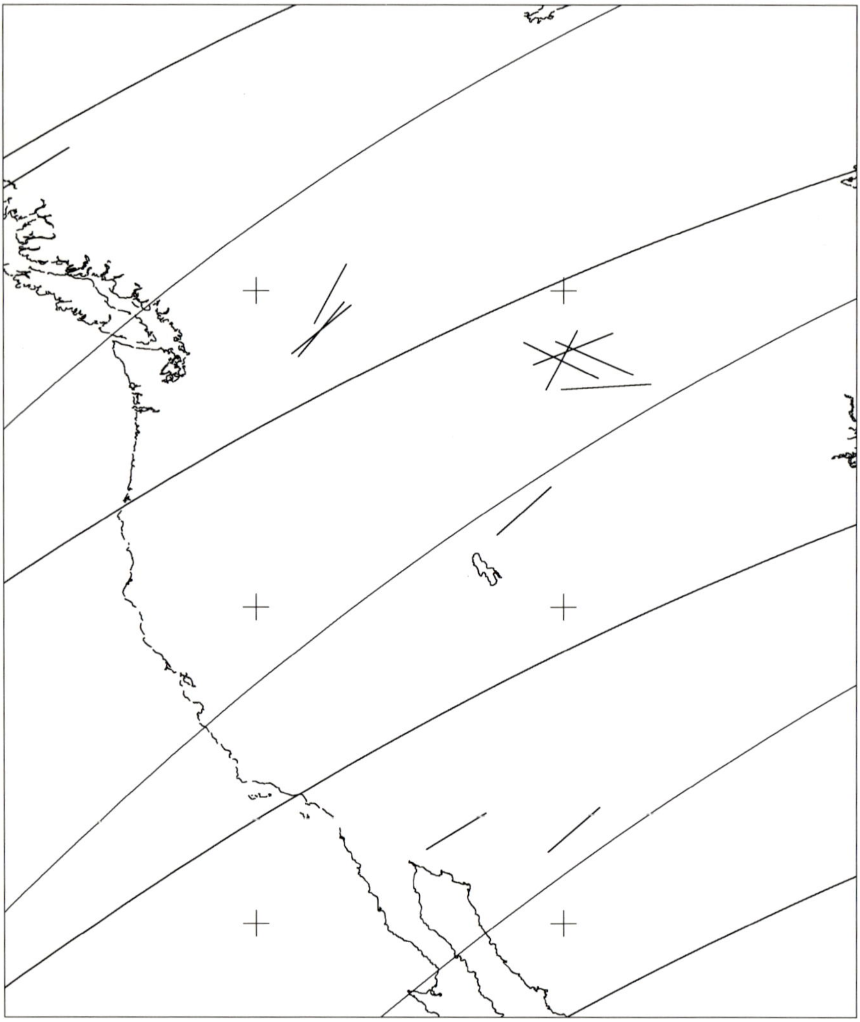

Fig. 8.11. North America: paleostresses and calculated flowlines in hotspot reference frame, 50–55 Ma. Bold flowline 50 Ma.

8.3.12
North America: 55–60 Ma

Only two data points are available for the 55–60 Ma interval (Fig. 8.12). One, from Idaho, trends east-west and does not correspond well with the calculated flowlines. The second data point is from East Texas, and, while it corresponds fairly well with the flowlines, it is poorly constrained in age (and included in older intervals as well).

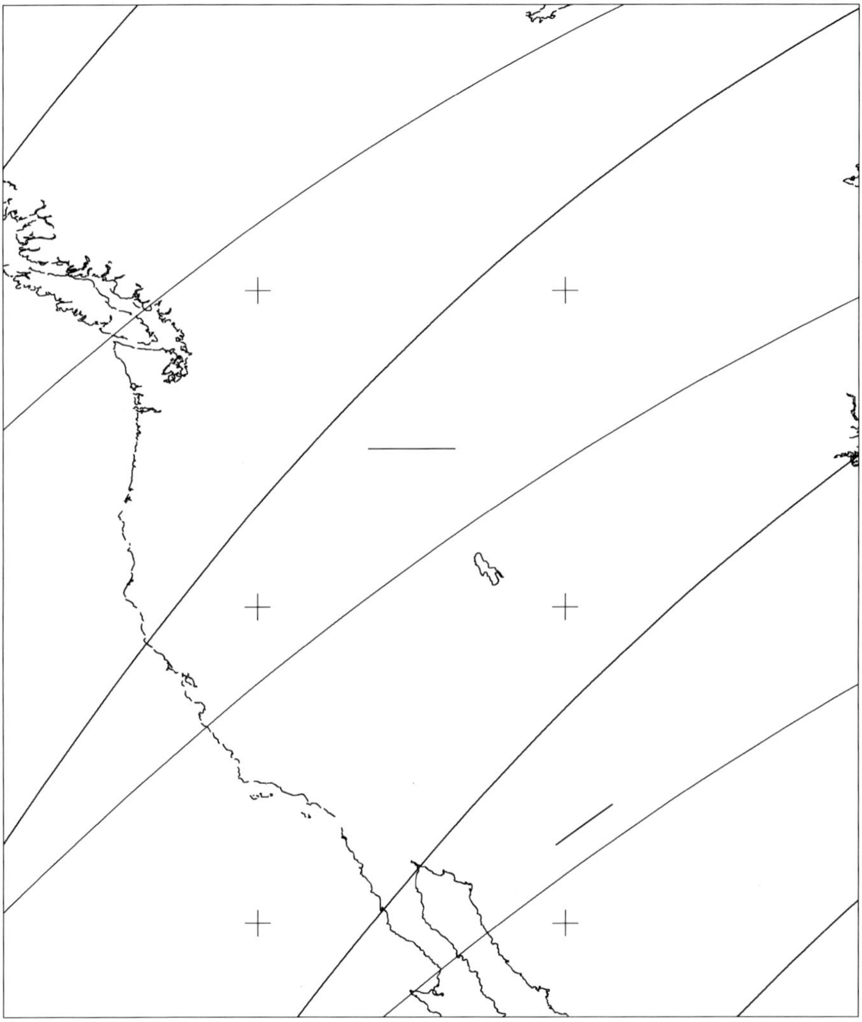

Fig. 8.12. North America: paleostresses and calculated flowlines in hotspot reference frame, 55–60 Ma. Bold flowline 60 Ma.

8.3.13
North America: 60–65 Ma

A few observations in the 60–65 Ma interval, from Colorado, New Mexico, and Arizona trend southwest-northeast (Fig. 8.13). Similarly, the calculated flowlines are so oriented. The data point from East Texas, included in the 55–60 Ma interval, also, is close to the flowline orientation.

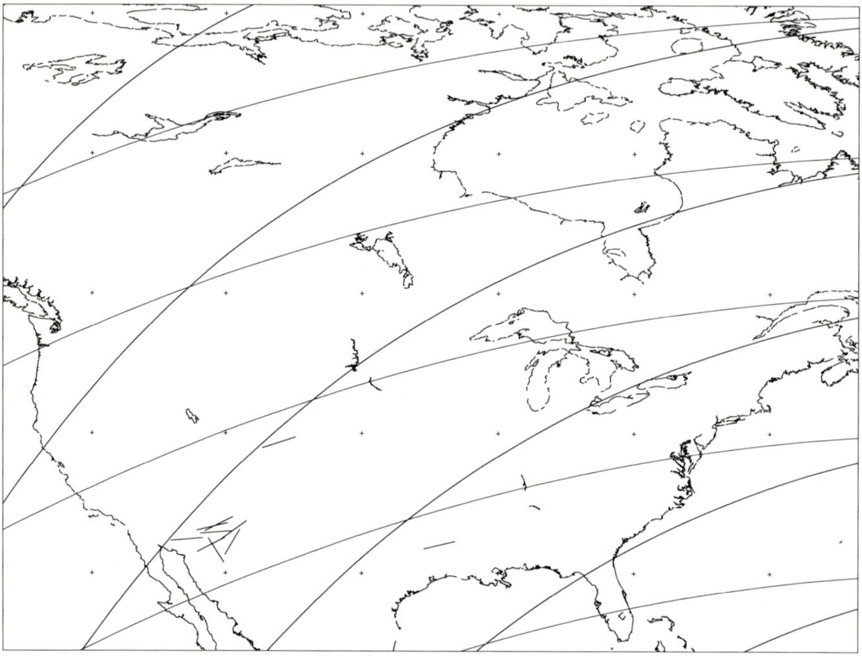

Fig. 8.13. North America: paleostresses and calculated flowlines in hotspot reference frame, 60–65 Ma. Bold flowline 60 Ma.

8.3.14
North America: 65–70 Ma

A number of σ_1 observations are available from the southern U.S. Cordillera for the 65–70 Ma interval (Fig. 8.14). Those from Arizona and New Mexico trend from southwest-northeast to east-west, corresponding with the calculated flowlines. One observation from Colorado is close to southwest-northeast, while a second observation is divergent from the flowlines, as are several measurements from Nevada, within the Sevier belt and the later Basin Range province.

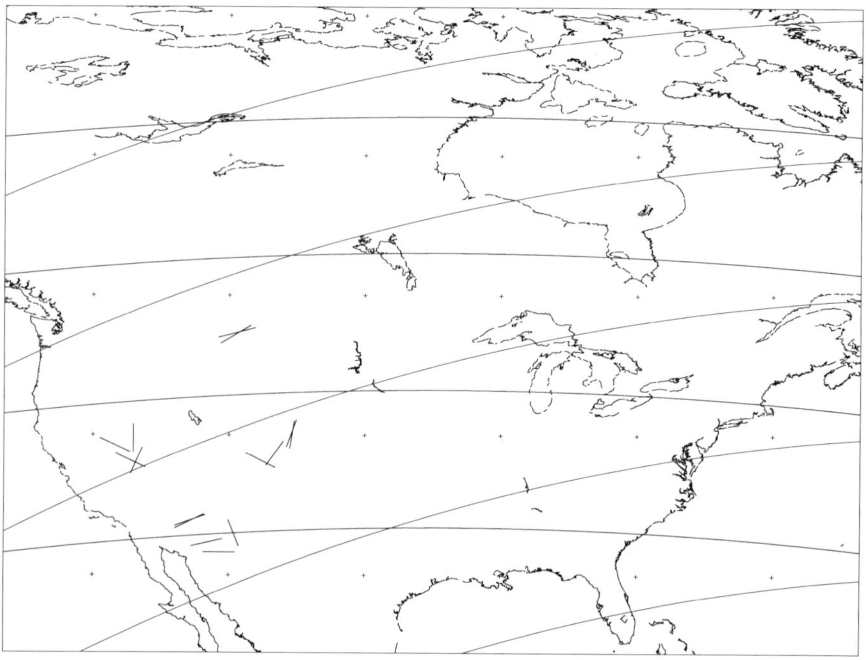

Fig. 8.14. North America: paleostresses and calculated flowlines in hotspot reference frame, 65–70 Ma. Bold flowline 70 Ma.

8.3.15
North America: 70–75 Ma

Southwesterly oriented σ_1 measurements are available in southern Arizona and New Mexico between 70 and 75 Ma (Fig. 8.15), with one similar measurement from Nevada. The calculated flowline for 70 Ma is close to these measurements.

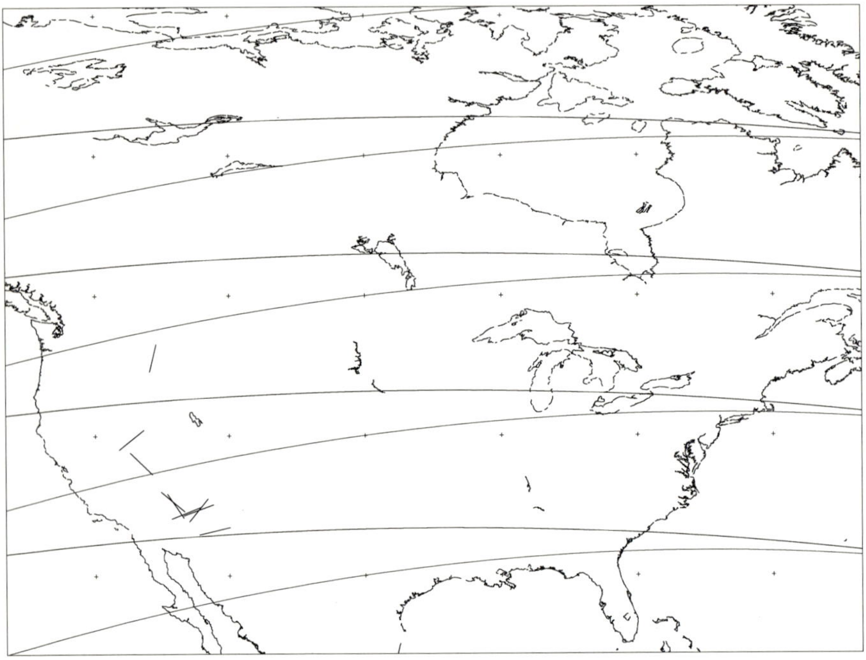

Fig. 8.15. North America: paleostresses and calculated flowlines in hotspot reference frame, 70–75 Ma. Bold flowline 70 Ma.

8.3.16
North America: 75–80 Ma

Two east-west oriented σ_1 measurements are available for the interval of 75–80 Ma (Fig. 8.16) from the southern United States Cordillera. Calculated flowlines for 75 and 80 Ma show a similar orientation.

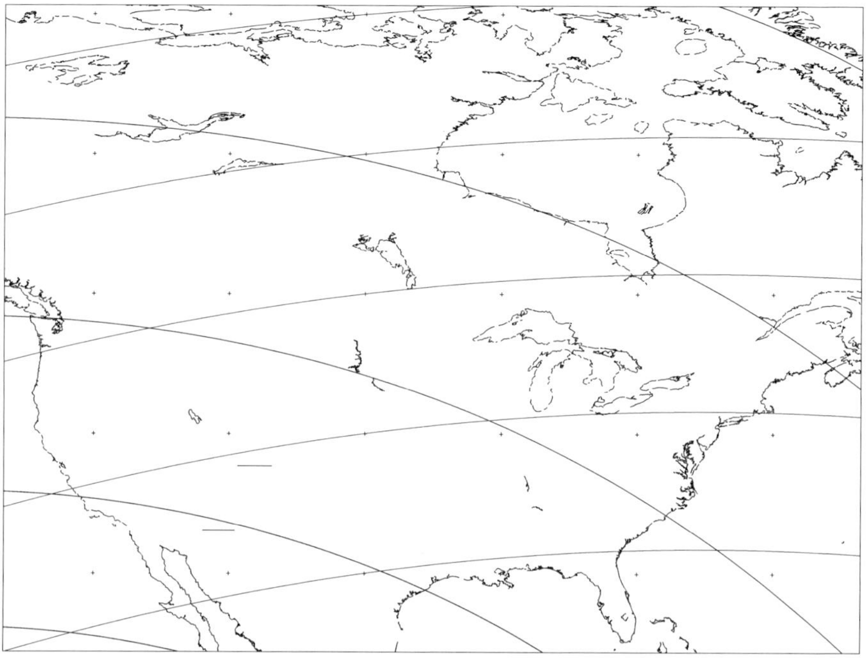

Fig. 8.16. North America: paleostresses and calculated flowlines in hotspot reference frame, 75–80 Ma. Bold flowline 80 Ma.

8.3.17
North America: 80–85 Ma

No σ_1 observations are available for the time interval between 80 and 85 Ma. If present, they would be expected to have an east-west to northeast-southwest orientation, based on calculated flowlines (Fig. 8.17).

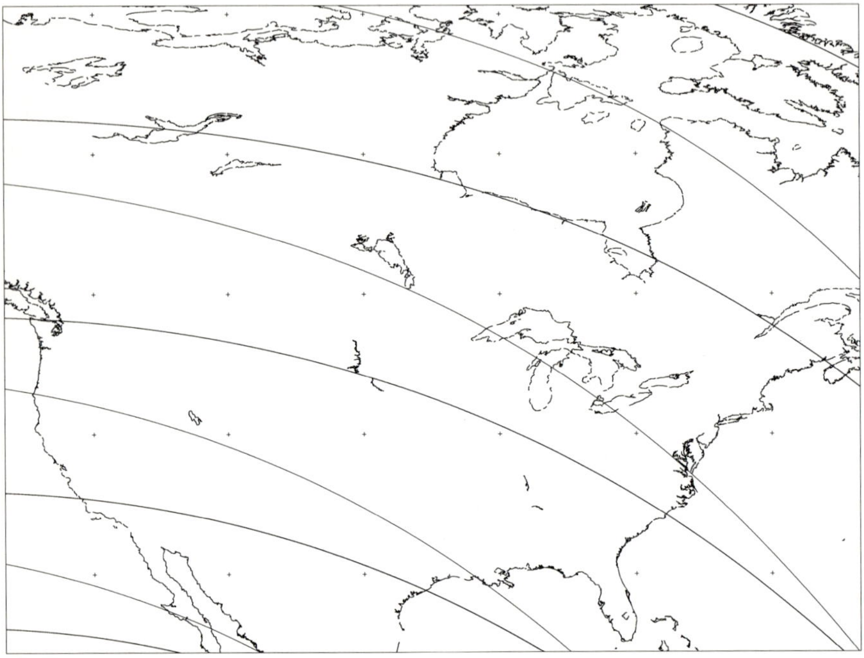

Fig. 8.17. North America: calculated flowlines in hotspot reference frame, 80–85 Ma. Bold flowline 80 Ma.

8.3.18
North America: 85–90 Ma

A number of east-west to northwest-southeast oriented σ_1 measurements are available from the Cordillera and southcentral United States (Fig. 8.18) for the 85–90 Ma interval. Calculate flowlines show a similar orientation. The two measurements from South Texas are stratigraphically constrained and, therefore, their ages are not as precise as those from dikes.

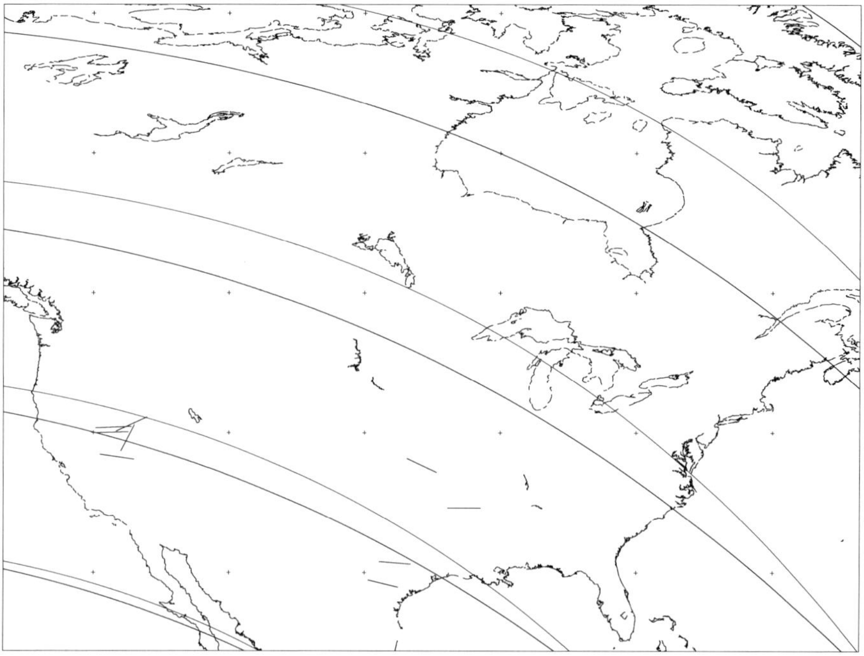

Fig. 8.18. North America: paleostresses and calculated flowlines in hotspot reference frame, 85–90 Ma. Bold flowline 90 Ma.

8.3.19
North America: 90–95 Ma

Several σ_1 observations from the south-central United States are oriented east-west to northwest-southeast for the interval 90–95 Ma (Fig. 8.19). The calculated flowline for 90 Ma has a similar orientation. Measurements from within the Cordilleran Sevier belt are quite diverse.

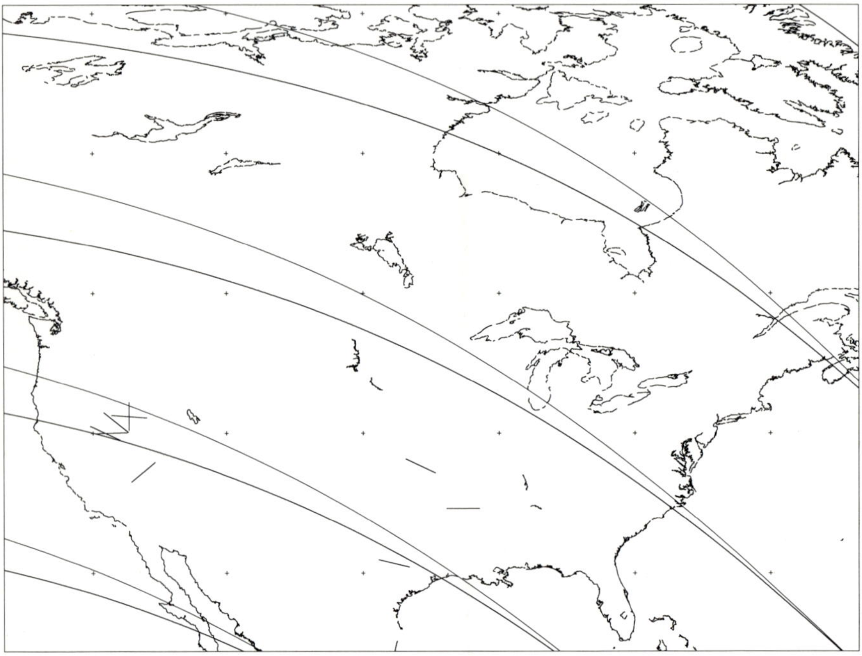

Fig. 8.19. North America: paleostresses and calculated flowlines in hotspot reference frame, 90–95 Ma. Bold flowline 90 Ma.

8.3.20
North America: 95–100 Ma

One σ_1 measurement, from New England, trends west-northwest–east-southeast for the interval 95–100 Ma (Fig. 8.20). Calculated flowlines for this interval correspond with the two measurements.

8.3.21
North America: 100–105 Ma

One paleostress measurement from North America is available for the time interval between 100 and 105 Ma (Fig. 8.21). Predicted flowlines trend from northwest-southeast. The single measurement trends east of north.

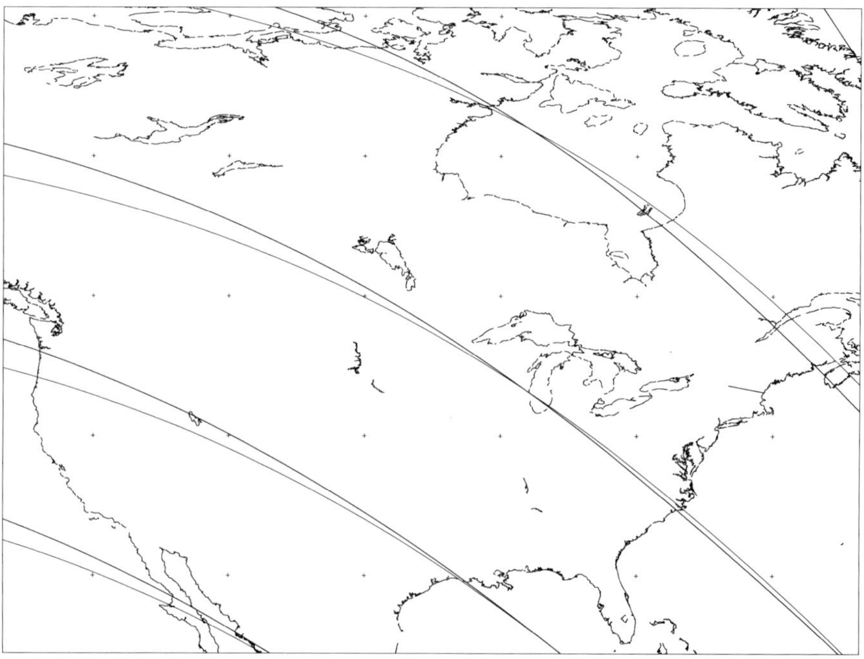

Fig. 8.20. North America: paleostress and calculated flowlines in hotspot reference frame, 95–100 Ma. Bold flowline 100 Ma.

8.3.22
North America: 105–110 Ma

Only two σ_1 measurements are available for the time interval from 105 to 110 Ma (Fig. 8.22). Both measurements, from Nevada, have orientations close to the predicted flowlines: west-northwest–east-southeast.

8.3.23
North America: 110–115 Ma

Calculated flowlines trend northwest-southeast for the interval 110–115 Ma (Fig. 8.23). No paleostress measurements are available for this interval.

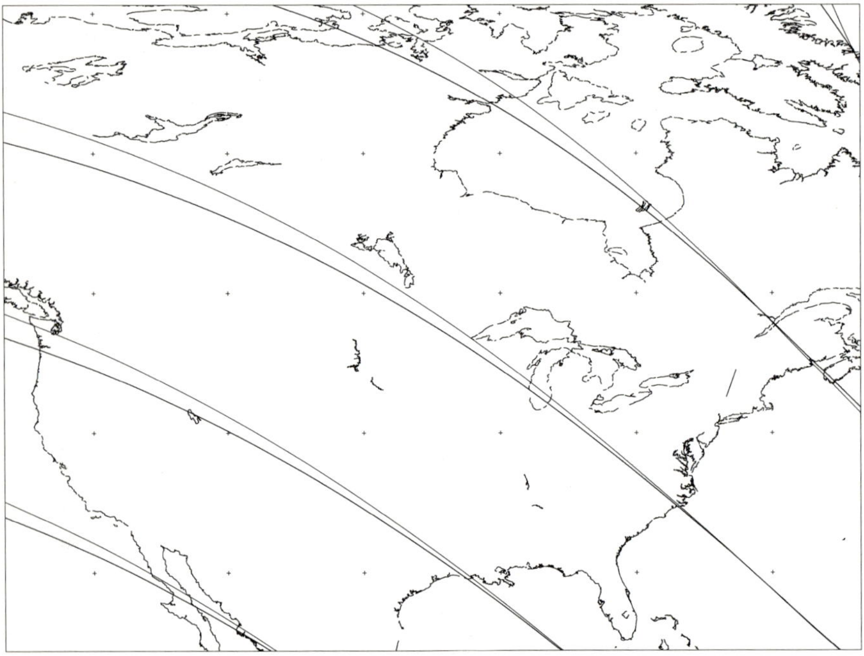

Fig. 8.21. North America: paleostresses and calculated flowlines in hotspot reference frame, 100–105 Ma. Bold flowline 100 Ma.

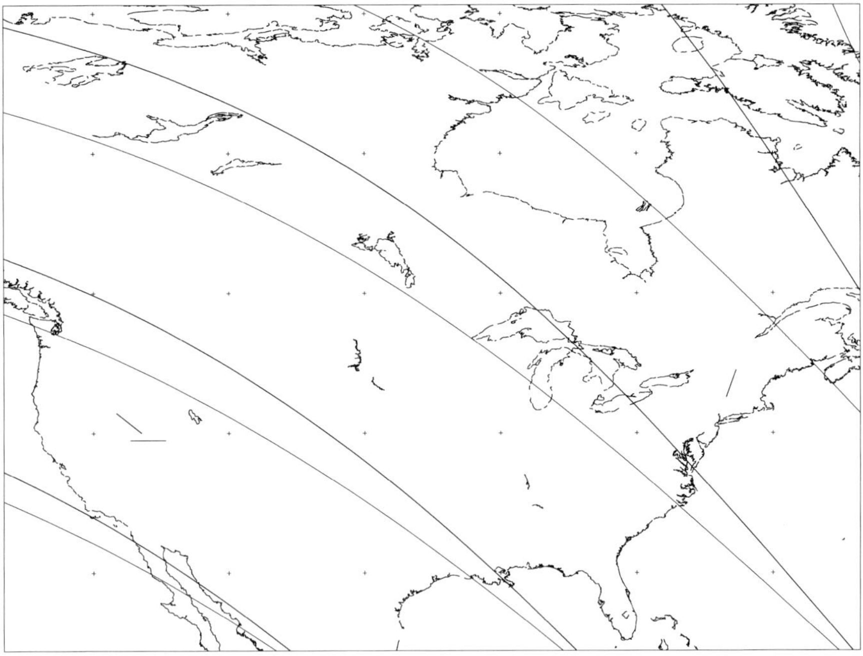

Fig. 8.22. North America: paleostresses and calculated flowlines in hotspot reference frame, 105–110 Ma. Bold flowline 110 Ma.

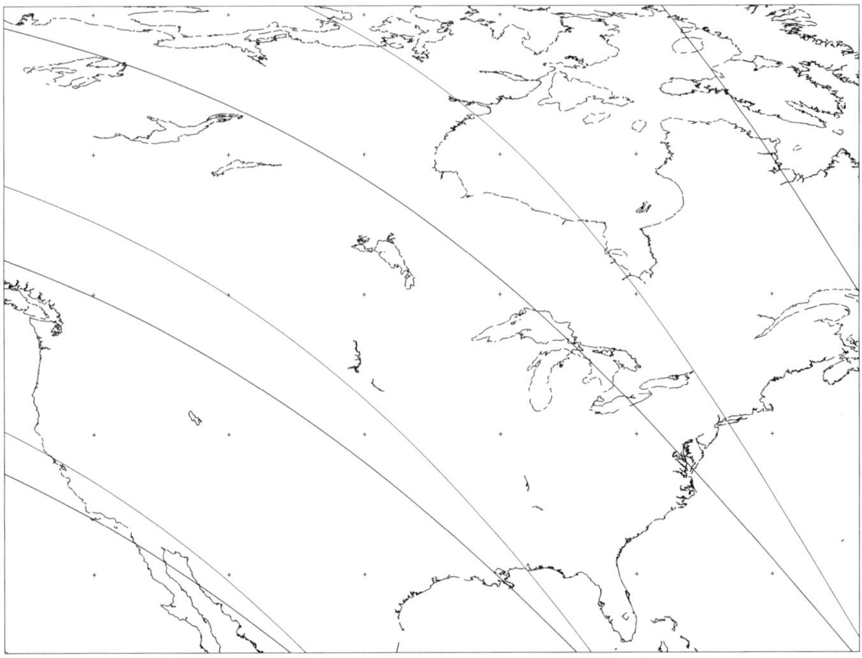

Fig. 8.23. North America: paleostresses calculated flowlines in hotspot reference frame, 110–115 Ma. Bold flowline 110 Ma.

8.3.24 North America: 115–120 Ma

A single σ_1 measurement is available for the 115–120 Ma interval (Fig. 8.24). Calculated flowlines would predict a more northwest-southeasterly orientation. However, the single measurement is within the actively deforming Sevier belt at this time.

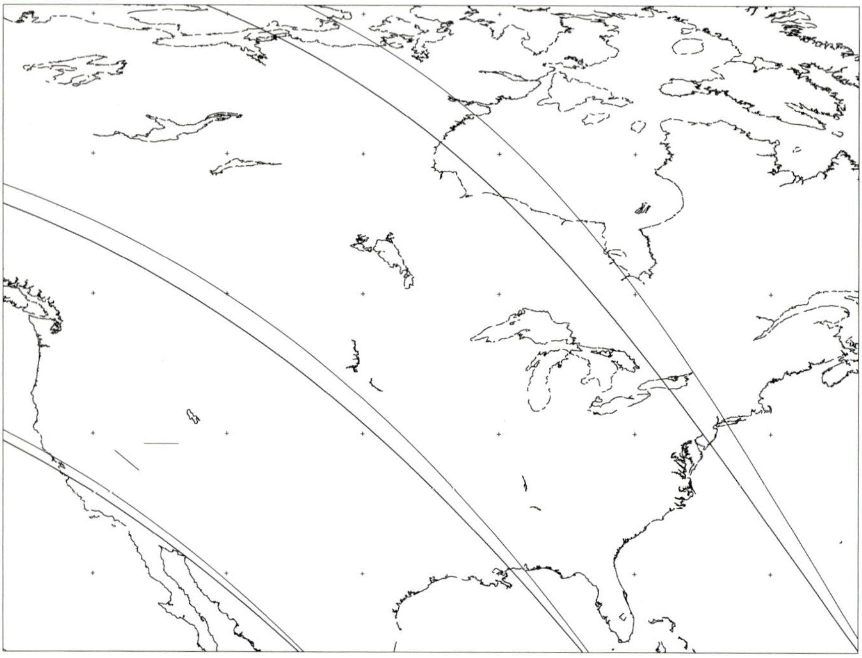

Fig. 8.24. North America: paleostresses and calculated flowlines in hotspot reference frame, 115–120 Ma. Bold flowline 120 Ma.

8.3.25
North America: 120–125 Ma

Two observations, from New England, are available for the interval from 120–125 Ma (Fig. 8.25). Both trend close to east-west, while calculated flowlines are oriented more northwest-southeasterly. The dikes have an orientation close to that of the portion of the northern Appalachians in which they occur.

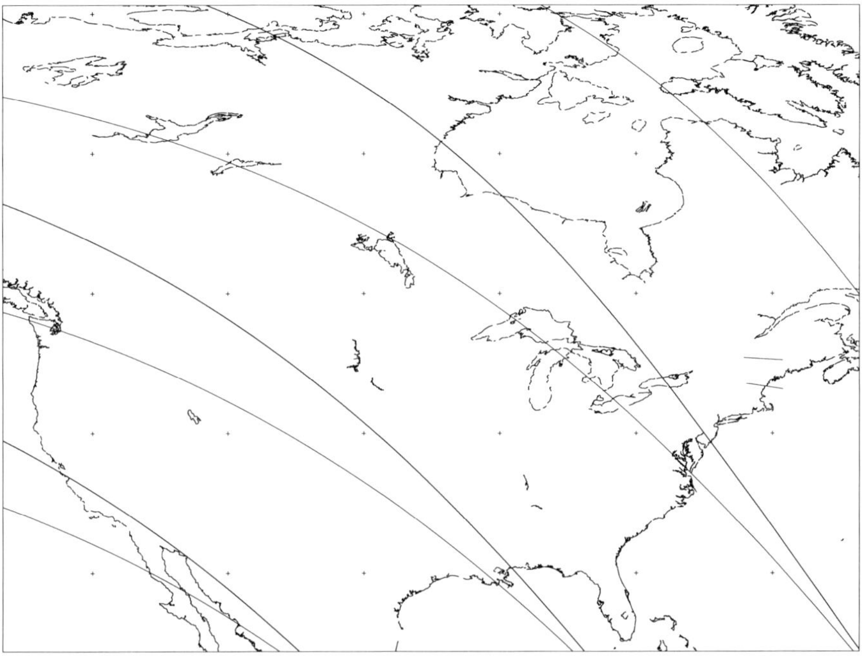

Fig. 8.25. North America: paleostresses and calculated flowlines in hotspot reference frame, 120–125 Ma. Bold flowline 120 Ma.

8.3.26
North America: 125–130 Ma

Three σ_1 measurements are available for the interval of 125–130 Ma (Fig. 8.26). Calculated flowlines would predict a northwest-southeast to east-west orientation. Only one paleostress measurement corresponds with the calculated flowlines.

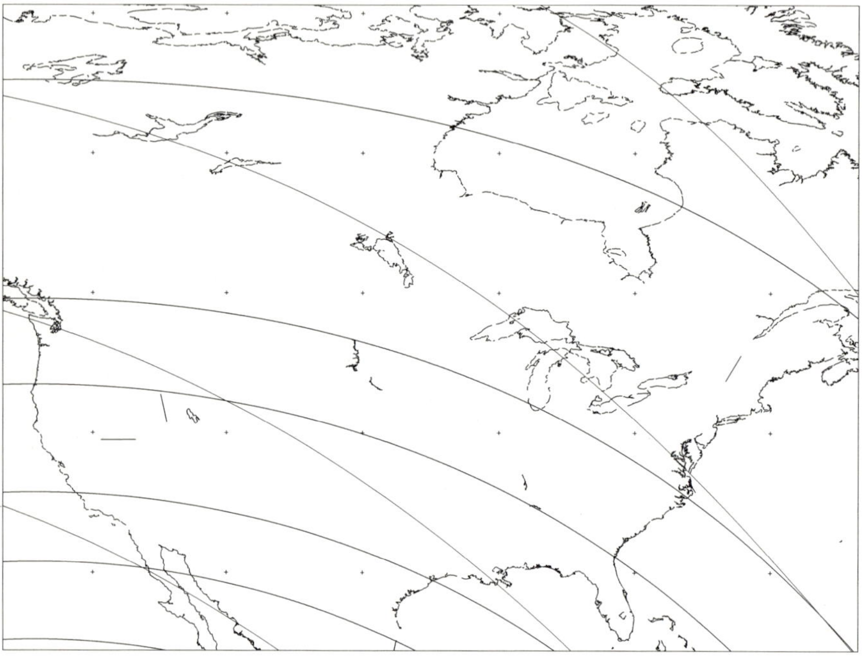

Fig. 8.26. North America: paleostresses and calculated flowlines in hotspot reference frame, 125–130 Ma. Bold flowline 130 Ma.

8.4
Europe: Paleostresses

For Europe, some contemporary stresses and young paleostresses are included in the first illustration (≤5 Ma). The remaining illustrations are at 5-my increments back to 100 Ma. Calculated streamlines are based on the global hotspot model of Gripp and Gordon (1991) and the Atlantic–Indian Ocean model of Müller et al. (1993) extended to Eurasia using the methods presented in this volume.

There is a possibility that the Atlantic–Indian Ocean model is not applicable to Eurasia. A separate reference frame may well exist (e.g., Norton, 2000). In each of the age intervals discussed below, this possibility needs to be kept in mind, whether explicitly mentioned or not.

8.4.1 Western Europe: 0–5 Ma

Paleostress measurements of σ_1 inWestern Europe trend from north-south to north-northwest–south-southwest in the time interval of 0–5 Ma (Fig. 8.27). Calculated streamlines for 0 and 5 Ma are almost 90° away from the observed measurements. However, at least one contemporary hotspot model (Minster et al. 1974) provides a very different streamline prediction. Almost all contemporary models have a large uncertainty in the location of the instantaneous rotation poles, so the discrepancy between calculated motion directions and observed σ_1 measurements are not surprising. Further, there is the possibility that a separate reference frame underlines Eurasia, so that calculated motions in the Atlantic–Indian Ocean frame may not be relevant.

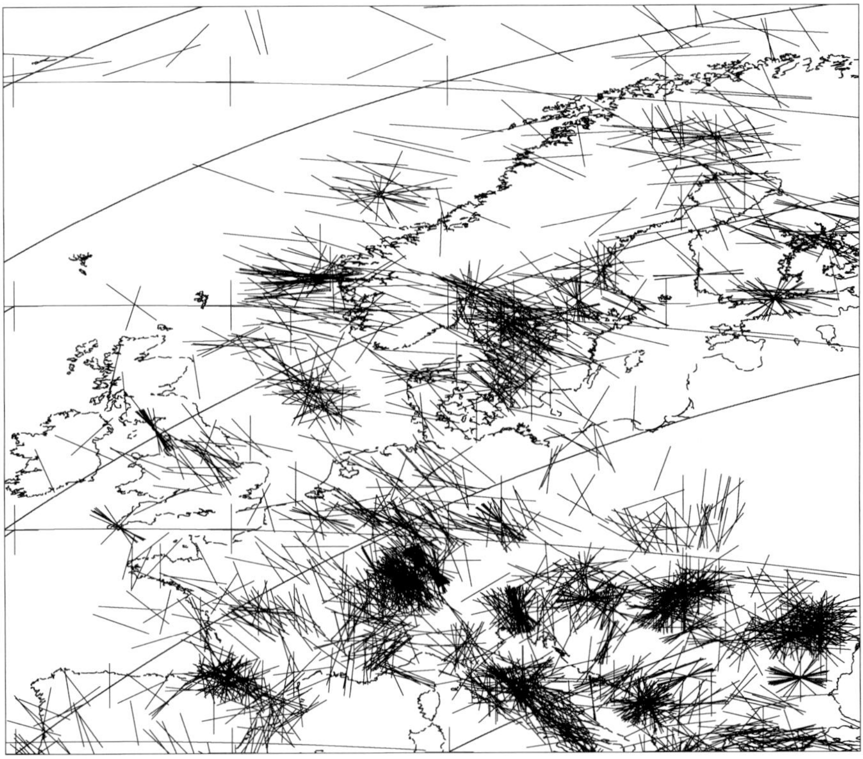

Fig. 8.27. Eurasia: paleostresses and calculated flowlines in hotspot reference frame, 0–5 Ma. Bold flowline 0 Ma.

8.4.2
Western Europe: 5–10 Ma

For the interval 5–10 Ma, observed σ_1 paleostress measurements trend between west-northwest and west-southwest, while calculated streamlines in the Atlantic–Indian hotspot frame trend similarly trend just north of west to west-southwesterly (Fig. 8.28). The ages of the paleostresses are stratigraphically constrained and therefore somewhat uncertain.

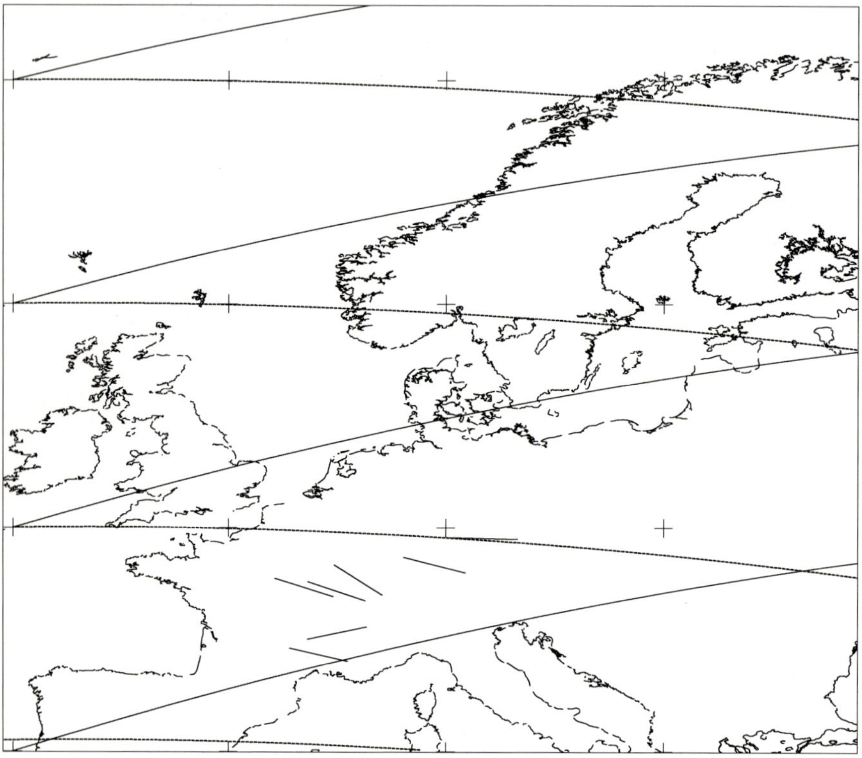

Fig. 8.28. Eurasia: paleostresses and calculated flowlines in hotspot reference frame, 5–10 Ma. Bold flowline 10 Ma.

8.4.3
Western Europe: 10–15 Ma

No paleostress measurements are available for the interval of 10–15 Ma. Calculated streamlines in the hotspot model trend southwesterly (Fig. 8.29).

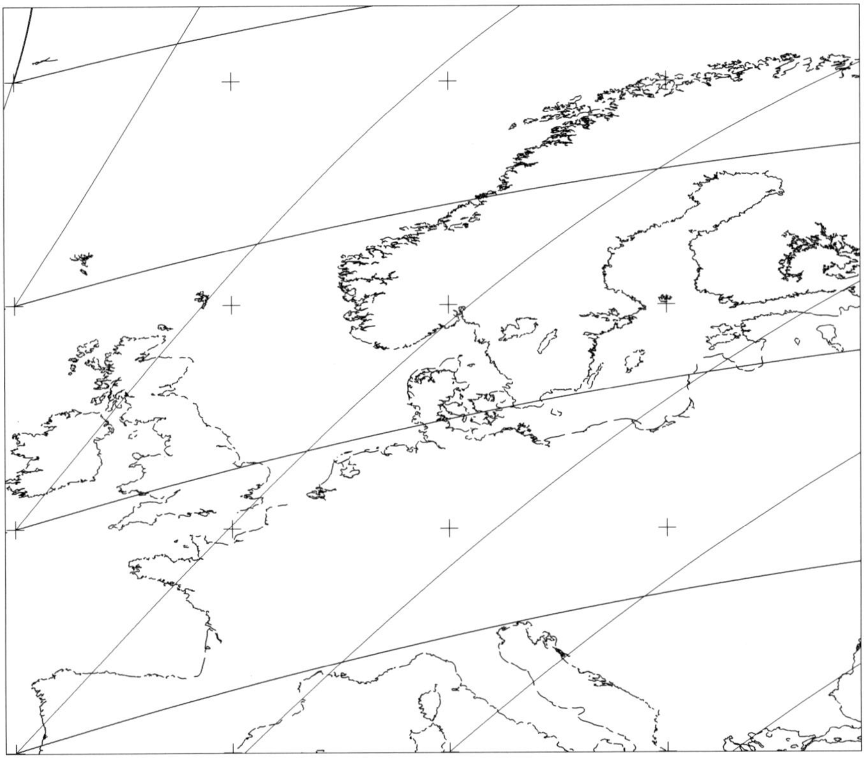

Fig. 8.29. Eurasia: calculated flowlines in hotspot reference frame, 10–15 Ma. Bold flowline 10 Ma.

8.4.4
Western Europe: 15–20 Ma

A number of paleostress measurements are available for the interval of 15–20 Ma, with southwesterly orientations (Fig. 8.30). Calculated streamlines for 15 Ma correspond with the σ_1 orientations, but the streamlines for 20 Ma are more south-southwesterly oriented.

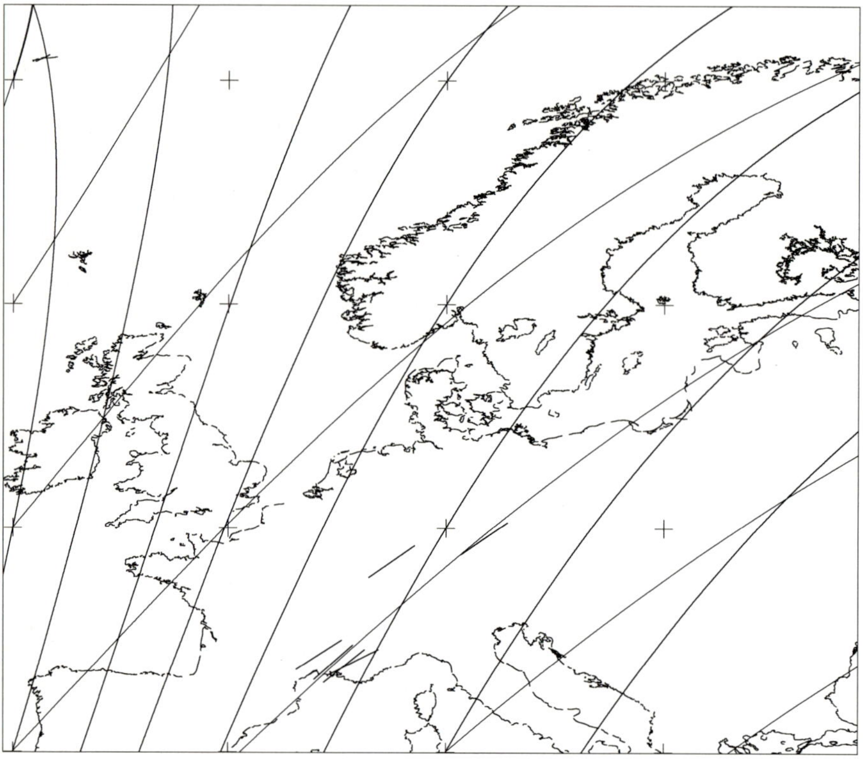

Fig. 8.30. Eurasia: paleostresses and calculated flowlines in hotspot reference frame, 15–20 Ma. Bold flowline 20 Ma.

8.4.5 Western Europe: 20–25 Ma

The measurements of σ_1 between 20 and 25 Ma have a similar orientation, southwesterly, as the 15–20 Ma interval (Fig. 8.31); in fact, they are the same measurements dated at 20 Ma from stratigraphic constraints. The southwesterly orientation of the 20-Ma streamlines are slightly oblique to the observed paleostresses, as are the streamlines for 25 Ma, with a high curvature.

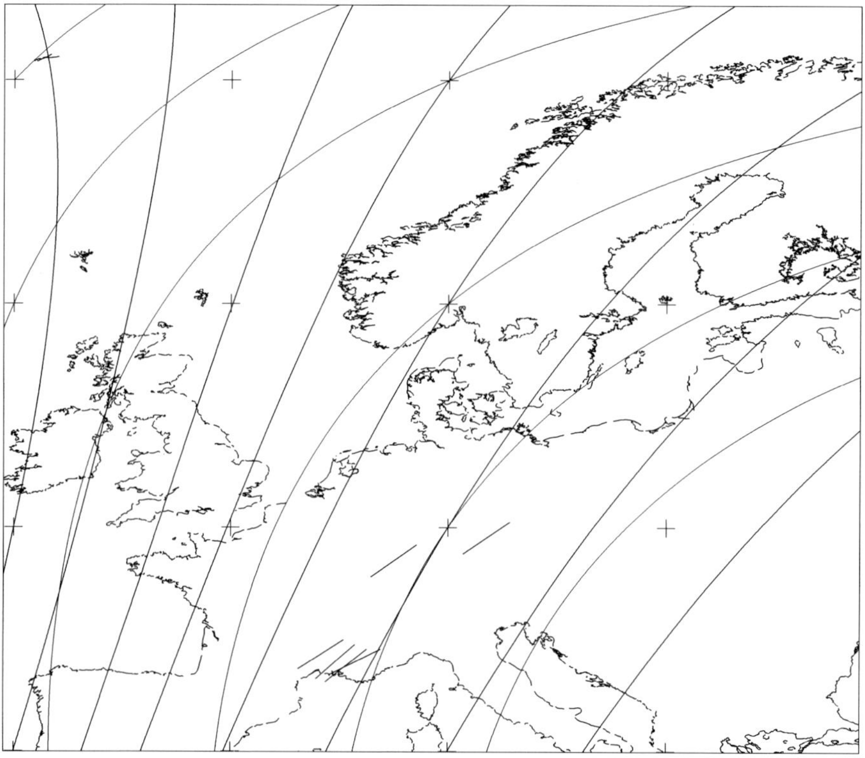

Fig. 8.31. Eurasia: paleostresses and calculated flowlines in hotspot reference frame, 20–25 Ma. Bold flowline 20 Ma.

8.4.6
Western Europe: 25–30 Ma

A single σ_1 observation is available for the 25–30 Ma time interval, in southern France and with a north-south orientation (Fig. 8.32). Streamlines for this interval range in orientation between northwesterly and northeasterly. Conceivably, a transitional kinematic orientation could correspond with the observed paleostresses.

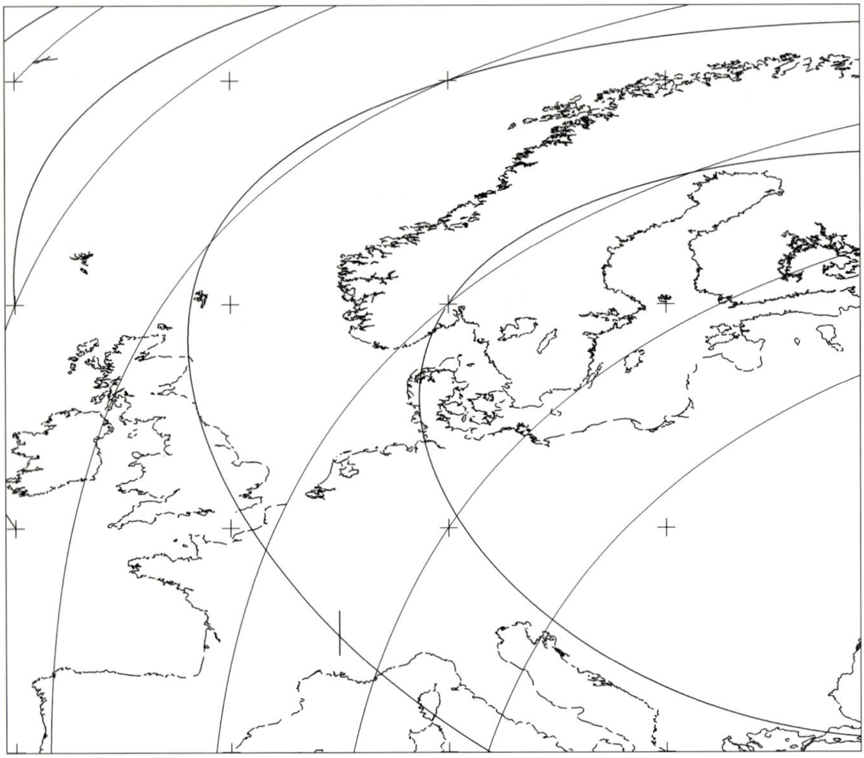

Fig. 8.32. Eurasia: paleostress and calculated flowlines in hotspot reference frame, 25–30 Ma. Bold flowline 30 Ma.

8.4.7
Western Europe: 30–35 Ma

Paleostress orientations in the interval 30-35 Ma exhibit both northerly and west-northwesterly orientations (Fig. 8.33). Calculated streamlines are oriented northwesterly.

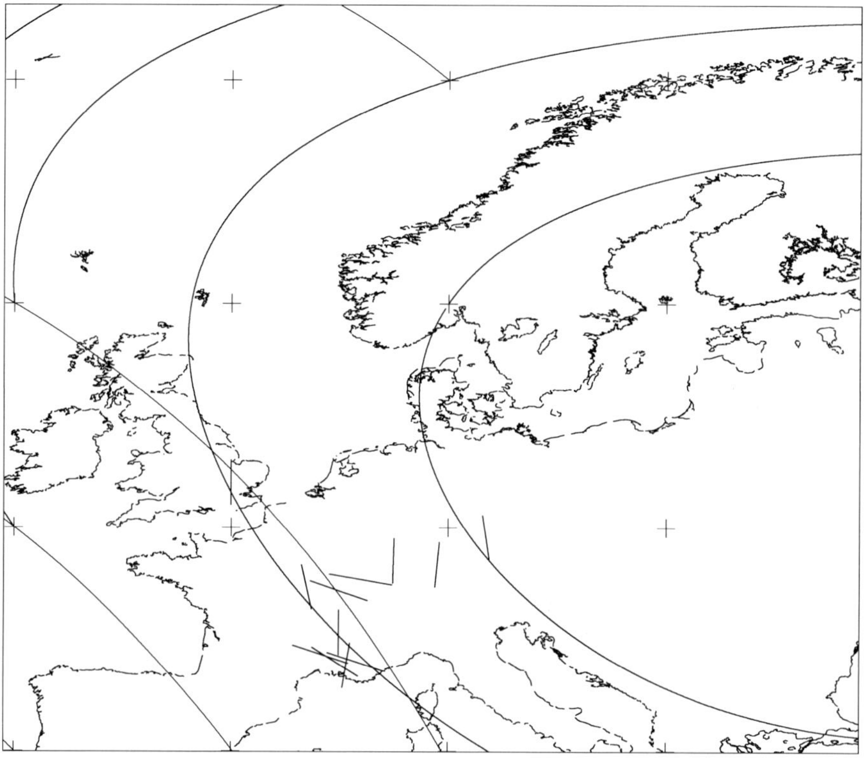

Fig. 8.33. Eurasia: paleostresses and calculated flowlines in hotspot reference frame, 30–35 Ma. Bold flowline 30 Ma.

8.4.8
Western Europe: 35–40 Ma

Available paleostress measurements of σ_1 from 35 to 40 Ma trend north-south (Fig. 8.34). Calculated streamlines for trend north-northwesterly.

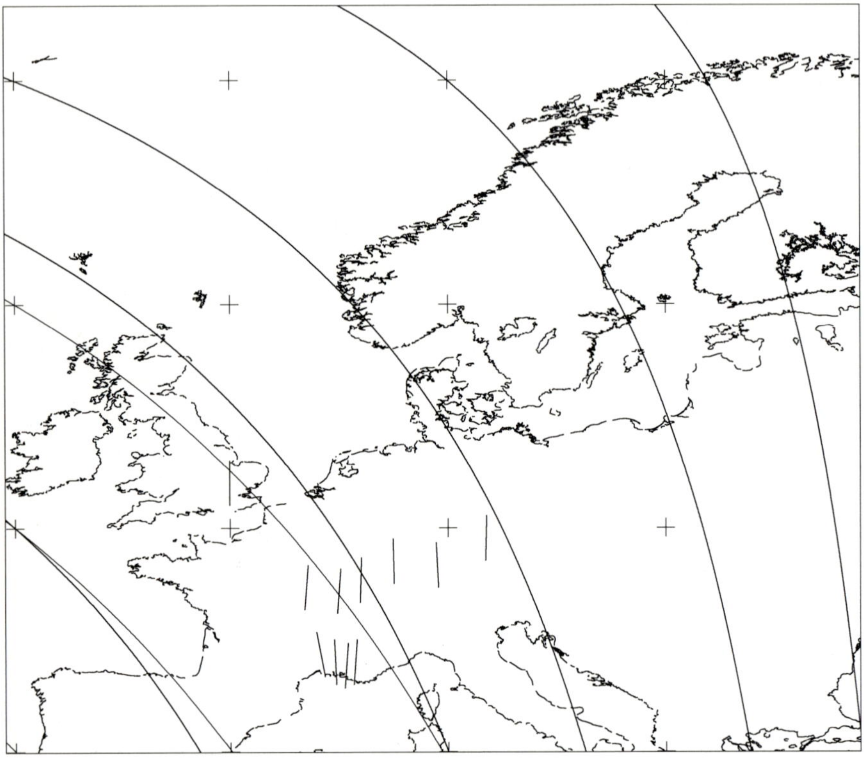

Fig. 8.34. Eurasia: paleostresses and calculated flowlines in hotspot reference frame, 35–40 Ma. Bold flowline 40 Ma.

8.4.9
Western Europe: 40–45 Ma

No paleostress measurements are available for the time interval 40–45 Ma (Fig. 8.35). Calculated streamlines range from north-northeasterly to north-westerly.

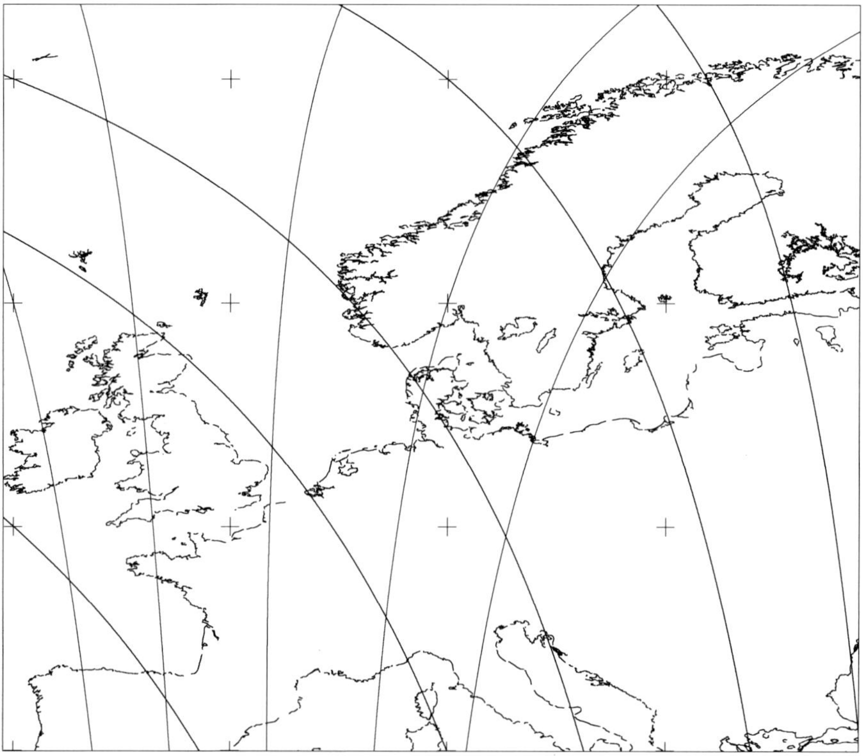

Fig. 8.35. Eurasia: calculated flowlines in hotspot reference frame, 40–45 Ma. Bold flowline 40 Ma.

8.4.10
Western Europe: 45–50 Ma

For the time interval 45–50 Ma, paleostress measurements show σ_1 orientations from west-northwesterly to west-southwesterly (Fig. 8.36). Calculated streamlines for 50 Ma are just south of west, while the 45-Ma streamlines trend nearly north-south.

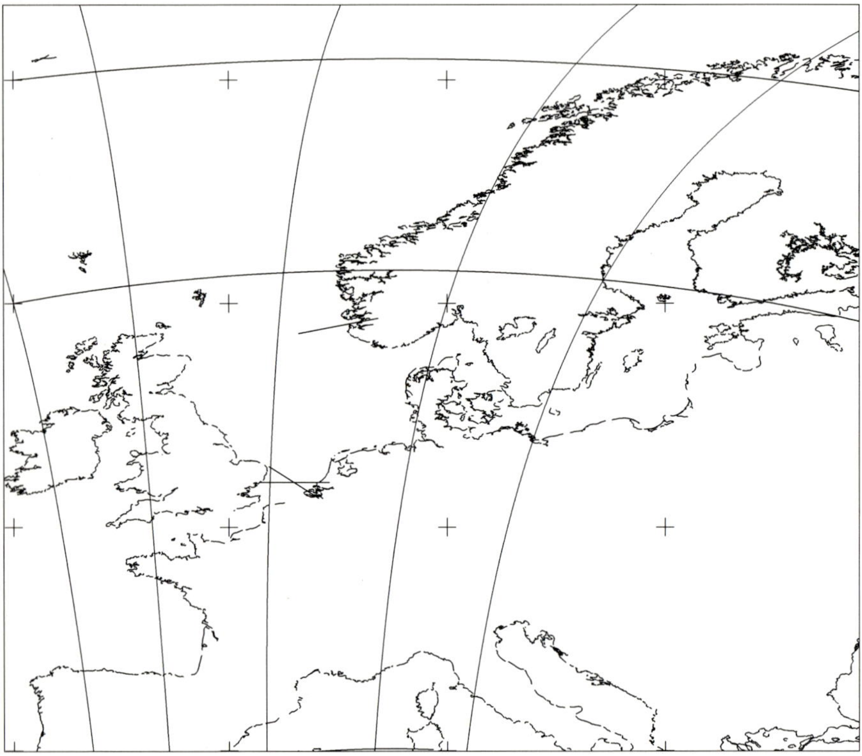

Fig. 8.36. Eurasia: paleostresses and calculated flowlines in hotspot reference frame, 45–50 Ma. Bold flowline 50 Ma.

8.4.11
Western Europe: 50–55 Ma

No stress measurements are available for the interval 50–55 Ma (Fig. 8.37). Streamlines trend east-west to northwest.

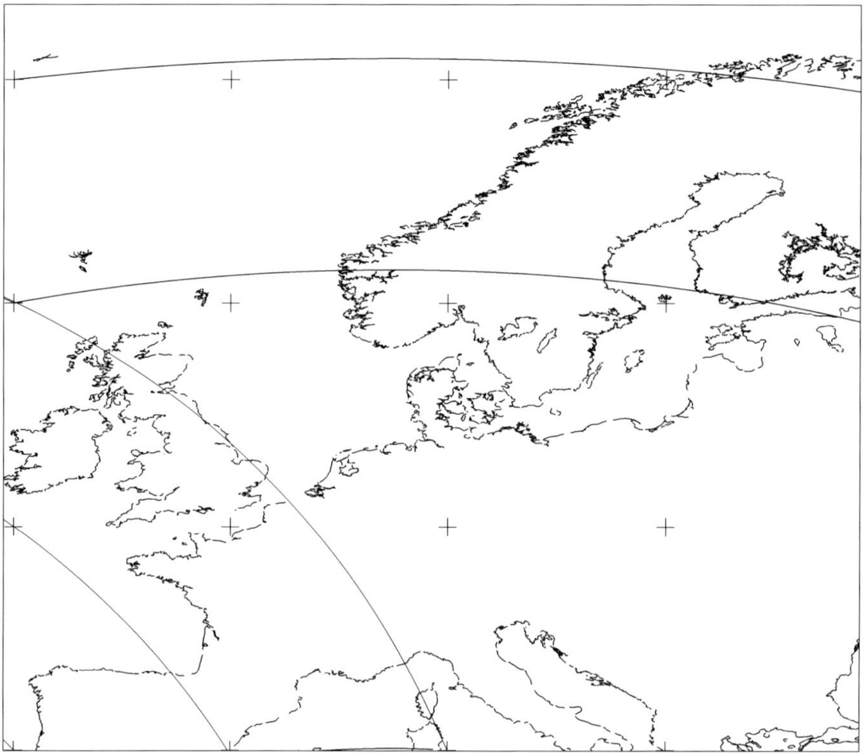

Fig. 8.37. Eurasia: calculated flowlines in hotspot reference frame, 50–55 Ma. Bold flowline 50 Ma.

8.4.12
Western Europe: 55–60 Ma

A number of dated dikes from the northern British Isles trend west-northwesterly (Fig. 8.38). Calculated streamlines for the region trend parallel with the dikes.

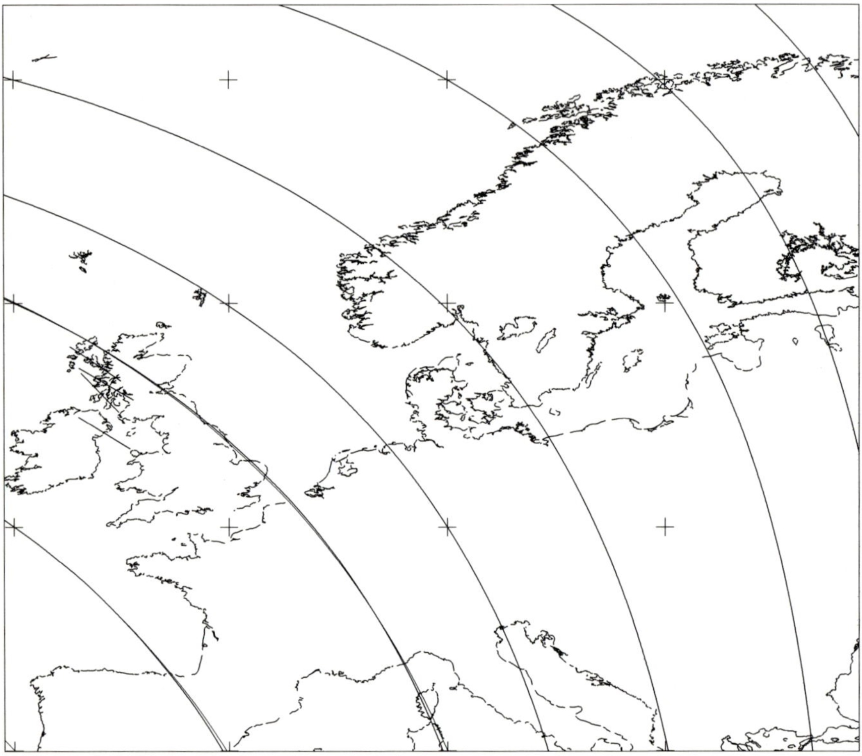

Fig. 8.38. Eurasia: paleostresses and calculated flowlines in hotspot reference frame, 55–60 Ma. Bold flowline 60 Ma.

8.4.13
Western Europe: 60–65 Ma

For the interval 60 to 65 Ma, observed σ_1 measurements include several dikes in the northern British Isles (Fig. 8.39). Calculated streamlines also trend northwesterly.

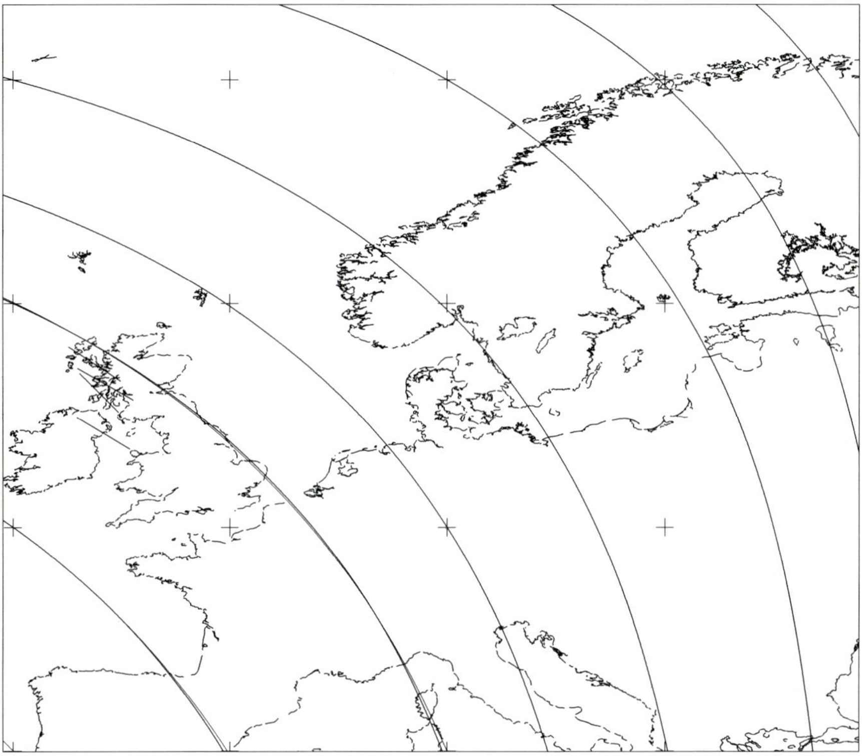

Fig. 8.39. Eurasia: paleostresses and calculated flowlines in hotspot reference frame, 60–65 Ma. Bold flowline 60 Ma.

8.4.14
Western Europe: 65–70 Ma

From 65–70 Ma, observed σ_1 measurements include two from faults in offshore Netherlands trending northwest and southwest (Fig. 8.40). Calculated streamlines trend parallel with the northwest oriented indicator.

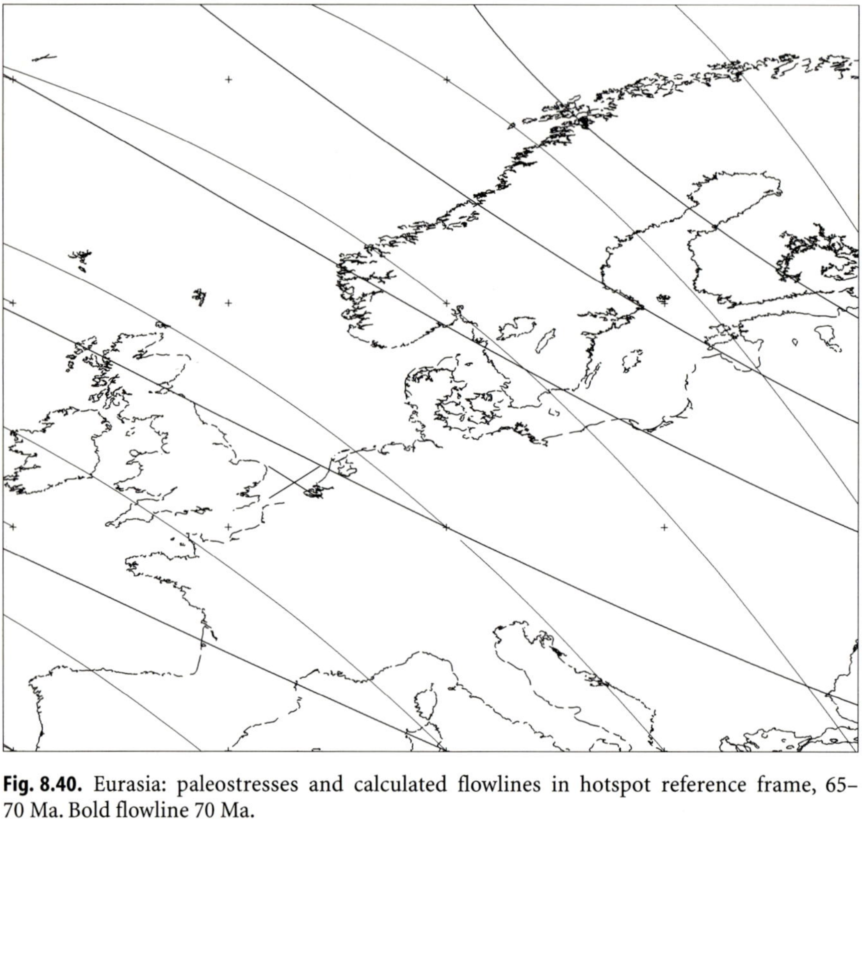

Fig. 8.40. Eurasia: paleostresses and calculated flowlines in hotspot reference frame, 65–70 Ma. Bold flowline 70 Ma.

8.4.15
Western Europe: 70–75 Ma

Between 70 and 75 Ma, all available σ_1 measurements trend north-south (Fig. 8.41). Calculated streamlines trend northwesterly.

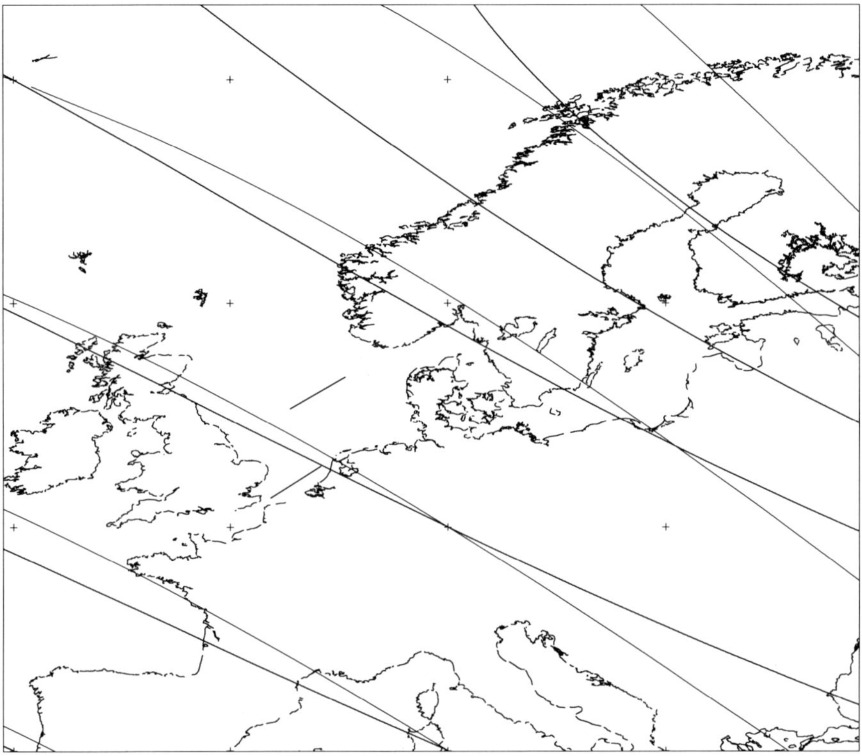

Fig. 8.41. Eurasia: paleostresses and calculated flowlines in hotspot reference frame, 70–75 Ma. Bold flowline 70 Ma.

8.4.16
Western Europe: 75–80 Ma

No paleostress measurements are available between 75 and 80 Ma. Calculated streamlines range from north-northeasterly to northwesterly (Fig. 8.42).

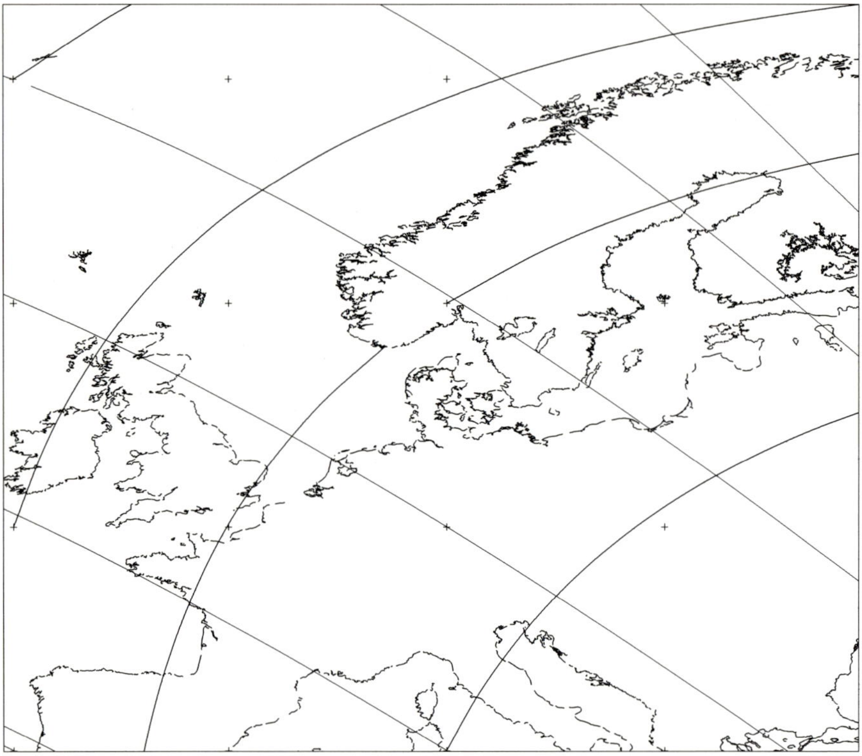

Fig. 8.42. Eurasia: calculated flowlines in hotspot reference frame, 75–80 Ma. Bold flowline 80 Ma.

8.5
East Africa: Paleostresses

All of the paleostress measurements in East Africa are from isotopically dated dikes. Calculated streamlines of African motion in the Atlantic–Indian Ocean hotspot frame based on the models of DeMets et al. (1990) and Muller et al. (1993) interpolated using the methods of this volume show less variability than those of Eurasia (see Sect. 8.4).

8.5.1
East Africa: 0–5 Ma

A single young dike in the 0–5 Ma age range trends slightly west of north, while contemporary stresses are diverse (Fig. 8.43). The contemporary streamline trends west-southwesterly while the 5 Ma streamline trends close to that of the dike.

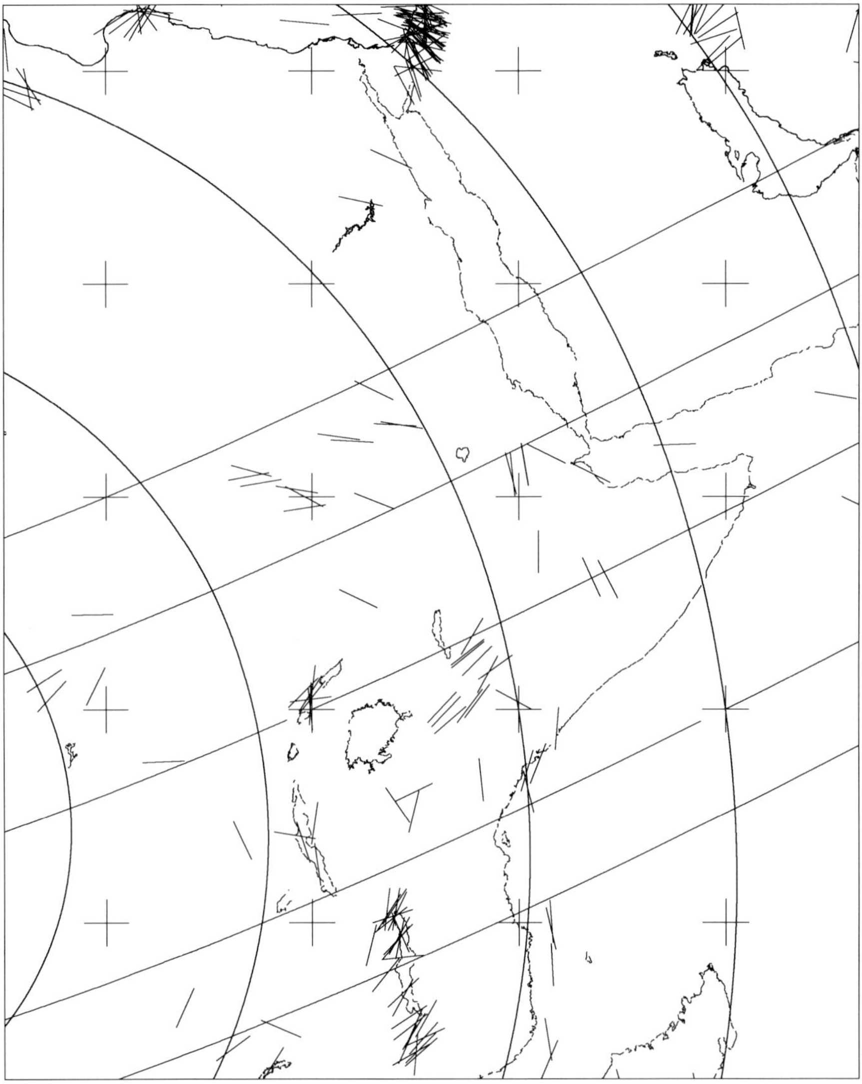

Fig. 8.43. East Africa: paleostresses and calculated flowlines in hotspot reference frame, 0–5 Ma. Bold flowline 0 Ma.

8.5.2
East Africa: 5–10 Ma

No paleostresses are available for 5–10 Ma (Fig. 8.44). Calculated streamlines trend east-northeasterly.

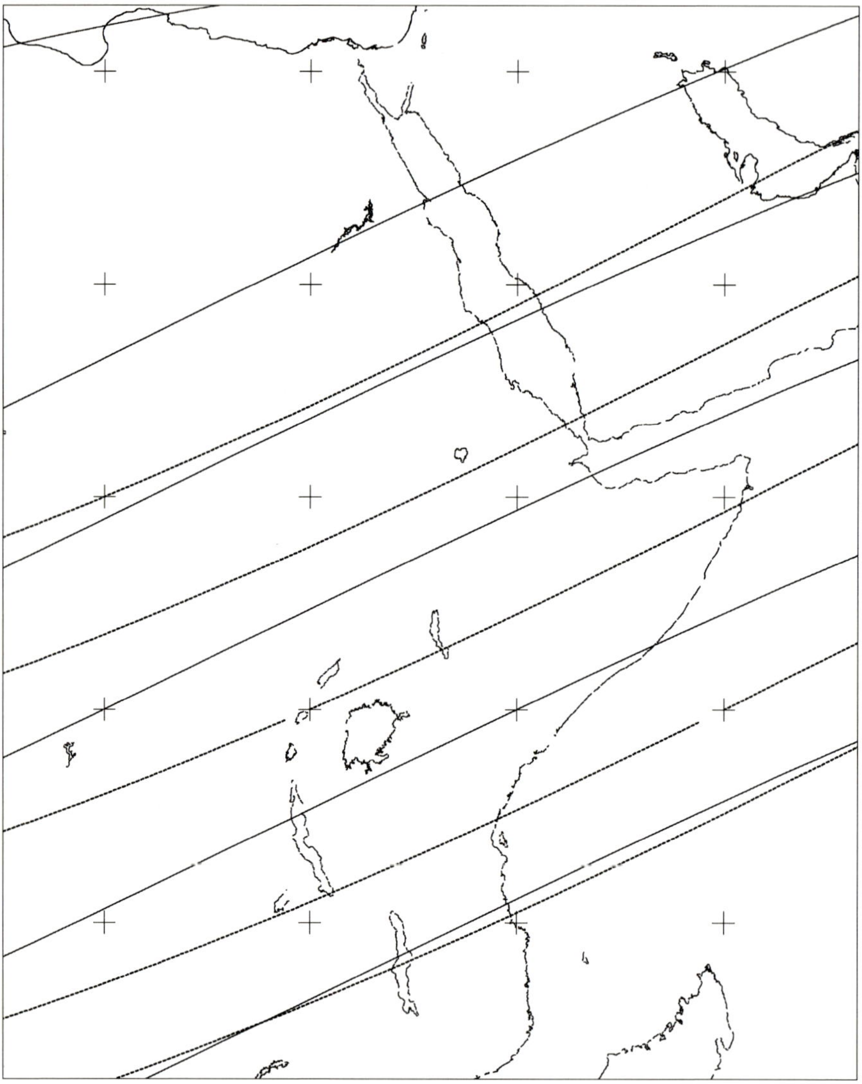

Fig. 8.44. East Africa: paleostresses and calculated flowlines in hotspot reference frame, 5–10 Ma. Bold flowline 10 Ma.

8.5.3
East Africa: 10–15 Ma

One dike trends northwesterly, three trend northerly, and a number of dikes trend northeasterly in the interval 10–15 Ma (Fig. 8.45). Calculated streamlines trend from east-northeasterly to north-northeasterly.

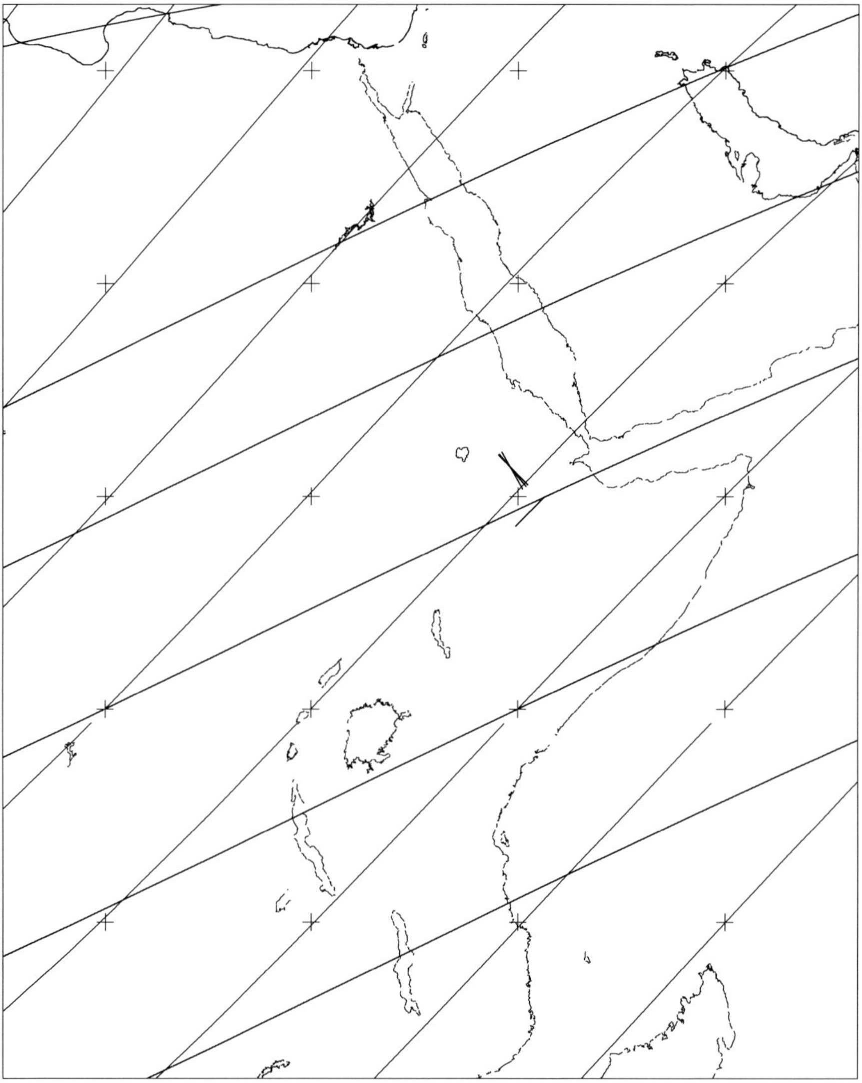

Fig. 8.45. East Africa: paleostresses and calculated flowlines in hotspot reference frame, 10–15 Ma. Bold flowline 10 Ma.

8.5.4
East Africa: 15–20 Ma

Dikes in the age range 15–20 Ma trend from northerly to northeasterly (Fig. 8.46). Calculated streamlines trend slightly north of northeast.

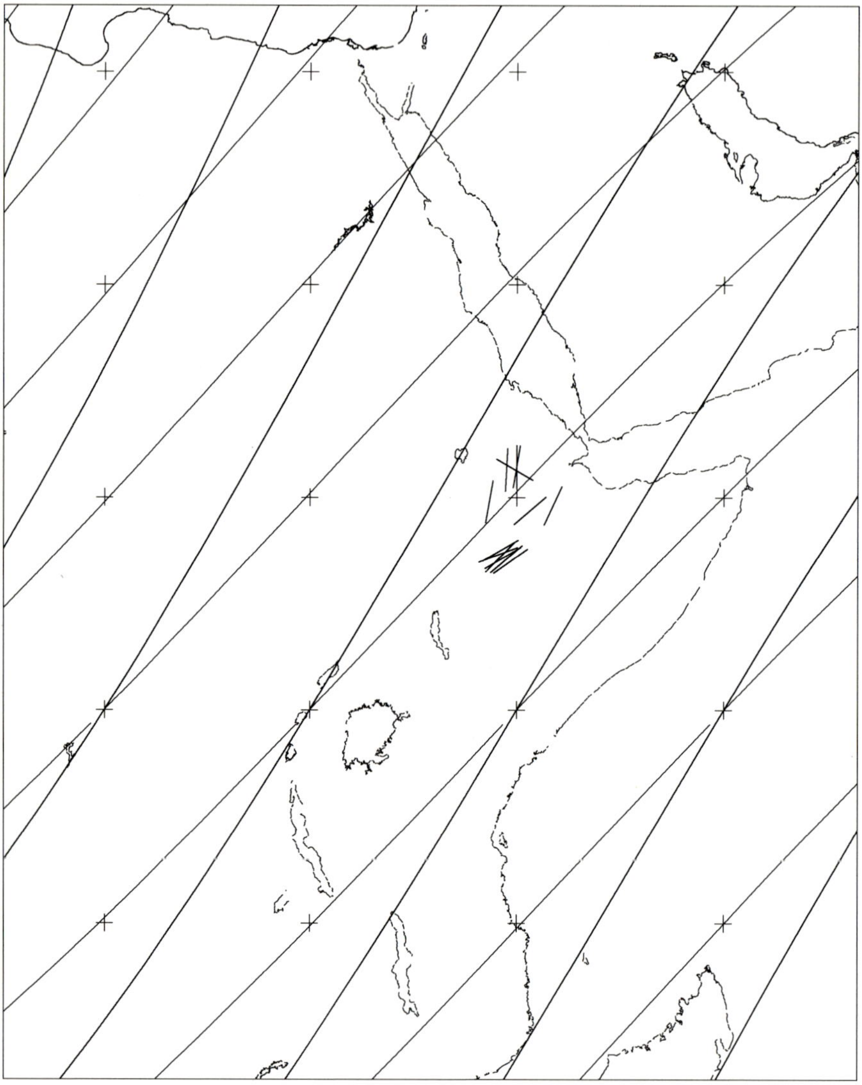

Fig. 8.46. East Africa: paleostresses and calculated flowlines in hotspot reference frame, 15–20 Ma. Bold flowline 20 Ma.

8.5.5 East Africa: 20–25 Ma

A single dike, trending west of north is present in the time range 20–25 Ma (Fig. 8.47). Calculated flowlines trend north-northeast.

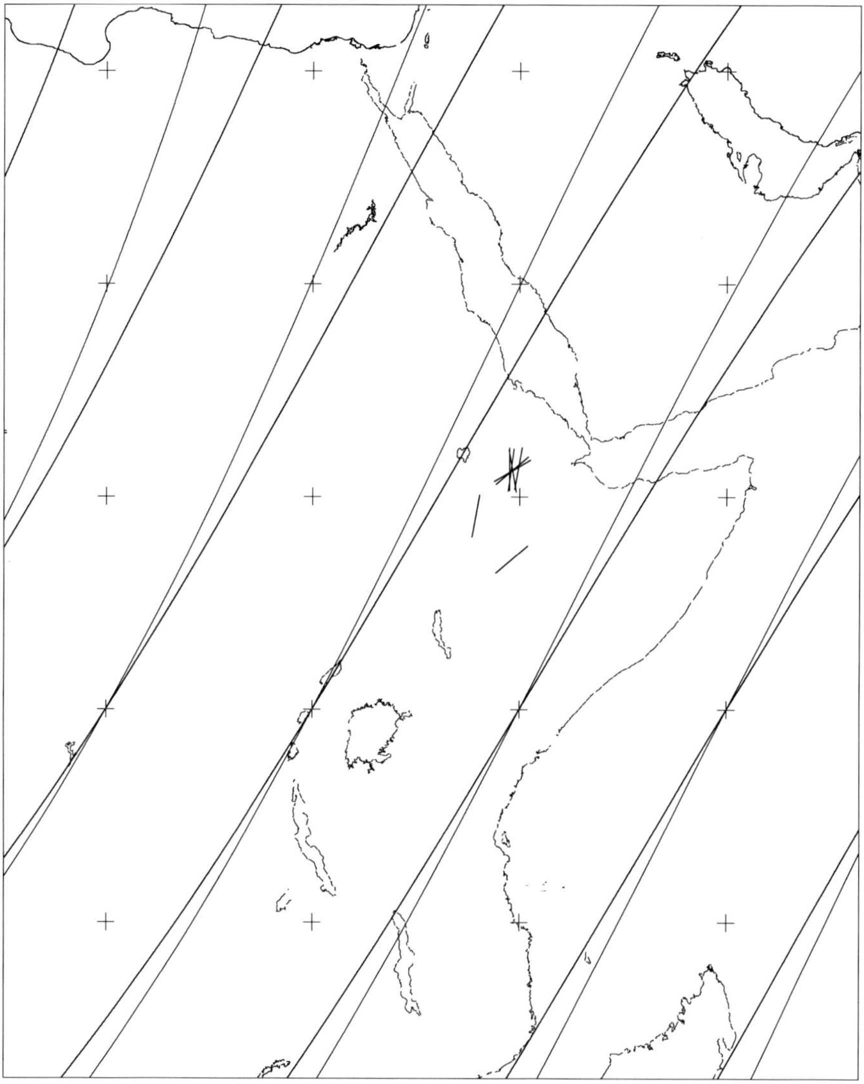

Fig. 8.47. East Africa: paleostresses and calculated flowlines in hotspot reference frame, 20–25 Ma. Bold flowline 20 Ma.

8.5.6
East Africa: 25–30 Ma

In the time interval 25–30 Ma, three dikes are oriented northwesterly, while three dikes have a northeasterly orientation (Fig. 8.48). Calculated flowlines trend slightly east of northeast.

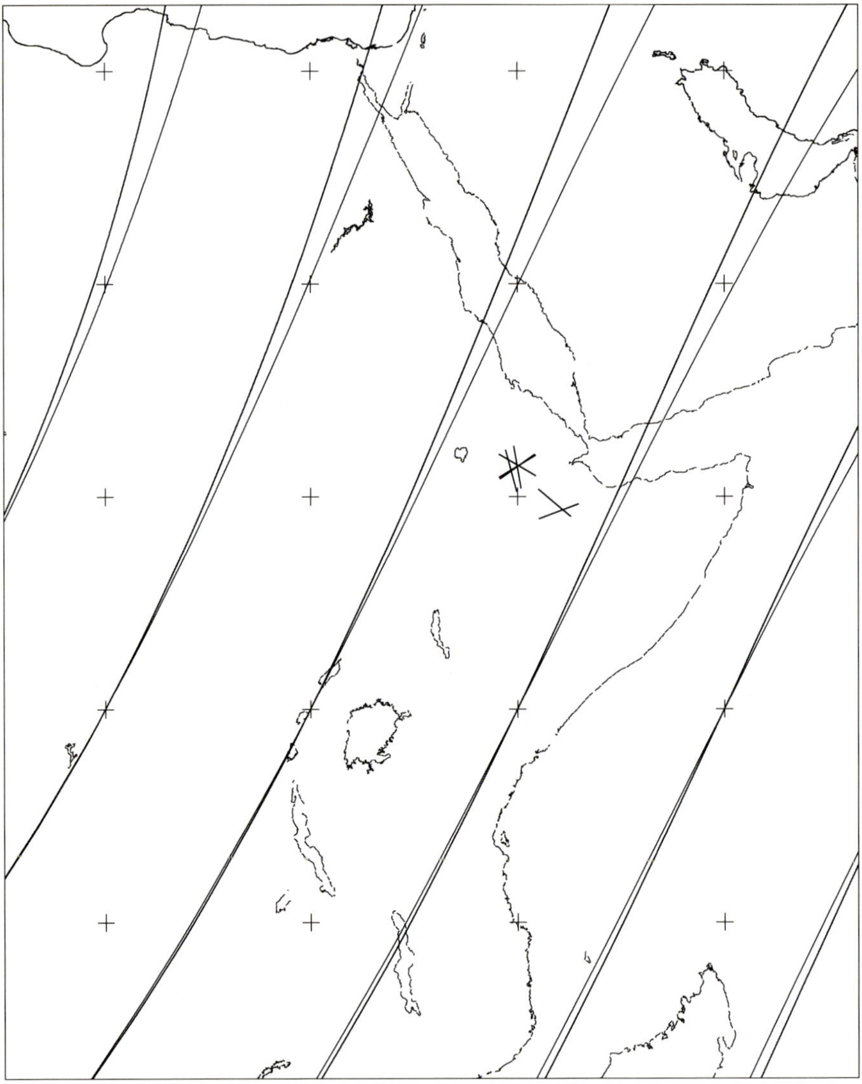

Fig. 8.48. East Africa: paleostresses and calculated flowlines in hotspot reference frame, 30–35 Ma. Bold flowline 30 Ma.

8.5.7 East Africa: 30–35 Ma

No paleostresses are available in the interval from 30 to 35 Ma (Fig. 8.49). Flowlines for the interval trend between east-northeasterly to north-northeasterly.

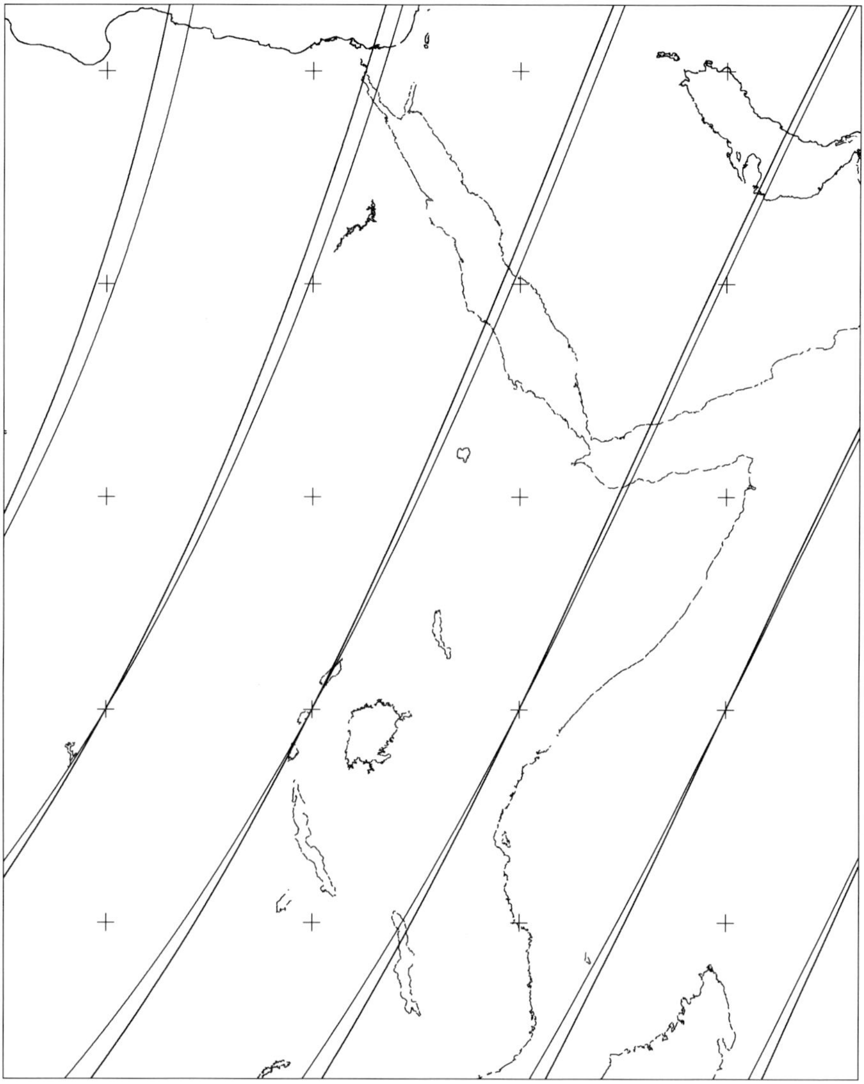

Fig. 8.49. East Africa: calculated flowlines in hotspot reference frame, 35–40 Ma. Bold flowline 30 Ma.

8.5.8
East Africa: 35–40 Ma

For 35–40 Ma no paleostresses are available (Fig. 8.50). Flowlines are calculated to trend northeasterly to north-northeasterly.

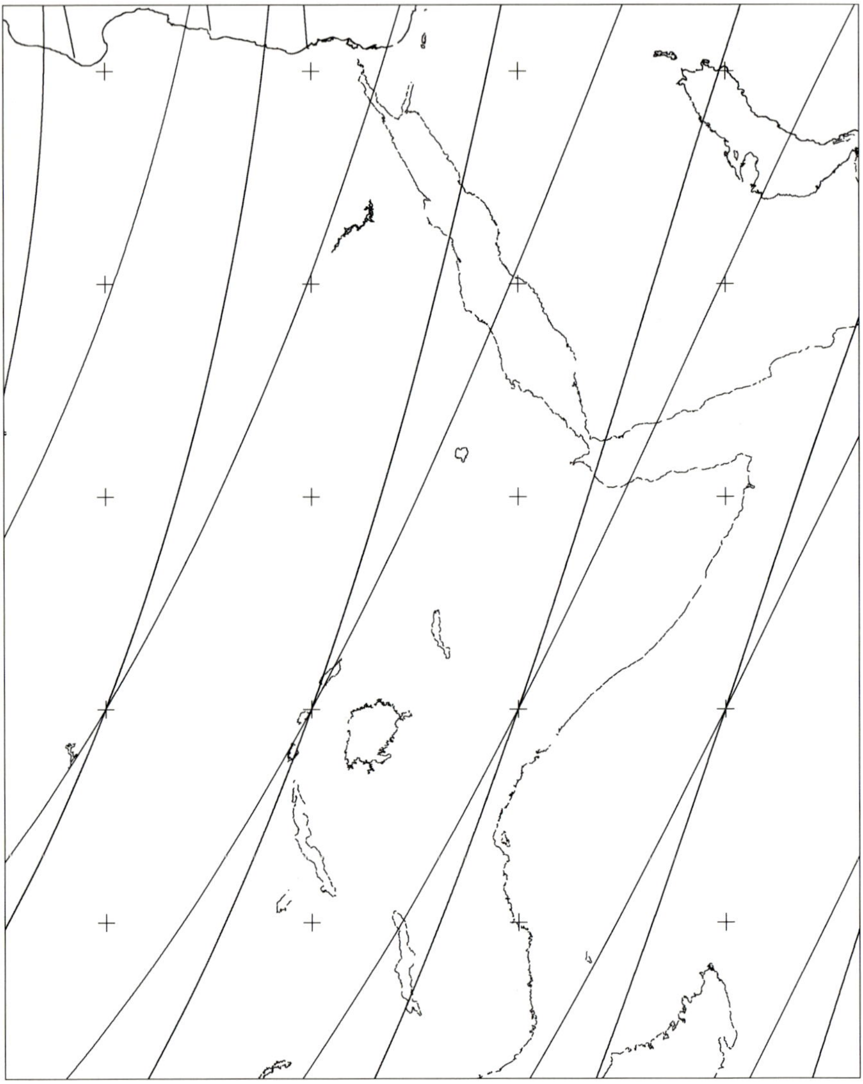

Fig. 8.50. East Africa: calculated flowlines in hotspot reference frame, 40–45 Ma. Bold flowline 40 Ma.

8.5.9
East Africa: 40–45 Ma

No paleostresses are available in the time range 40–45 Ma (Fig. 8.51). Flowlines trend north-northeasterly.

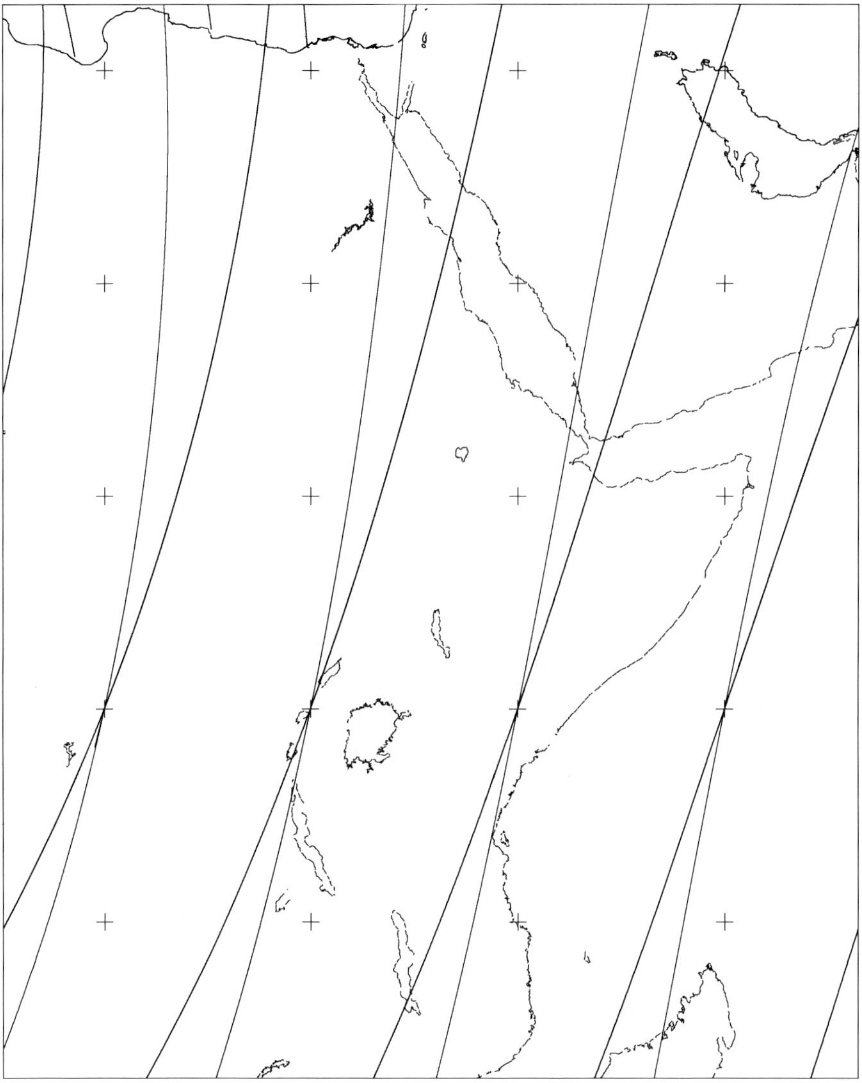

Fig. 8.51. East Africa: calculated flowlines in hotspot reference frame, 45–50 Ma. Bold flowline 50 Ma.

8.5.10
East Africa: 45–50 Ma

No dikes are available in the 45–50 Ma time range. Calculated flowlines trend north-northeasterly (Fig. 8.52).

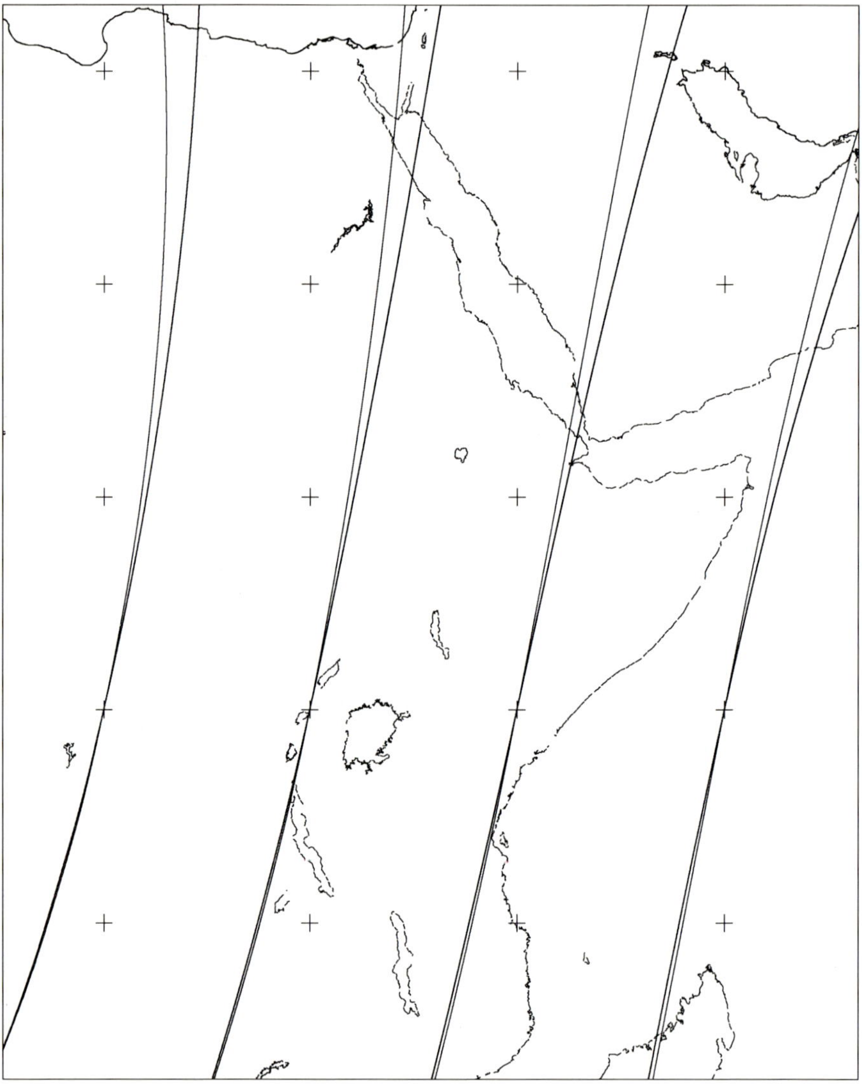

Fig. 8.52. East Africa: calculated flowlines in hotspot reference frame, 50–55 Ma. Bold flowline 50 Ma.

8.5.11
East Africa: 50–55 Ma

No dikes are available in the 50–55 Ma time range. Calculated flowlines trend north-northeasterly (Fig. 8.53).

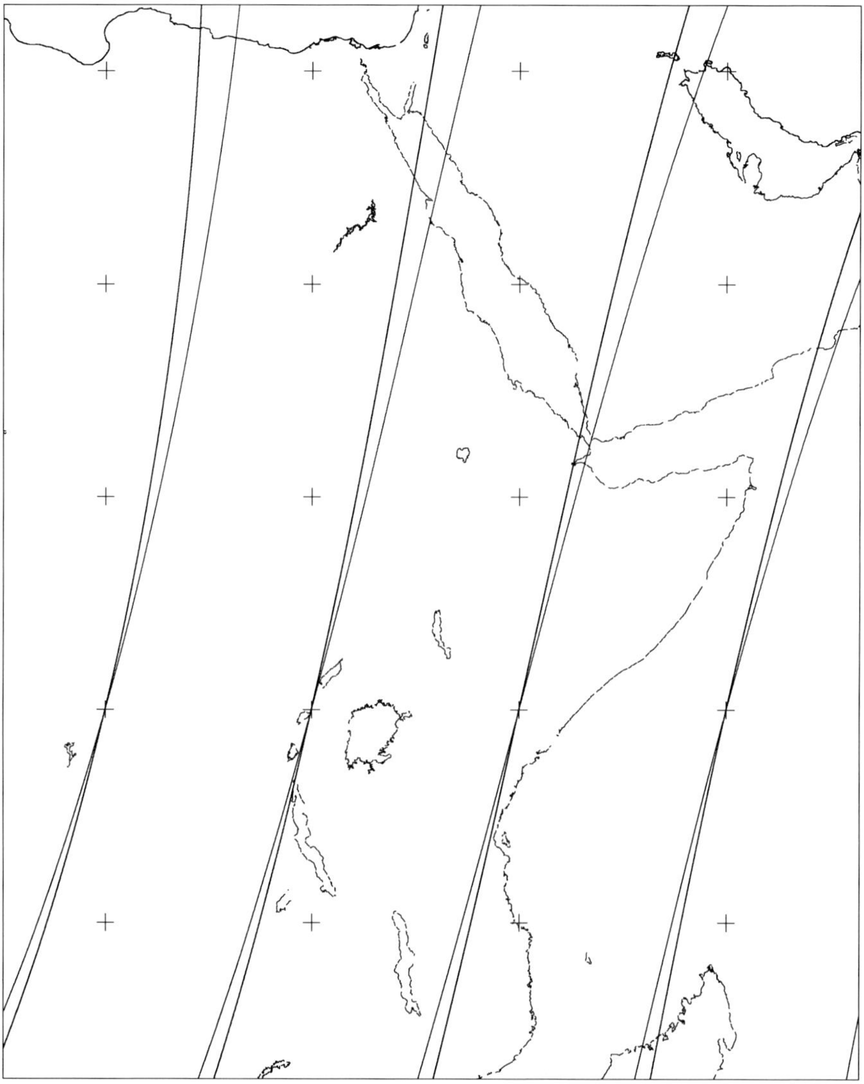

Fig. 8.53. East Africa: calculated flowlines in hotspot reference frame, 55–60 Ma. Bold flowline 60 Ma.

8.5.12
East Africa: 55–60 Ma

No dikes are available in the 55–60 Ma time range. Calculated flowlines trend north-northeasterly (Fig. 8.54).

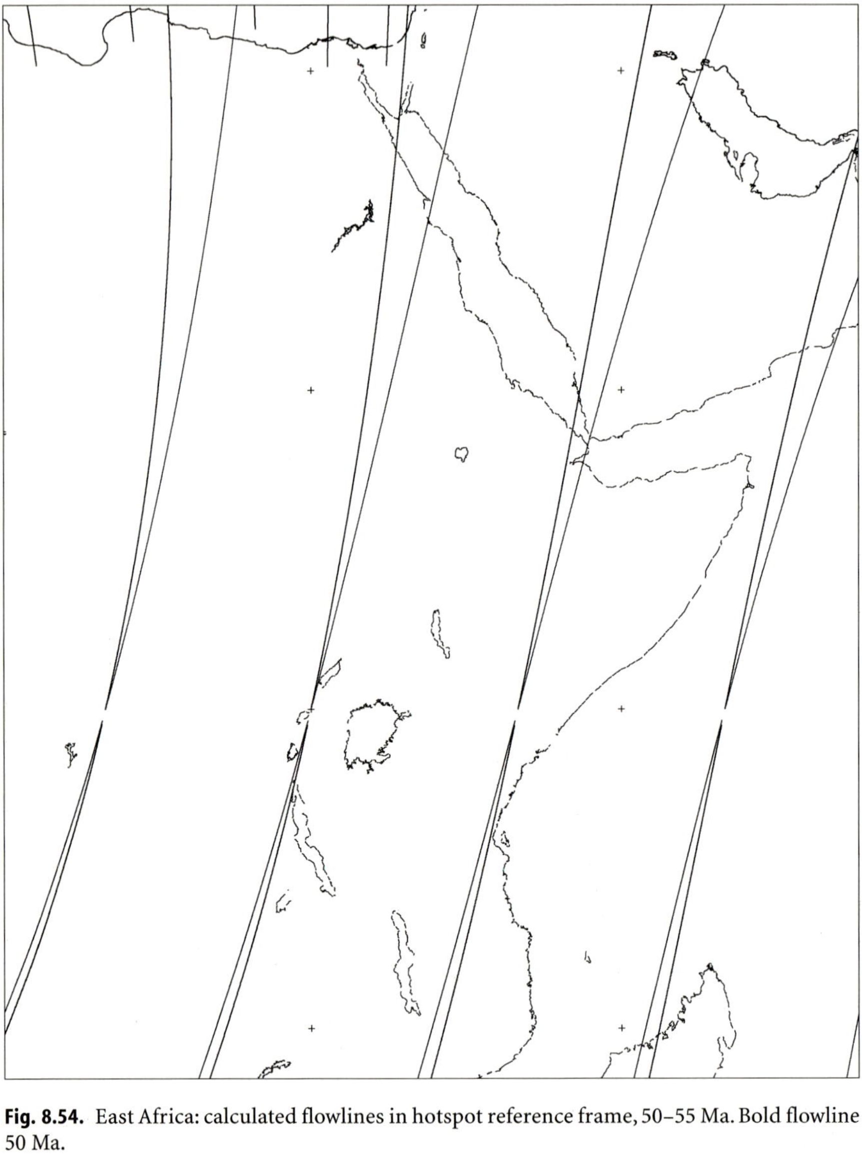

Fig. 8.54. East Africa: calculated flowlines in hotspot reference frame, 50–55 Ma. Bold flowline 50 Ma.

8.6 East Australia: Paleostresses

8.6.1 East Australia: 0–5 Ma

Two paleostress observations are available from Queensland (Fig. 8.55), with northeasterly orientations. Contemporary plate motion in the Atlantic–Indian Ocean hotspot reference frame is slightly east of north, while motion at 5 Ma was north-northeasterly.

8.6.2 East Australia: 5–20 Ma

Calculated flowlines in the hotspot frame range from north-northeasterly to almost northeasterly between 5 and 15 Ma (Figs. 8.56–8.58). No paleostress measurements are available for this time range.

8.6.3 East Australia: 20–25 Ma

Calculated flowlines in the hotspot frame range from north-south to north (Fig. 8.59). No paleostress measurements are available for this time range. A single paleostress measurement from the East Australian Highlands also trends close to north-south; the offshore measurement is from the trend of the Tasman seamount chain.

8.6.4 East Australia: 25–35 Ma

Calculated flowlines in the hotspot frame range from north-south to northeasterly between 25 and 35 Ma (Figs. 8.60, 8.61). No paleostress measurements are available for this time range.

8.6.5 East Australia: 35–55 Ma

Calculated flowlines in the hotspot frame range from north-northeasterly to almost northeasterly between 35 and 55 Ma (Figs. 8.62–8.65). No paleostress measurements are available for this time range.

8.6.6 East Australia: 55–60 Ma

Calculated flowlines in the hotspot frame range from northeasterly to northwesterly (Fig. 8.66) in the interval from 55 to 60 Ma. A single paleostress measurement from the East Australian Highlands trends north-northwesterly.

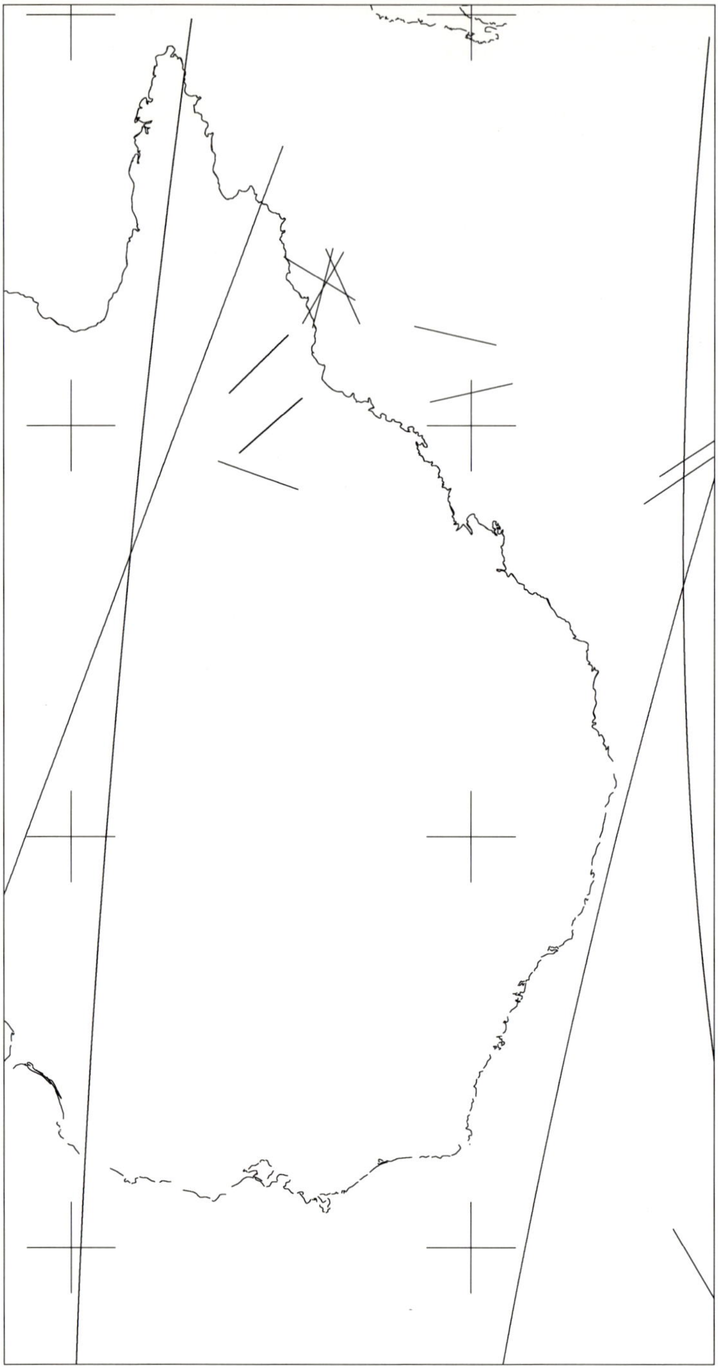

Fig. 8.55. East Australia: contemporary stresses, paleostresses and calculated flowlines in hotspot reference frame, 0–5 Ma. Bold flowline 0 Ma.

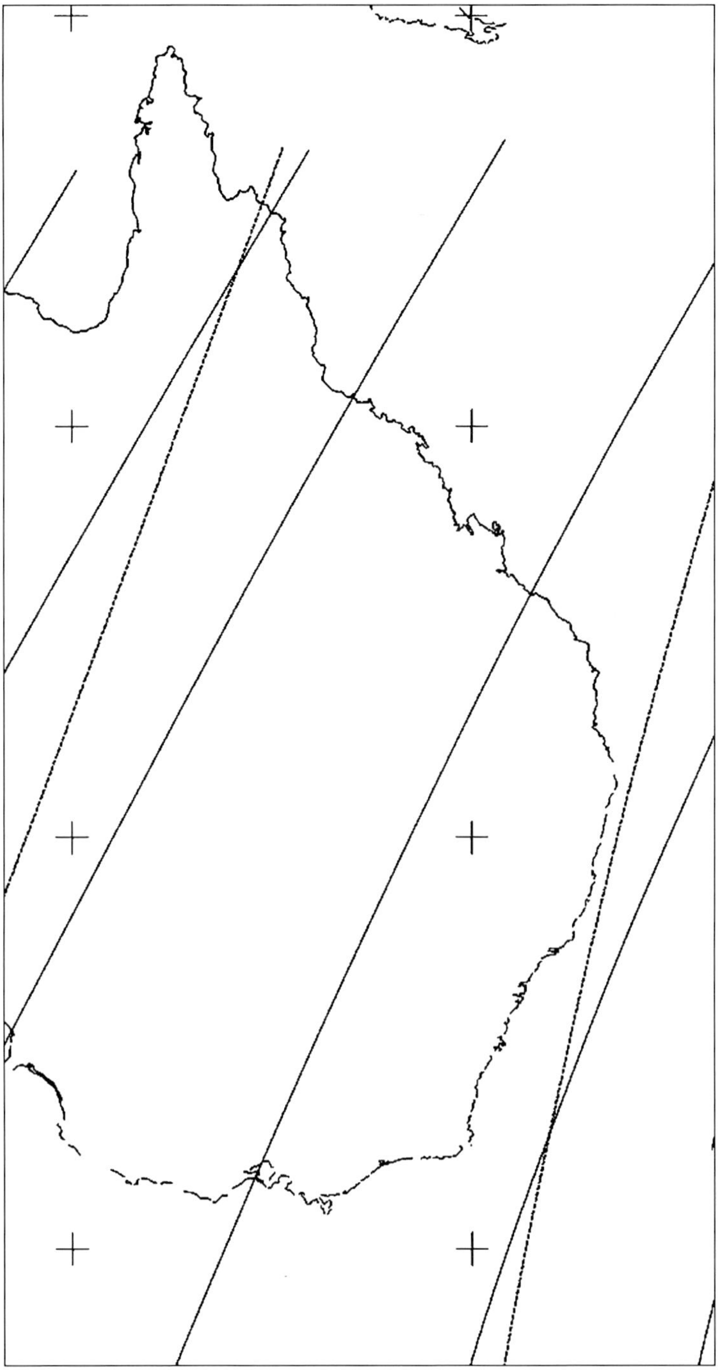

Fig. 8.56. East Australia: calculated flowlines in hotspot reference frame, 5–10 Ma. Bold flowline 10 Ma.

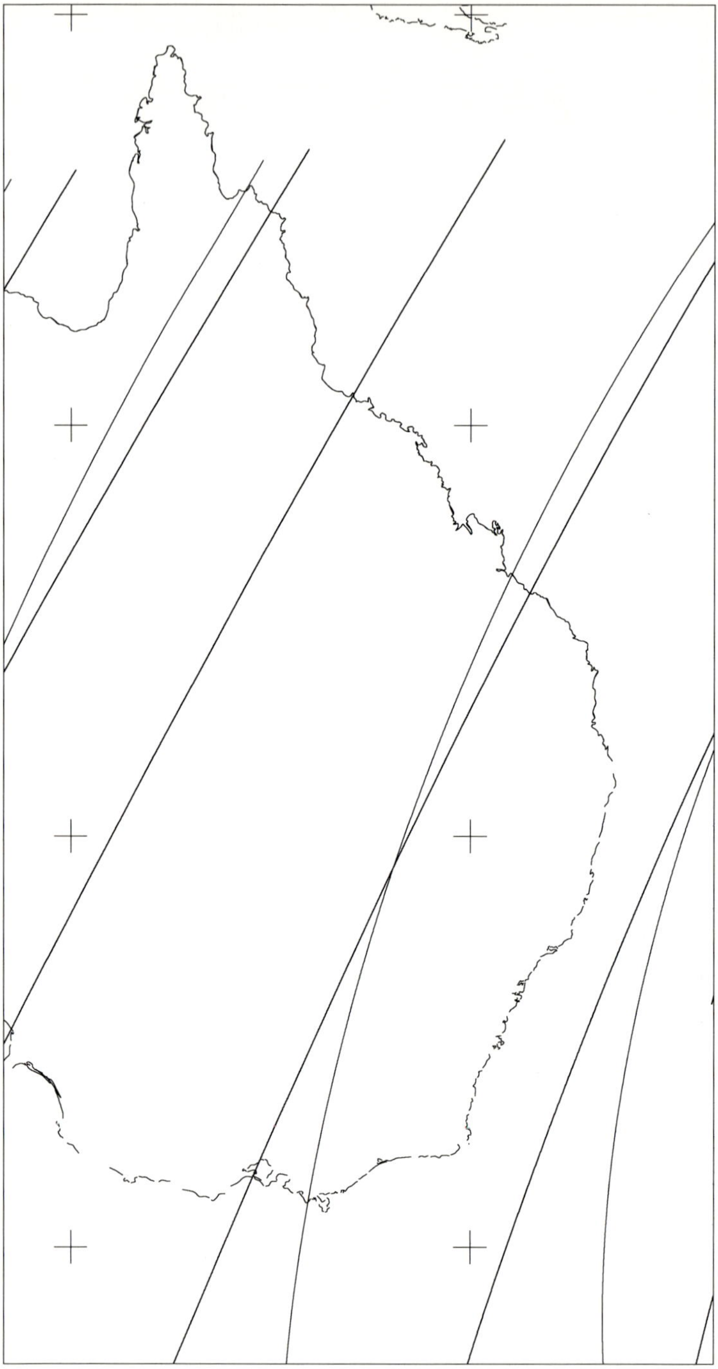

Fig. 8.57. East Australia: calculated flowlines in hotspot reference frame, 10–15 Ma. Bold flowline 10 Ma.

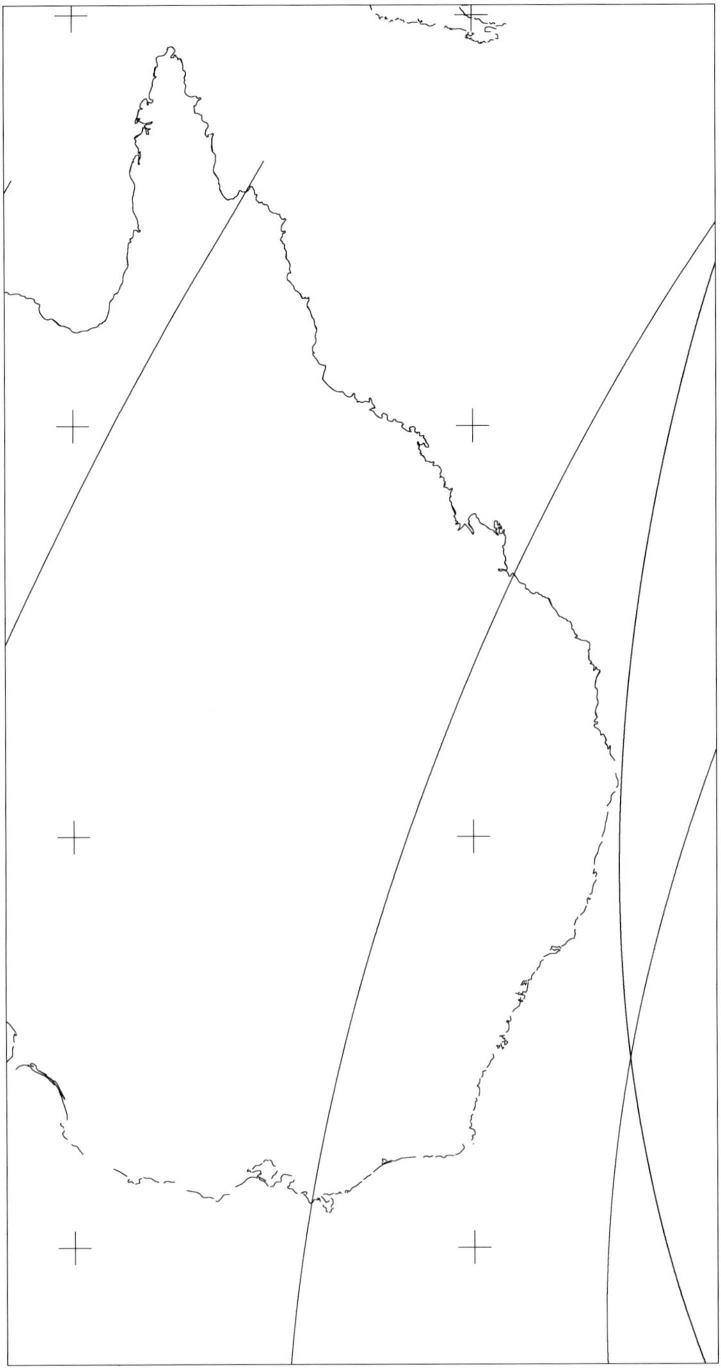

Fig. 8.58. East Australia: calculated flowlines in hotspot reference frame, 15–20 Ma. Bold flowline 20 Ma.

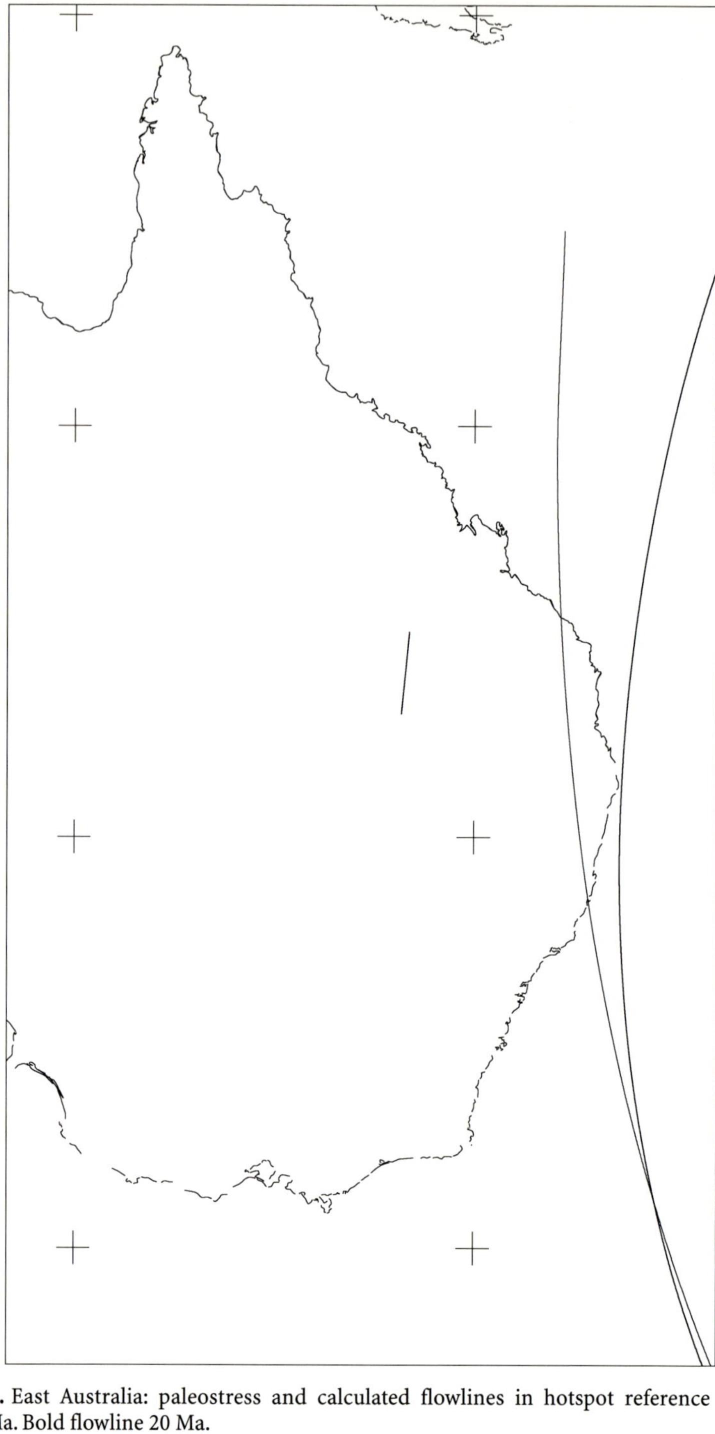

Fig. 8.59. East Australia: paleostress and calculated flowlines in hotspot reference frame, 20–25 Ma. Bold flowline 20 Ma.

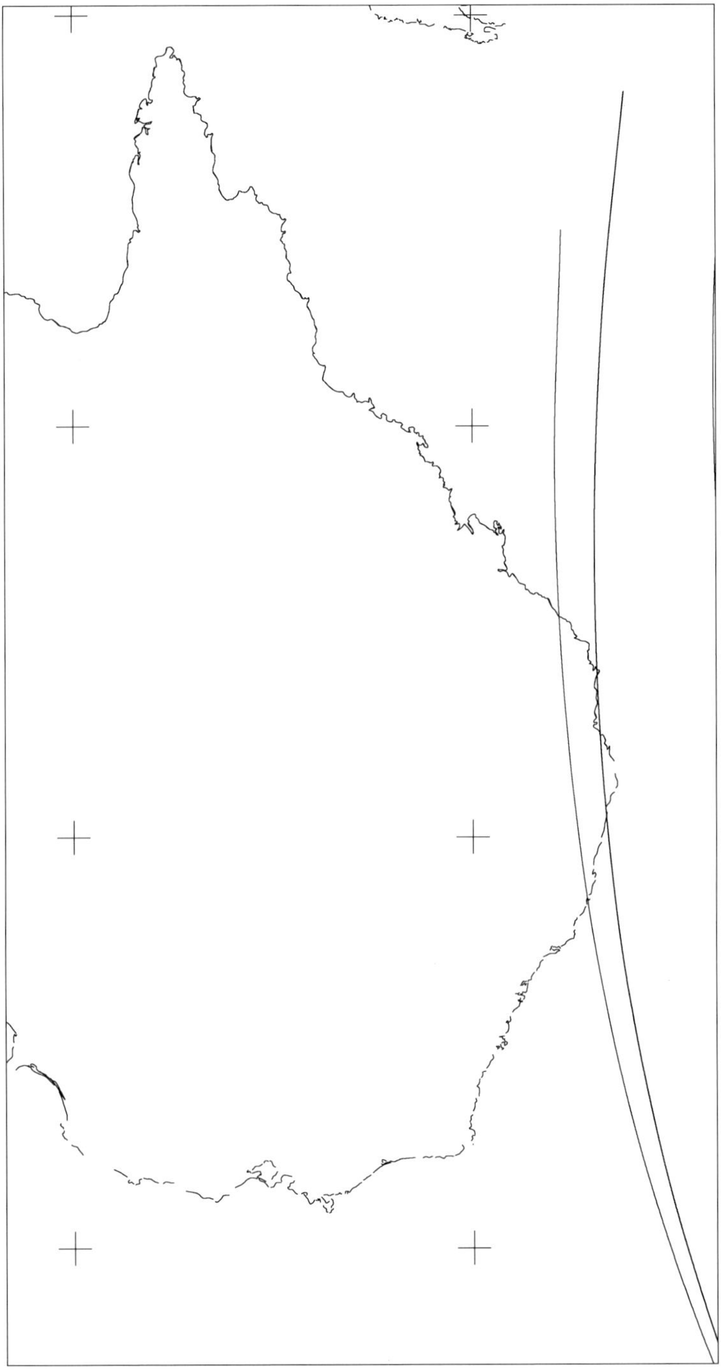

Fig. 8.60. East Australia: calculated flowlines in hotspot reference frame, 25–30 Bold flowline 30 Ma.

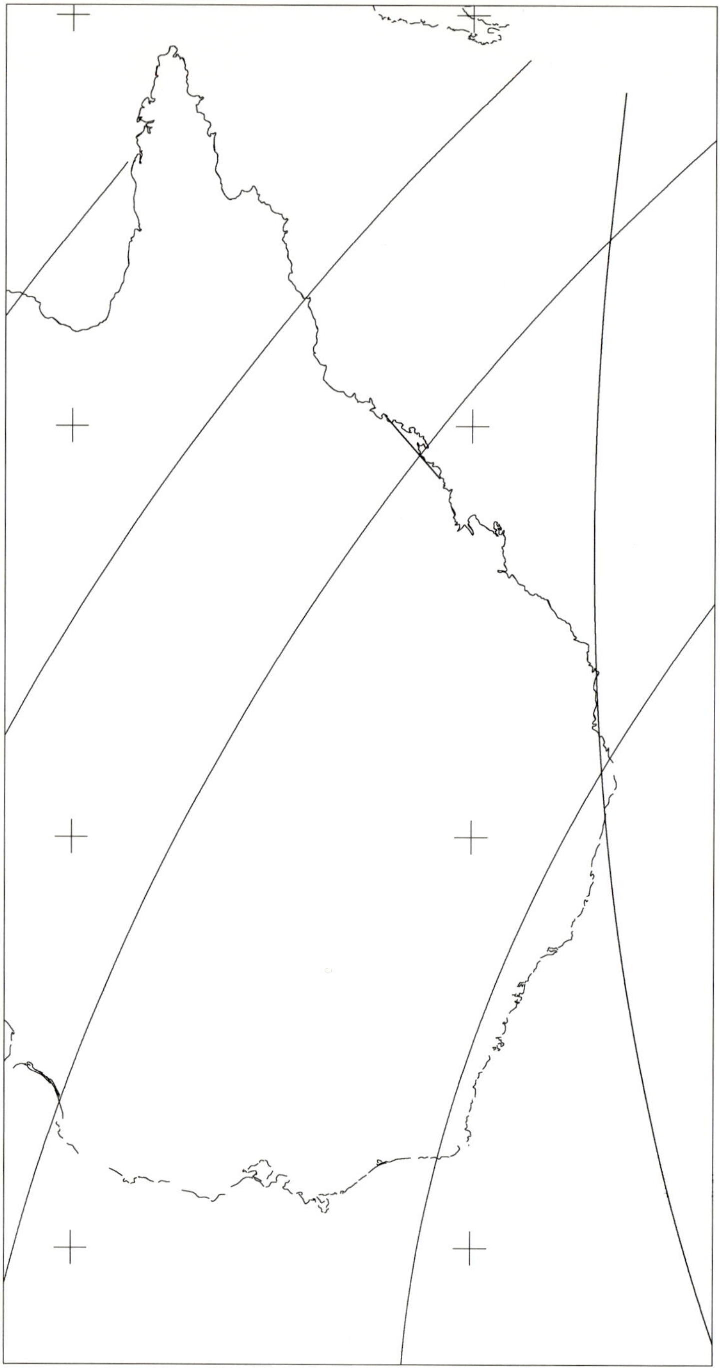

Fig. 8.61. East Australia: calculated flowlines in hotspot reference frame, 30–35 Ma. Bold flowline 30 Ma.

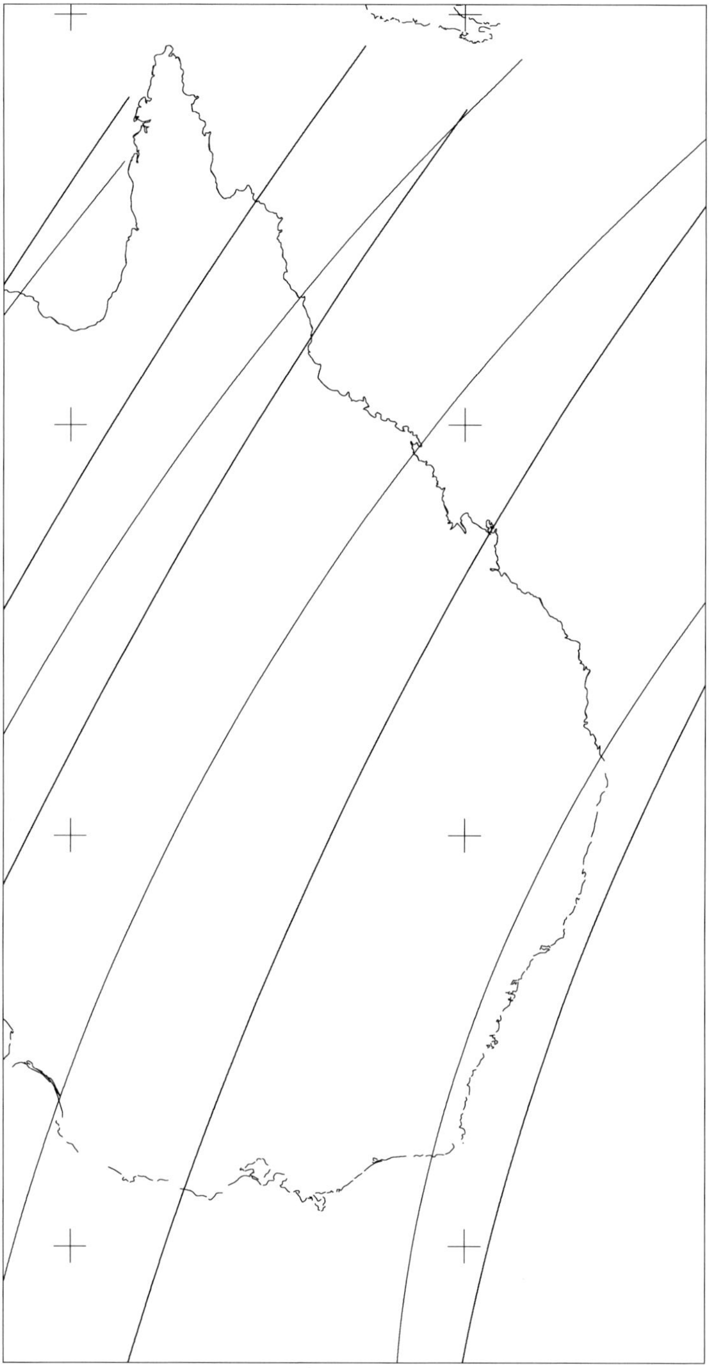

Fig. 8.62. East Australia: calculated flowlines in hotspot reference frame, 35–40 Ma. Bold flowline 40 Ma.

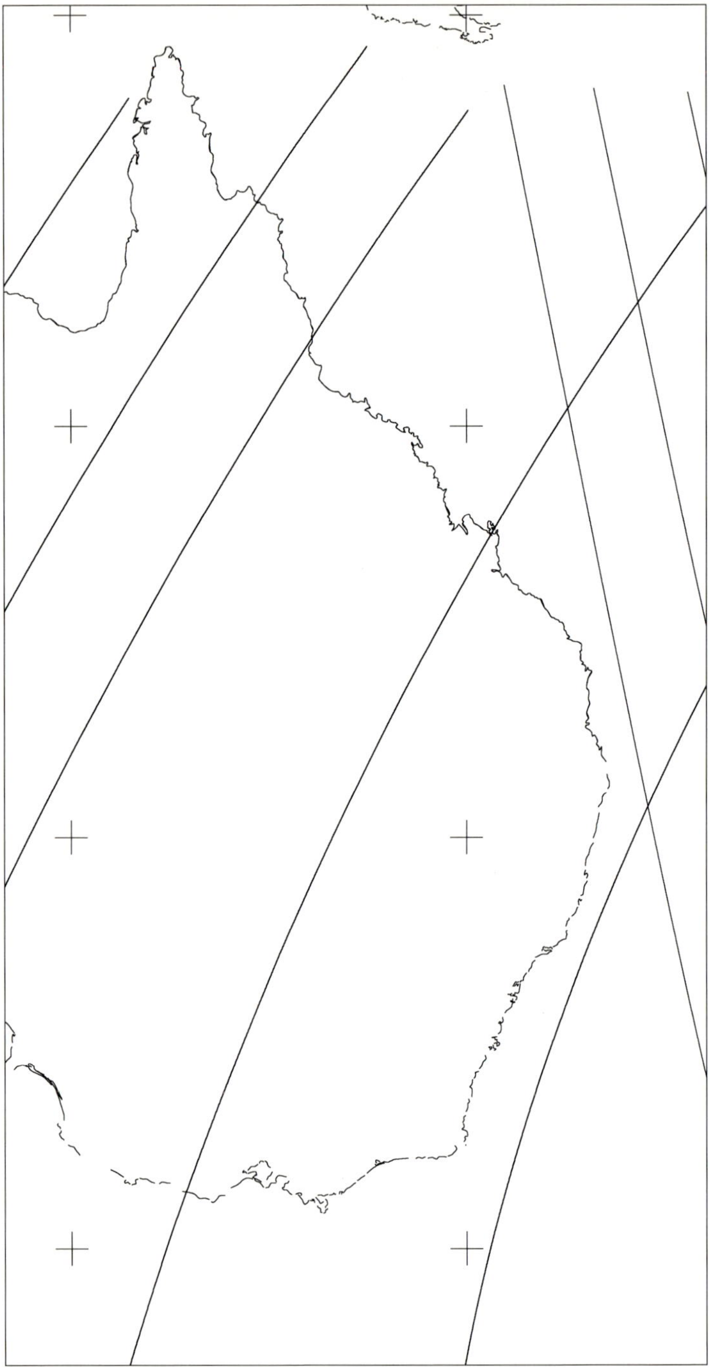

Fig. 8.63. East Australia: calculated flowlines in hotspot reference frame, 40–45 Ma. Bold flowline 40 Ma.

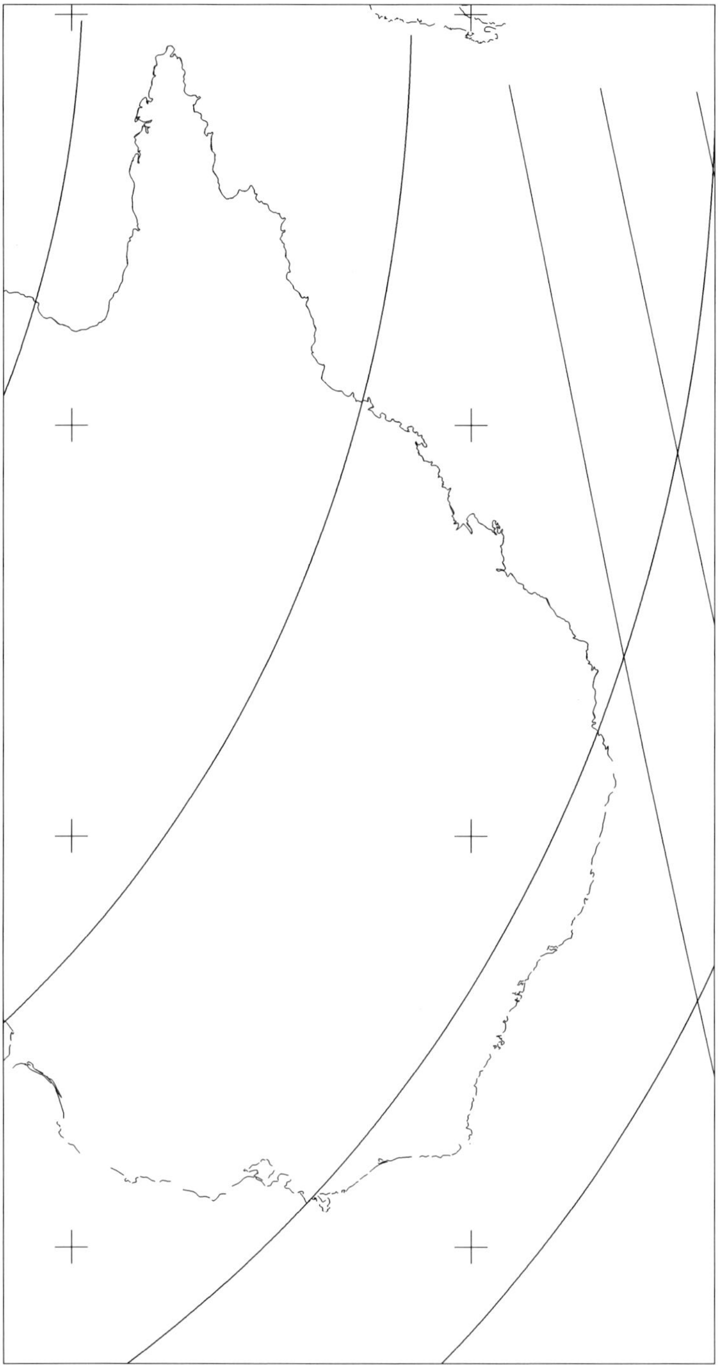

Fig. 8.64. East Australia: calculated flowlines in hotspot reference frame, 45–50 Ma. Bold flowline 50 Ma.

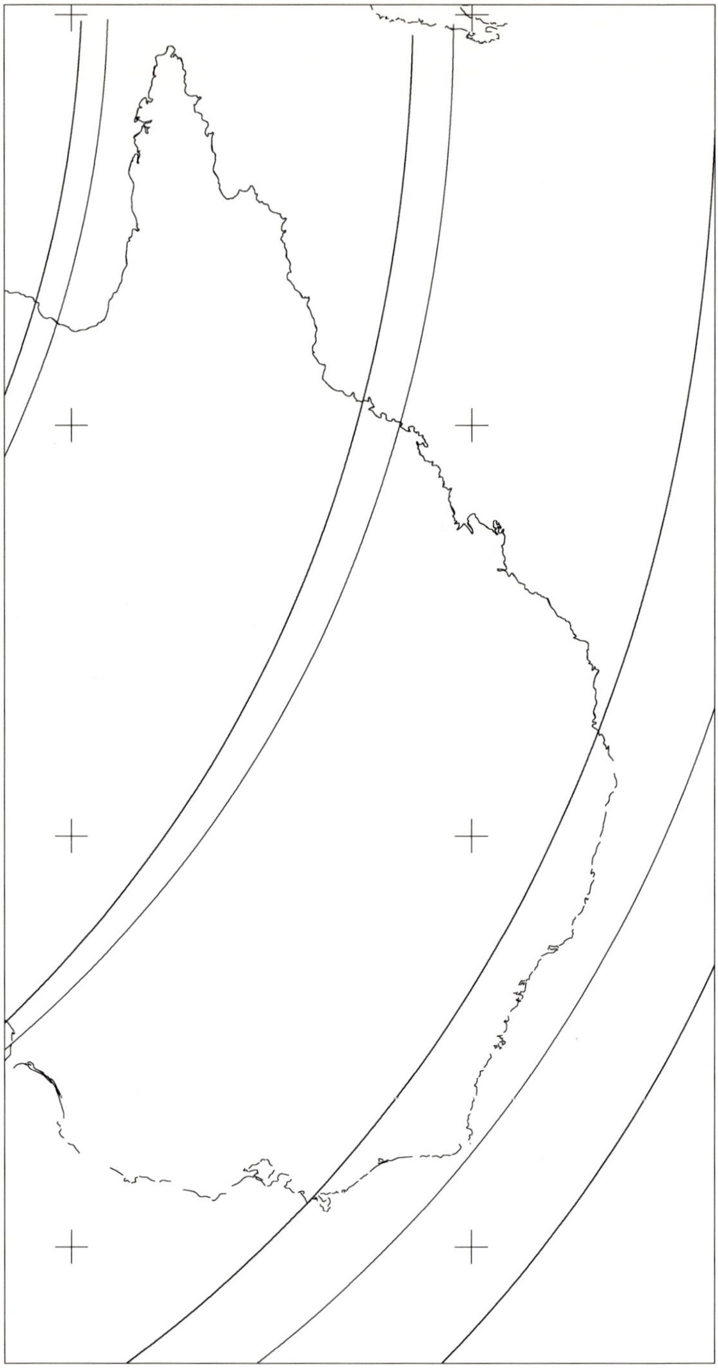

Fig. 8.65. East Australia: calculated flowlines in hotspot reference frame, 50–55 Ma. Bold flowline 50 Ma.

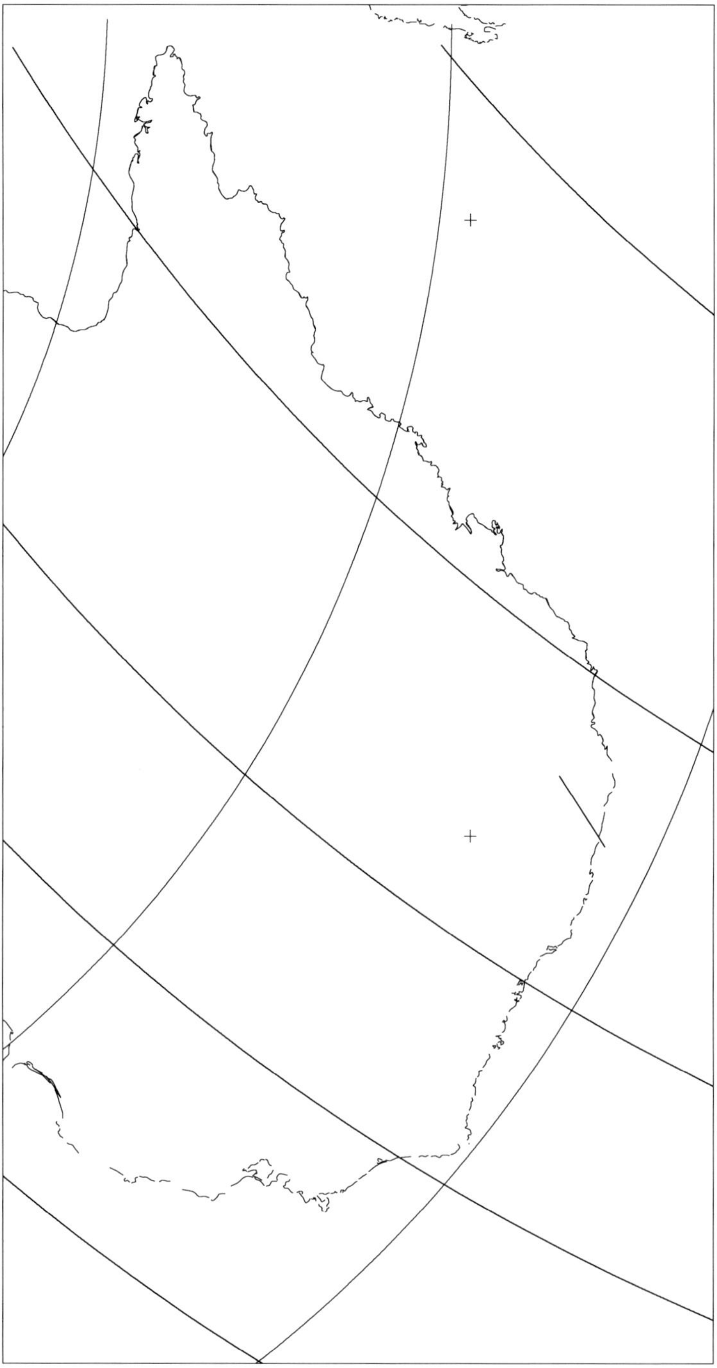

Fig. 8.66. East Australia: paleostress and calculated flowlines in hotspot reference frame, 55–60 Ma. Bold flowline 60 Ma.

CHAPTER 9

Paleomagnetic Reference Frame

9.1
Aside

Paleomagnetism, after morphology and geology, provided key evidence for the great precursor of plate tectonics – continental drift. One of the great hopes with plate tectonics was for paleomagnetism to provide additional kinematic constraints. Alas, the very processes apparently involved in generation of the geomagnetic field conspire against providing the kind of high-resolution information the kinematicist would want. The question of the relation of polar wander curves to hotspot reference frames is a corollary. Even though high resolution is not possible, it is still worthwhile comparing paleomagnetic observations with the hotspot frame.

9.2
Paleomagnetism

After morphological and geological arguments (Wegener 1912; du Toit 1937), paleomagnetic evidence provided the next great body of evidence for continental drift, prior to the advent of nascent plate tectonics (Hess 1962; Vine and Matthews 1963; Morley and Larochelle 1964; Wilson 1965b). The divergence of measured apparent polar wander curves between North America and Europe suggested progressive separation of the two continents during the Late Mesozoic and Cenozoic. With the much higher resolution evidence from oceanic lineations and fracture zone offsets, paleomagnetic data no longer provide significant constraints on the relative motion of continents and plates. Nevertheless, the dipole hypothesis as applied to paleomagnetic data provides a possible internal reference frame for plate motions, especially if the magnetic poles correspond with the Earth's rotation axis. (Actually, of course, even if the dipole does not correspond with the rotation axis paleomagnetism could still provide a reference frame. It seems more likely that the existence of a correspondence with the rotational pole would then imply greater tectonic significance.)

Prior to acceptance of plate tectonics (and, therefore, continental drift), one of the explanations for deviation of apparent polar wander curves from the rotational poles was "true polar wander"; at present there are still arguments being advanced in support of either significant shifts of the magnetic pole (or even the rotational and magnetic poles together) to explain large deviations between the apparent paleomagnetic polar wander curve and polar wander curves in the hotspot reference frame.

Without directly addressing the question of true polar wander, it seems of value to evaluate the correspondence (or lack thereof) of paleomagnetic data with plate motions in the hotspot reference frame(s). A great deal of well-dated paleomagnetic data has been acquired over the past 40 years. Fortunately, a publicly available compilation of the paleomagnetic data is available to examine possible correspondence with inferred plate motions in the hotspot reference frame.

From the isotopic age of a paleomagnetic datum, the measured pole can be rotated back to its location in the hotspot frame at the time the paleomagnetic field was inferred to be recorded in the rock/mineral sample. Using the IAGA (2001) Paleomagnetic Database (as served by the Norwegian Geological Survey), paleomagnetic poles for North and South America, Europe, Greenland, Africa, India, Australia, and Antarctica were extracted with the following criteria: (1) Published since: 1970; (2) terrane: all; (3) rock type: all; (4) demagnetization procedure: Stereonets and vector plots or more; with results that supersede previous, lower quality measurements. Further, data with age ranges less than 20% of the maximum age were filtered out.

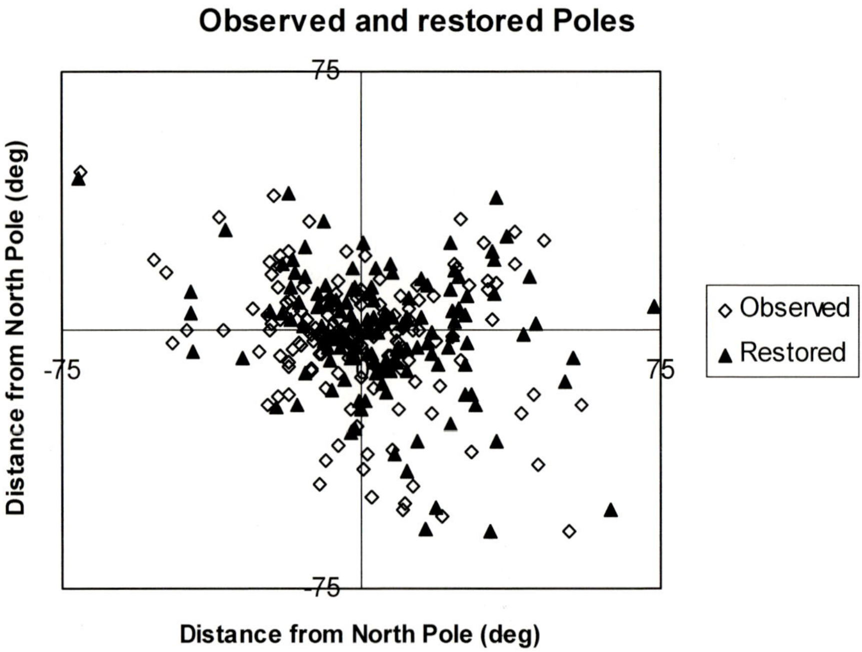

Fig. 9.1. Paleomagnetic poles from North America, Eurasia, Africa, India, Antarctica, and Australia. Observed (*diamonds*) and restored (*triangles*) to Atlantic–Indian Ocean hotspot reference frame (rotation parameters determined by isotopic age and method outlined in Chap. 3). Note the tighter clustering of the restored data compared with the present-day configuration.

Figure 9.1 illustrates the present-day apparent paleomagnetic poles and the same poles rotated back to their inferred time of formation relative to the Atlantic–Indian Ocean hotspot reference frame using the model of Müller et al. (1993) and the interpolation methods presented here. Note that the scatter in the data is significantly reduced from the observed distribution, as would be expected, since the hotspot reconstruction includes removal of continental divergence subsequent to freezing-in of the paleomagnetism. However, note also that the reconstructed clustering is, nevertheless, displaced somewhat from the rotational pole. The weighted-mean reconstructed pole location is 86.20°N, 5.93°E, which is still not too far from the rotational pole.

Comparison of predicted and observed apparent pole locations shows an interesting pattern (Fig. 9.2a, b). For all continents, there is only a broad correspondence in latitude, with significant dispersion (Fig. 9.2a). While there is limited correspondence in longitude (Fig. 9.2b), there is a much greater apparent dispersion (as would be expected give the proximity of the magnetic pole to the geographic pole as well as the dispersion in latitude). Note that predicted latitudes tend to be much higher than observed latitudes (Fig. 9.2a).

A comparison of observed and predicted poles as a function of age further amplifies the generalization regarding polar latitudes (Fig. 9.3a–c). By further separating the comparison on the basis of continent, it is apparent which continents show the most striking trend in observed and predicted latitude - North America and Europe, which also have the greatest number of data points (Figs. 9.3–9.5).

In his elaboration of the fixed hotspot/plume hypothesis, Morgan (1972a) recognized that paleomagnetic data did not correspond with inferred plate motions relative to the hotspots according to the model he produced for the Atlantic and Indian Oceans. While subsequent hotspot reference frame models of the Atlantic–Indian Ocean plates have been able to improve their fit to the hotspots, it is apparent that there is still a discrepancy between the hotspot model and the paleomagnetic reference frame, in the same sense as Morgan originally observed.

Figure 9.11 summarizes the difference between observed paleomagnetic and predicted hotspot-model poles. The amount of dispersion in paleomagnetic observations is discouraging insofar as the derivation of a precise polar wander curve and a kinematic model-reconstruction model are concerned. It would be desirable to be able to calculate the relative motion of the paleomagnetic field and Atlantic–Indian Ocean and the Pacific Ocean hotspots. However, in addition to the scatter in paleomagnetic observations, there is the additional factor that only the magnetic pole location is sampled. Thus, a wide spectrum of kinematic models could be derived for motion of the lithosphere relative to the paleomagnetic reference frame.

To summarize, the Atlantic–Indian Ocean hotspot reference frame is separate from the paleomagnetic frame. However, the scatter in paleomagnetic observations limits the ability to precisely quantify the kinematics of the difference between the two reference frames.

The question might arise as to why weighted averaging of the reconstructed poles could not be undertaken. Certainly, there have been a number of previous

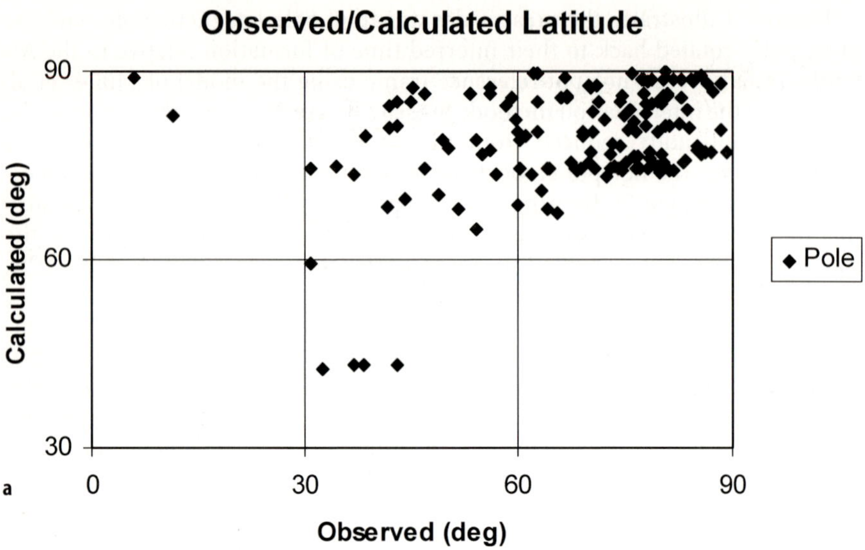

a

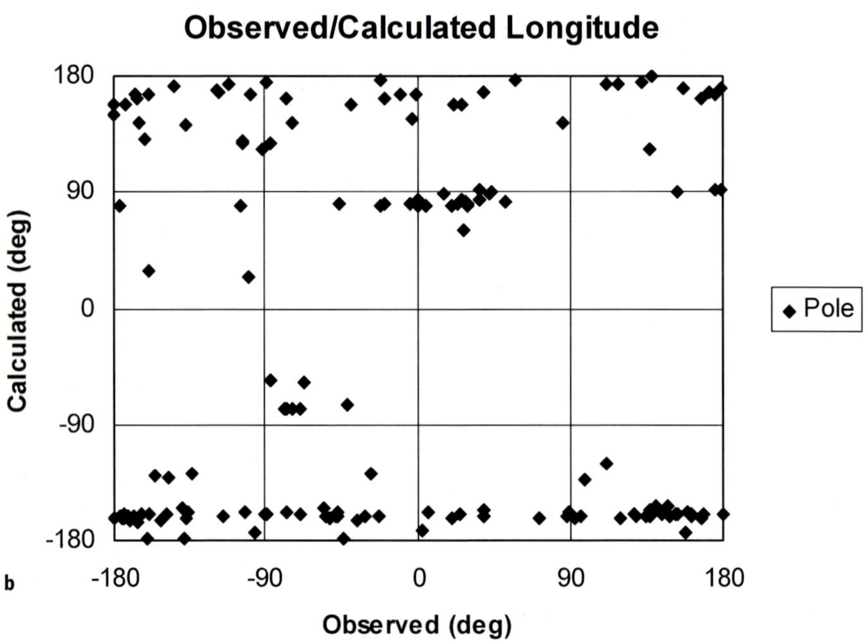

b

Fig. 9.2 a, b. Observed versus predicted paleomagnetic (a) latitude and (b) longitude of paleomagnetic poles from North America, Eurasia, Africa, India, Antarctica, and Australia.

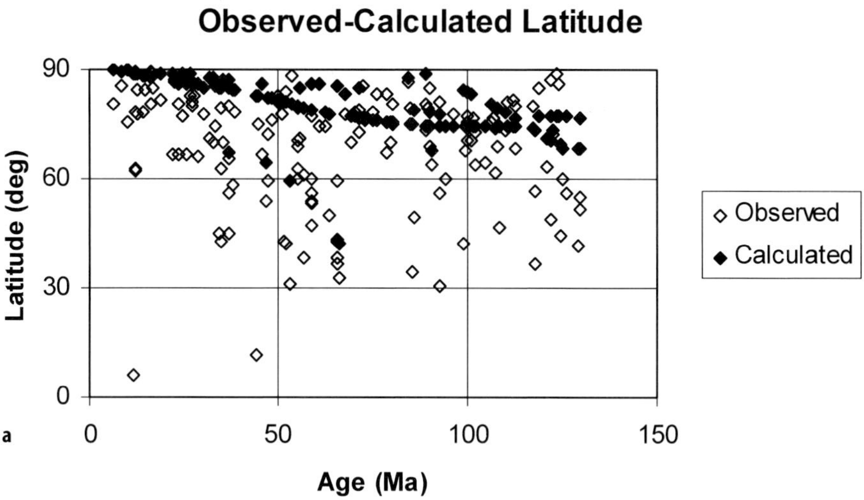

a

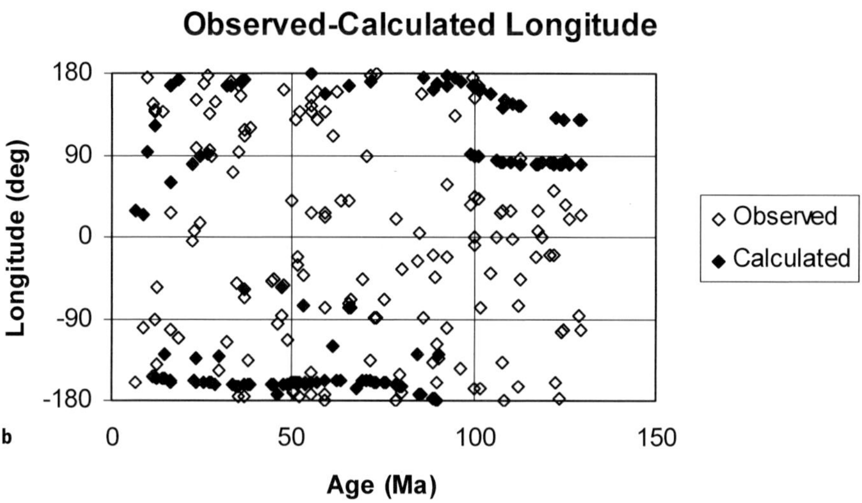

b

Fig. 9.3 a, b. Predicted versus observed paleomagnetic pole locations from North America, Europe, Africa, India, and Australia as a function of age. **a** Latitude; **b** longitude.

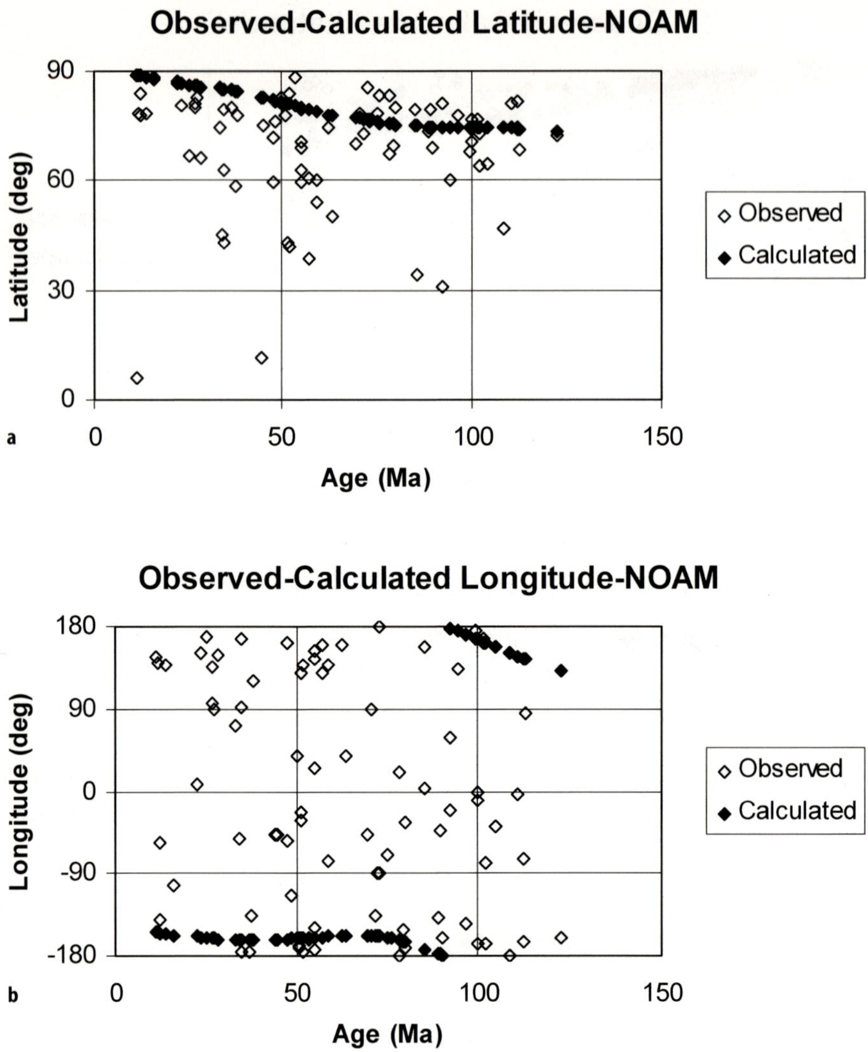

Fig. 9.4 a, b. Predicted and observed paleomagnetic pole locations versus age from North America. **a** Latitude; **b** longitude.

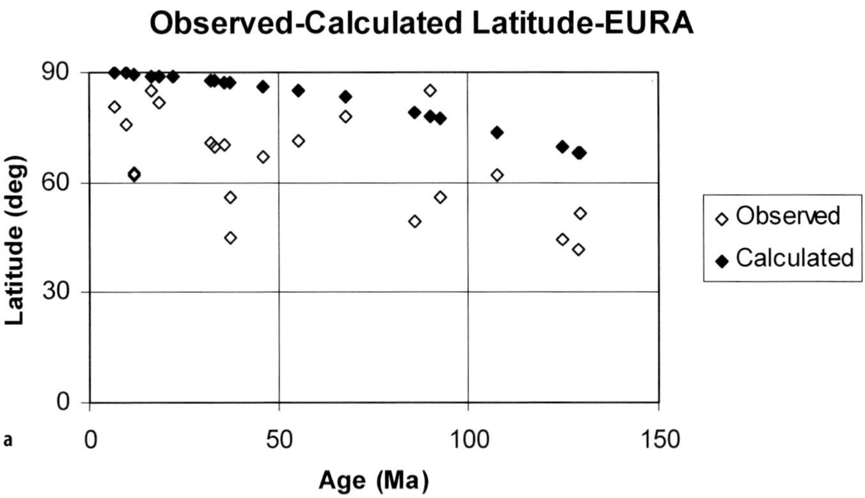

a

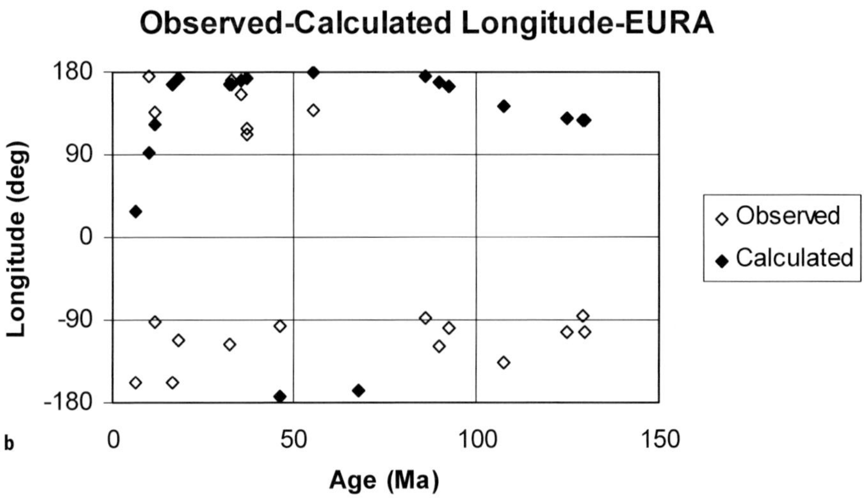

b

Fig. 9.5 a, b. Predicted and observed paleomagnetic pole locations versus age from Europe. **a** Latitude; **b** longitude.

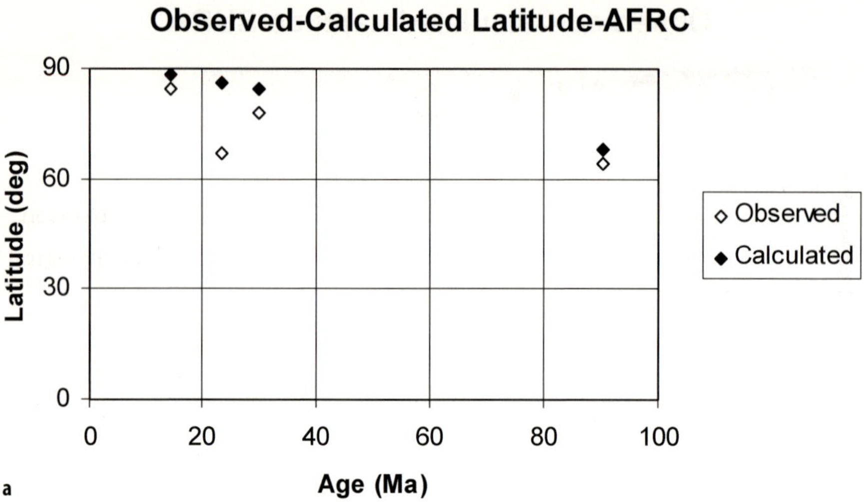

a

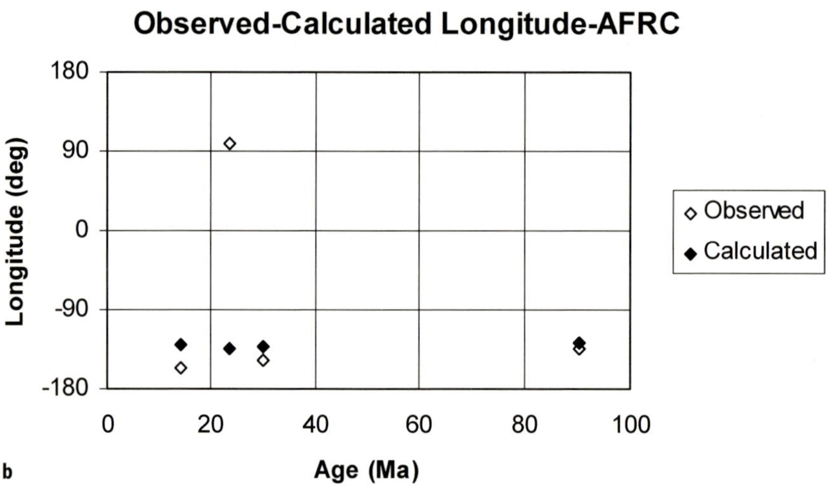

b

Fig. 9.6 a, b. Predicted and observed paleomagnetic pole locations versus age from Africa. **a** Latitude; **b** longitude.

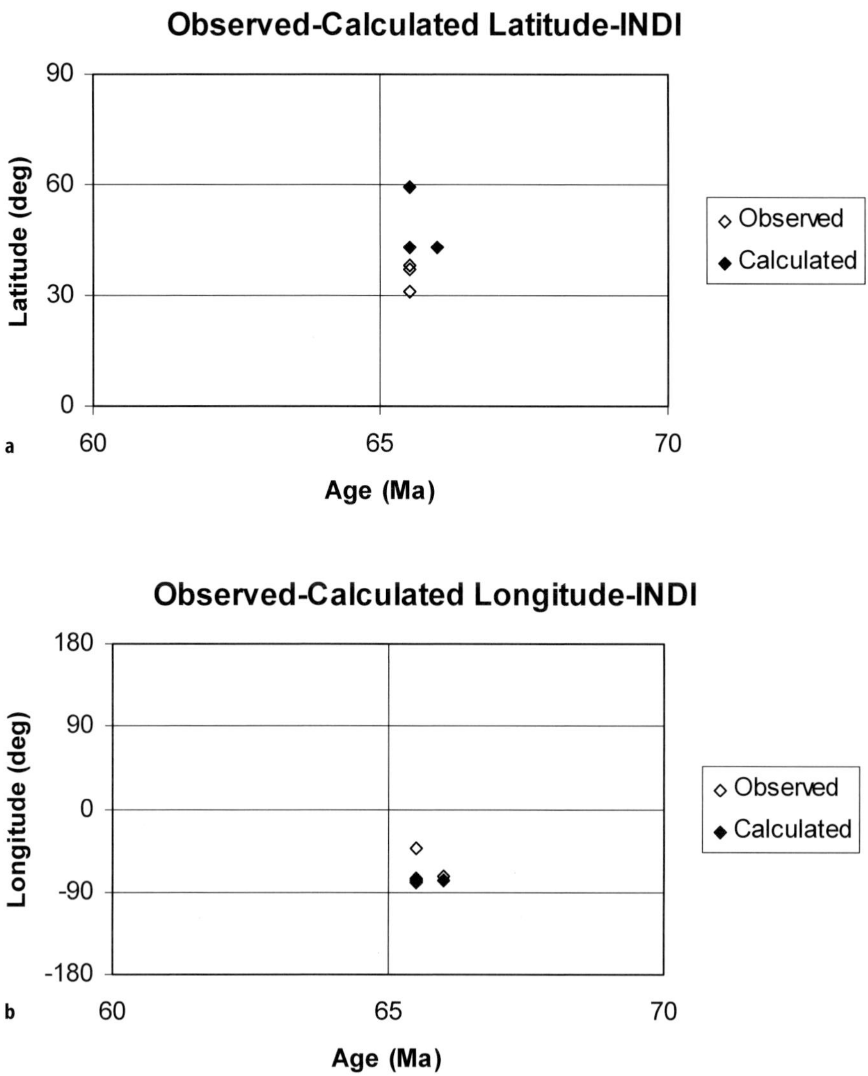

Fig. 9.7 a, b. Predicted and observed paleomagnetic pole locations versus age from India. **a** Latitude; **b** longitude.

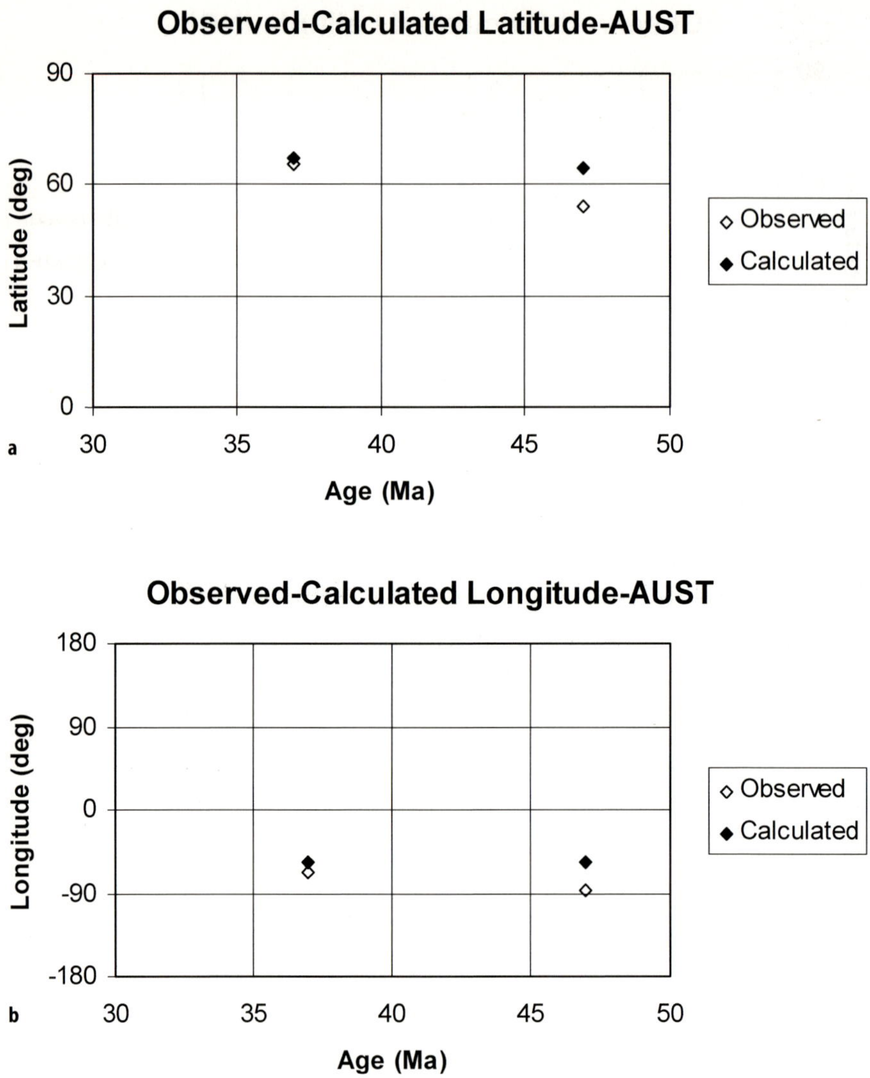

Fig. 9.8 a, b. Predicted and observed paleomagnetic pole locations versus age from Australia. **a** Latitude; **b** longitude.

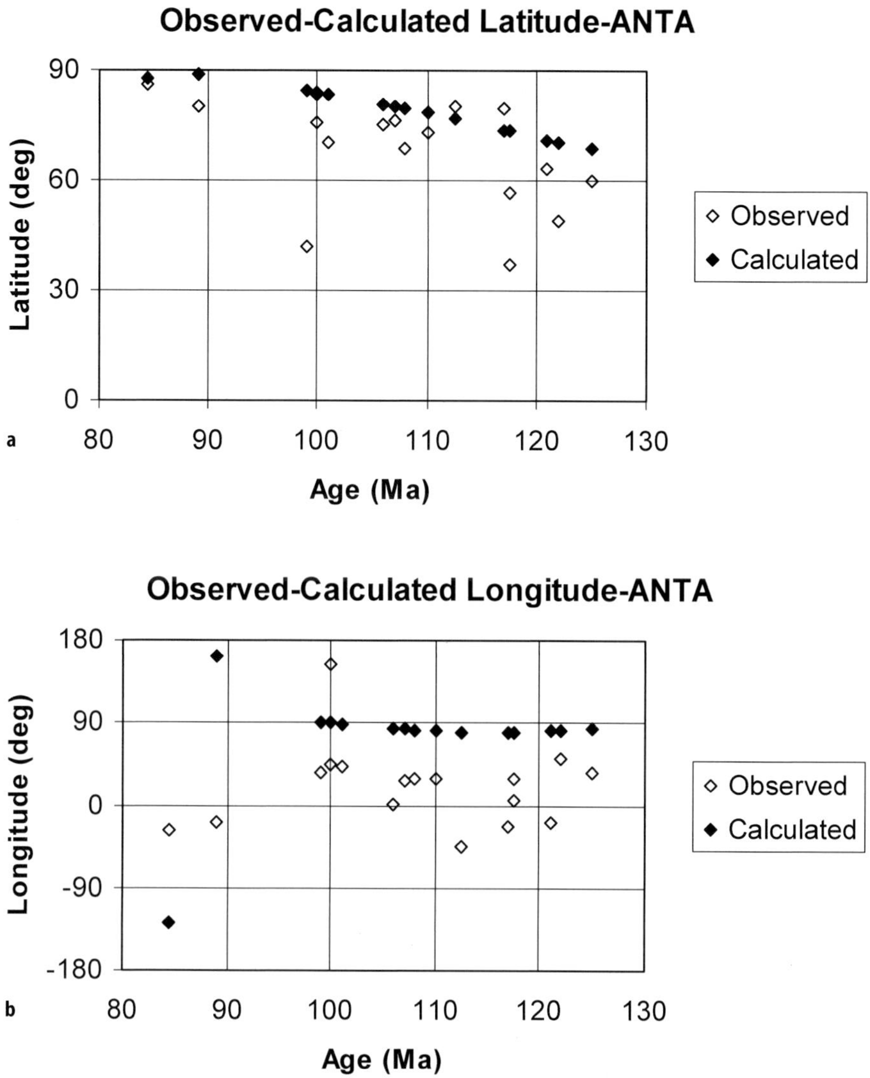

Fig. 9.9 a, b. Predicted and observed paleomagnetic pole locations versus age from Antarctica. **a** Latitude; **b** longitude.

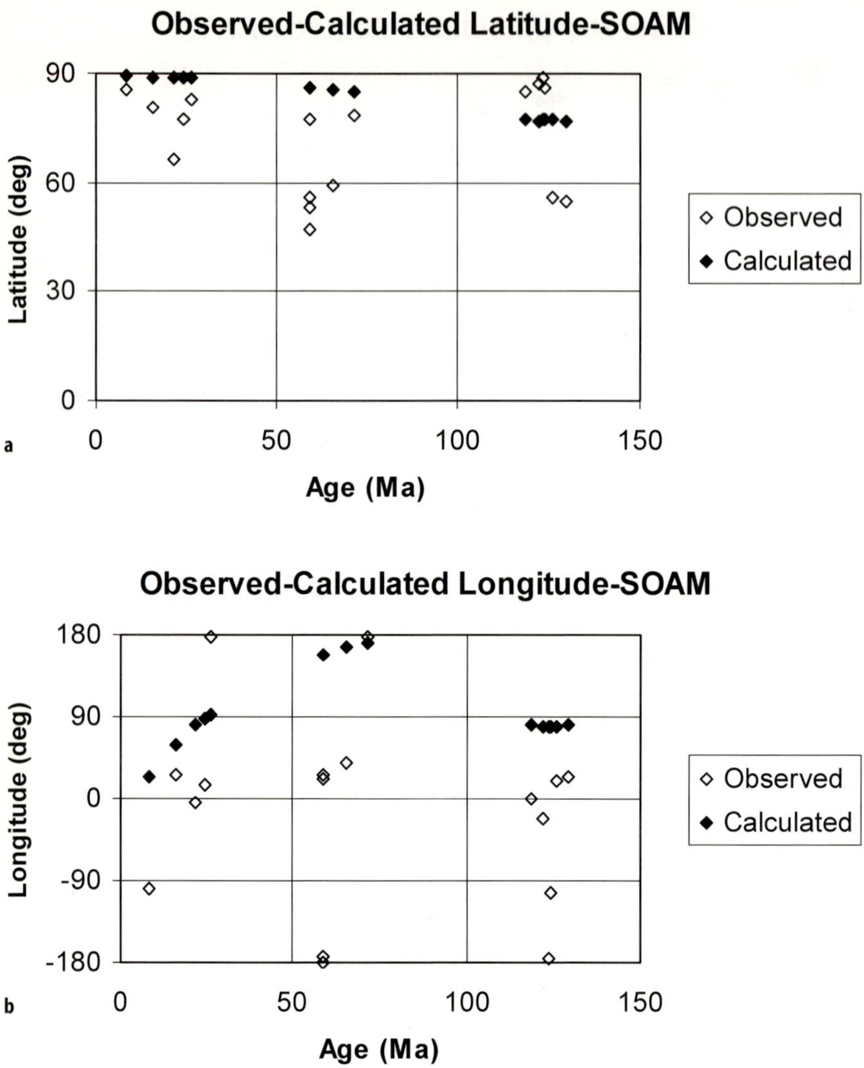

Fig. 9.10 a, b. Predicted and observed paleomagnetic pole locations versus age from South America. **a** Latitude; **b** longitude.

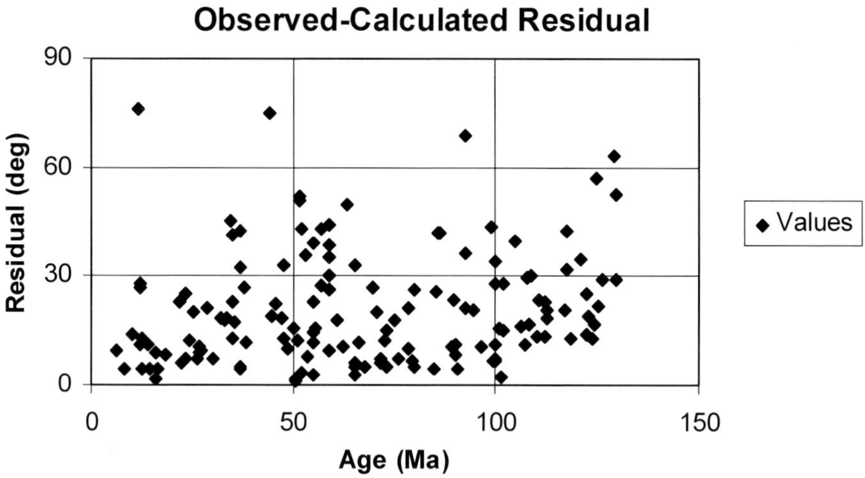

Fig. 9.11. Residual angles between observed and predicted paleomagnetic poles as a function of age.

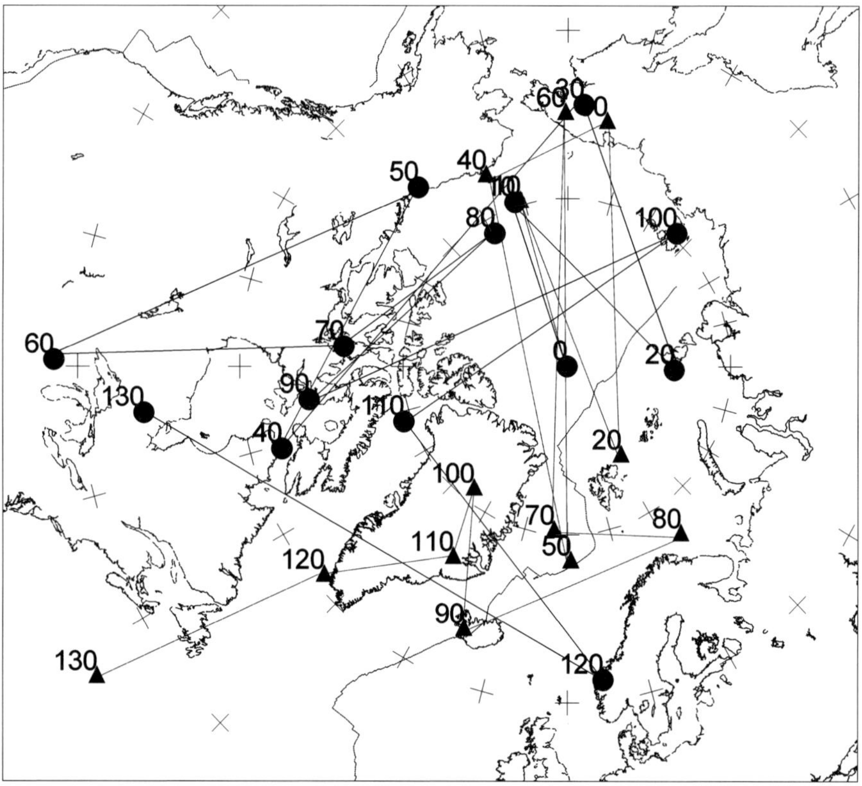

Fig. 9.12. Weighted-average of the paleomagnetic data rotated to the Atlantic–Indian hotspot frame using the methods presented in Chaps. 3 and 4. Circles-unrotated; triangles rotated.

such global averages (e.g., Irving 1980; Livermore et al. 1983). Figure 9.12 shows a weighted-average (1-my increments with a ±5-my moving window) of the paleomagnetic data rotated to the Atlantic–Indian hotspot frame using the methods presented in Chaps. 3 and 4. It would appear fruitless to attempt a kinematic parameterization of the average poles with the variation observed.

CHAPTER 10

Back-Arc Spreading and Proper Motion

10.1 Aside

Back-arc spreading was one of the big auxiliary surprises of second-generation plate tectonics. Tectonicists had visualized convergence of plates as resulting from converging convection cells with consequent compressive structures. As the geometric implications of plate tectonics began to be elaborated, the idea of convection cells tied to plate geometries evaporated. The documentation of back-arc spreading provided additional support for the decoupling of convection and plate kinematics. However, the mechanism for the origin of back-arc spreading has remained uncertain, although the ideas of Moberly were introduced shortly after spreading was demonstrated.

10.2 Back-Arcs

As support for his "absolute" motion model based on hotspots, Morgan (1972a) cited Moberly's (1972) idea that motion of the upper plate of a subduction zone controls the development of back-arc spreading. In the context of his contemporary kinematic model, indeed each active back-arc basin demonstrated motion relative to the subduction boundary that could "explain" back-arc divergence. In the context of the more recent contemporary instantaneous plate model of Gripp and Gordon (1991) and the Pacific and Atlantic–Indian Ocean hotspot models, those basins for which the motions can be predicted are considered.

10.2.1 Lau Basin

The Lau Basin in the Southwest Pacific (e.g., Zellmer and Taylor 2001) is the back-arc of the Tonga Trench, separating the Pacific and Australian plates (Fig. 10.1). Calculated motions in the contemporary global hotspot frame indicate that Australia is moving to the north. If the Tonga trench is stable (in the hotspot frame), then a component of extension is required because of the oblique divergence of the upper plate relative to the subduction zone.

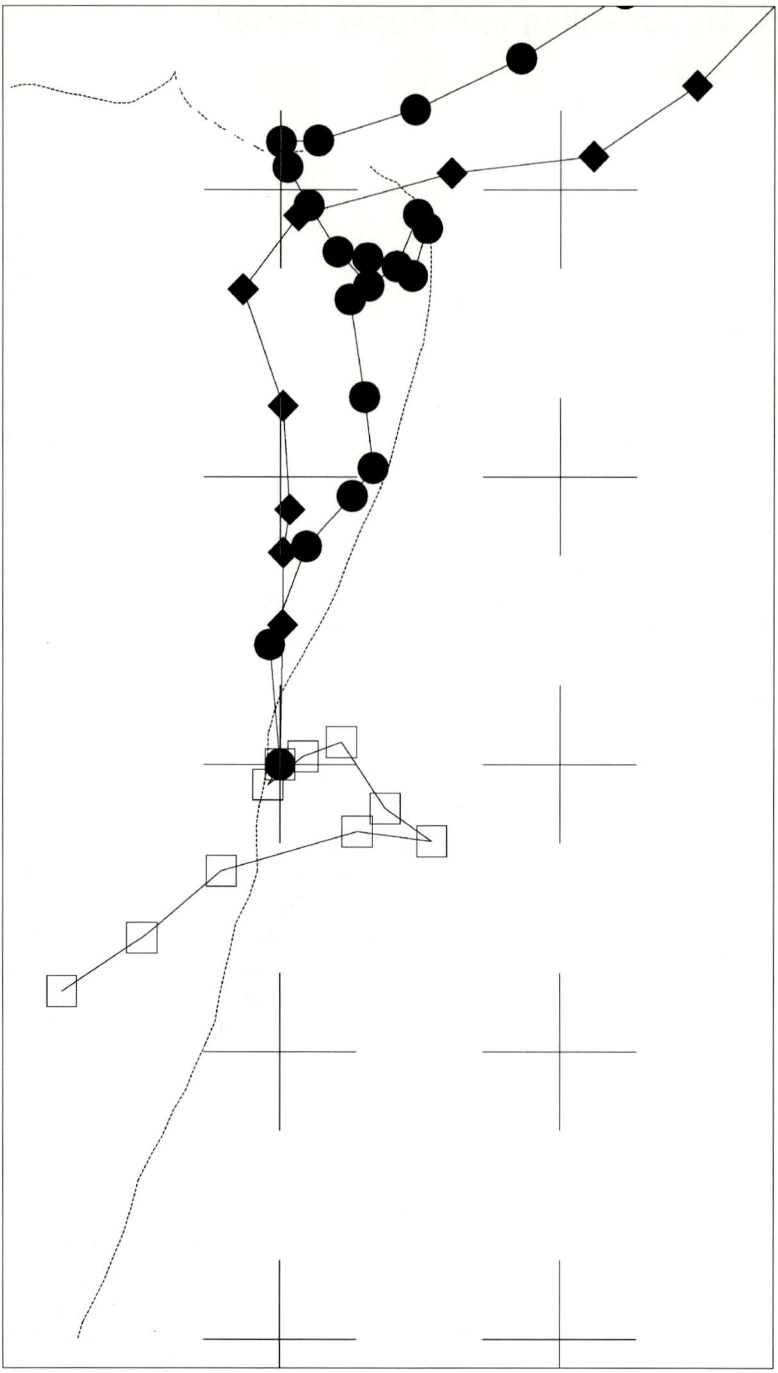

Fig. 10.1. Lau Basin circles: locus of Australian plate motion in Atlantic–Indian Ocean hotspot reference frame. Diamonds: Pacific Ocean hotspot reference frame. Squares: Pacific Ocean reference frame relative to Atlantic–Indian Ocean reference frame. Symbols are spaced at every 5 my.

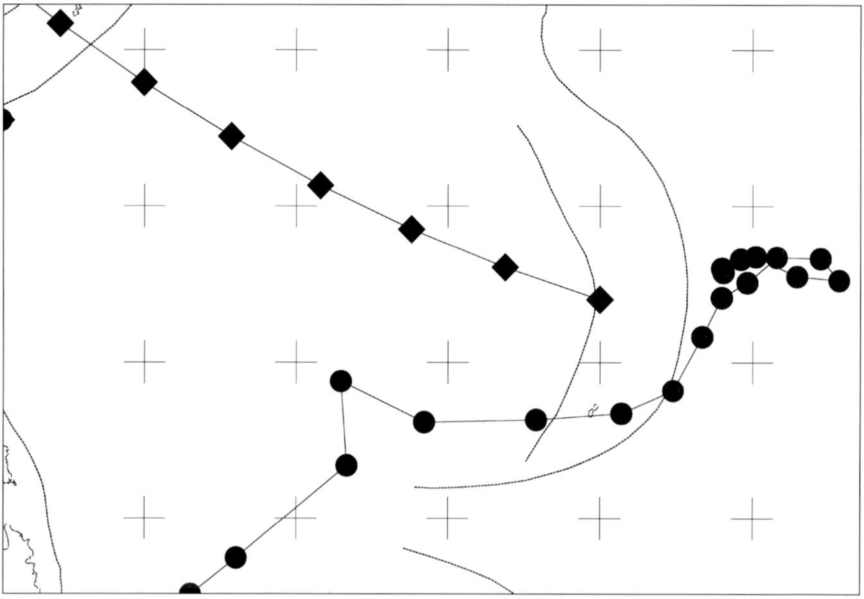

Fig. 10.2. Marianas diamonds: locus of Philipine plate motion in Pacific Ocean hotspot reference frame as in Fig. 10.1. Circles Pacific Ocean reference frame relative to Atlantic–Indian Ocean reference frame.

10.2.2 Marianas Arc

The Philippine plate's motion relative to the Pacific plate can be determined from seismic focal mechanisms. When combined with the global hotspot model, the Philippine plate is moving to the west (Fig. 10.2). If the Marianas trench is stable, then observed back-arc extension is consistent with the inferred motion of the upper plate.

10.2.3 Scotia Arc

The subduction zone beneath the Scotia arc (e.g., Livermore et al. 1999), if stable, enables back-arc extension as motion of the Scotia plate (Fig. 10.3). While poorly constrained, motion of the upper plate could be to the west in the contemporary hotspot frame when focal mechanisms of boundaries between the Scotia plate and the South American and Antarctic plates are taken into account.

10.2.4 Lesser Antilles Arc

There are suggestions that at various times extension has occurred behind the Lesser Antilles island arc (Fig. 10.4). Motions of the Caribbean plate relative to

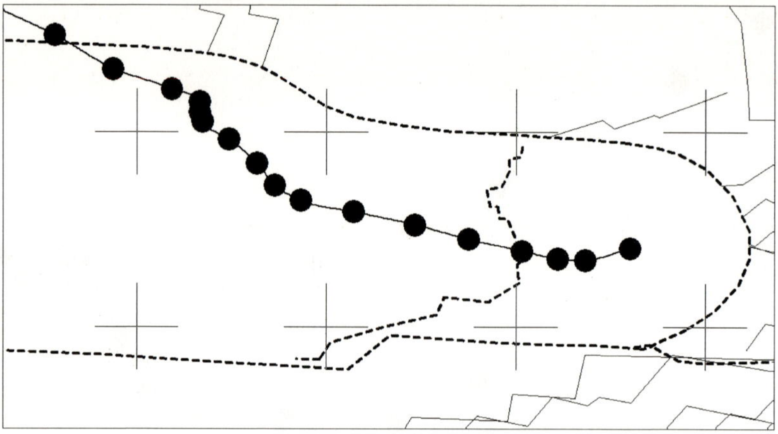

Fig. 10.3. Scotia: locus of South American plate motion in Atlantic–Indian Ocean hotspot reference frame as in Fig. 10.1.

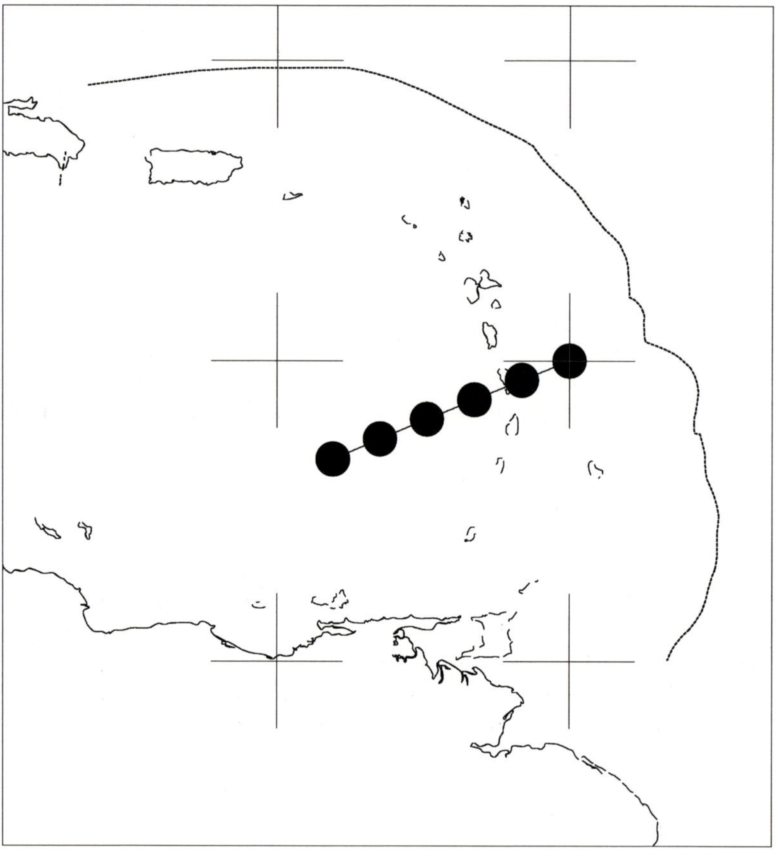

Fig. 10.4. Lesser Antilles: locus of Caribbean plate motion in Atlantic–Indian Ocean hotspot reference frame as in Fig. 10.1.

North America are poorly constrained (e.g., Müller et al. 1999), but, when combined with the hotspot model, they could suggest that the upper plate of the subduction zone has at times moved to the west relative to a stable subduction zone.

10.2.5 Sea of Japan

The Sea of Japan may be relatively inactive at present, but in much of the Cenozoic very probably was an active back-arc basin (Fig. 10.5). Calculated motions in the Atlantic–Indian Ocean hotspot frame suggest there were periods in which a relatively eastward motion of Eurasia relative to the Japan trench occurred. In particular, between 70 and 50 Ma, and between 15 and 5 Ma significant eastward motion of Eurasia in the hotspot frame is indicated. There is uncertainty, however, as to whether motion of Eurasia is best characterized by the Atlantic–Indian Ocean hotspot frame or a third reference frame. Takahashi and Saito (1997) propose a Miocene (20–10 Ma) age for the formation of the Sea of Japan by back-arc spreading.

10.2.6 Basin Range Province

Perhaps one of the more intriguing implications of the proper motion model of Müller et al. (1993) has to do with the tectonic evolution of western North America. The model implies that proper motion of the western U.S. was to the west northwest from 130 to 80 Ma (Figs. 8.16–8.26). As inferred by Cross and Pilger (1978a), a change in motion at about 80 Ma, with subsequent motion directed to

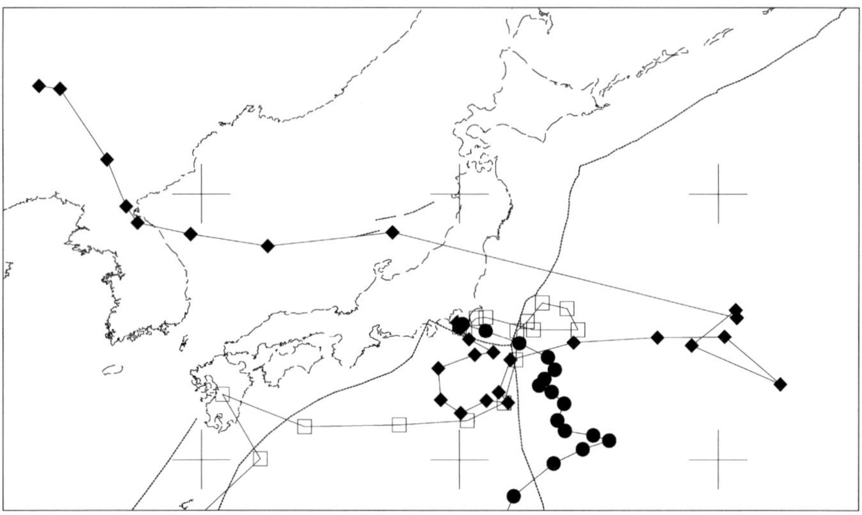

Fig. 10.5. Sea of Japan: Loci of Eurasian plate motion as in Fig. 10.1.

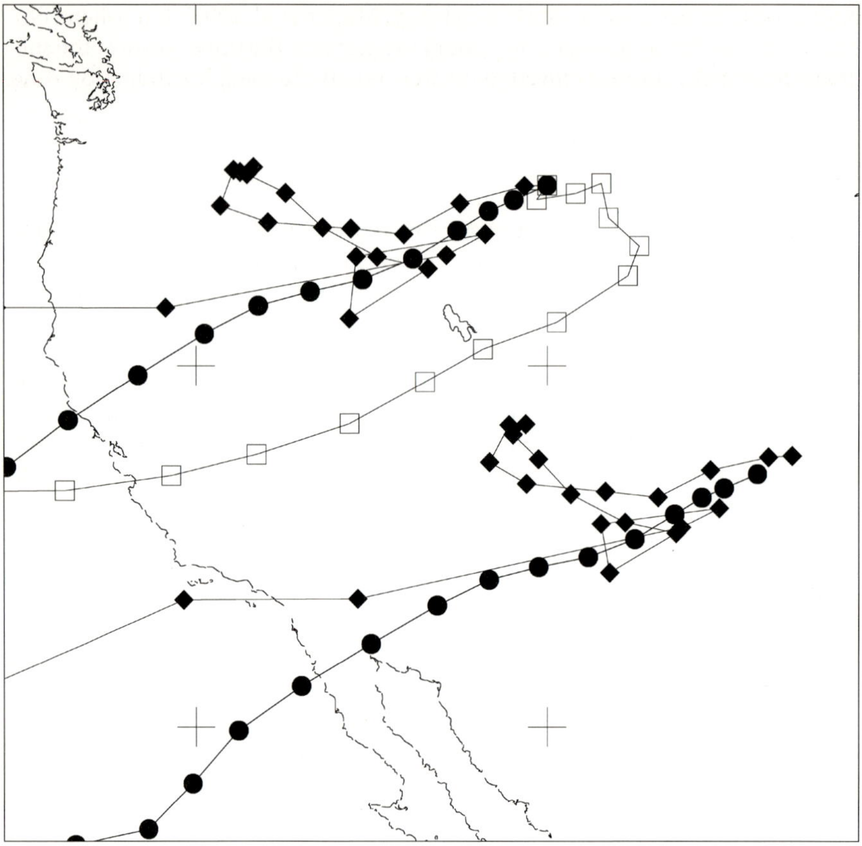

Fig. 10.6. Basin Range Province: loci of North American plate motion in Fig. 10.1.

the southwest (up until the present) corresponds with the cessation of the Sevier event in Utah and Nevada (and northwest-southeast directed thrusting) and onset of the Laramide event to the east (Cross 1986). The rate of North American plate motion immediately prior and subsequent to 80 Ma was fairly high. The change in character of tectonism, particularly the easterly shift in crustal deformation and volcanism, suggested to Cross and Pilger that extremely low-angle subduction of the Farallon (or another Pacific Ocean plate) occurred in the Early Cenozoic. The similar rates of plate motion during the Sevier and Laramide orogeny imply that an additional factor may have been involved, including either subduction of progressively younger (and therefore lower density) lithosphere or of an aseismic ridge (Cross and Pilger 1978b).

In the Late Cenozoic, the rate of North American proper plate motion decreased significantly, and was directed to the southwest, according to Müller et al.'s (1993) model (Fig. 10.6). The hypothesis of Cross and Pilger (1978a), that slow plate motion of North America may have facilitated back-arc extension resulting in Basin Range extension, is supported by this aspect of the proper motion model. Further, the apparently greatest amount of extension, over the full width of both the northern and southern parts of the province, occurs in those back-arc regions that are behind portions of the pale margin which trended highly oblique to the inferred direction of North American plate motion. It is the component of plate motion normal to the margin that is critical to the production of back-arc extension, according to the hypothesis of Moberly (1972).

Hotspot Reference Frames: The "Necessity" of Mesoplates

11.1 Aside

Part of the resistance to Plate Tectonics among earth scientists in the 1960s and 1970s involved the concept of rigidity. When advocates of the new global tectonics spoke of rigid plates, they encountered an intuitive misunderstanding of mechanical rigidity rather than kinematic rigidity. The idea of mesoplates as described below is likely to encounter a similar resistance. Is the domain of the mesoplate likely to be rigid? Isn't it more likely to behave plastically? Again, from a mechanical perspective, mesoplates, as defined below, may well consist of plastically deformable material at the pressures and temperatures at which it exists, if it is subjected to sufficient stresses. The key condition that allows for both lithoplates and mesoplates is the absence of pervasive stresses within their interiors. Deformation is restricted to the margins of plates – both lithoplates (of classic Plate Tectonics) and mesoplates.

11.2 Hotspot Systems

The progressive improvement of independent plate-hotspot models for the Atlantic–Indian Ocean system and for the Pacific Ocean system has been accompanied by parallel failure to reconcile the best models of the two hotspot systems. A number of workers have tested the relative fixity of the Pacific hotspot system relative to the Atlantic–Indian systems, beginning with Molnar and Atwater (1973). Despite improvements in relative plate reconstructions, especially in the South Pacific, and even recognition of a spreading plate boundary between East and West Antarctica plates (Cande et al. 1999), hotspot reconstructions tied to the Atlantic–Indian Ocean system (using the model of Müller et al. 1993) through Antarctica still do not correspond with inferred Pacific Ocean hotspot traces (e.g., Norton 2000; Raymond et al. 2000). In fact, the discrepancies observed in the latest models are quite similar to those recognized by Molnar and Atwater (1973), Molnar and Francheteau (1975), and Duncan (1981). What has become more apparent, however, is the remarkable internal consistency of each system taken separately, as noted by Norton (2000) and Raymond et al. (2000).

Those who want to explain the origin of hotspots in terms of deep-rooted mantle plumes are challenged by the existence of two different hotspot reference frames. It is difficult to explain how two different coherent sets of hotspots could be maintained while moving relative to each other. For example, Gaina et al. (2000) attempted to model seamount chains in the Tasman Sea by suggesting that one or more hotspot traces were derived in first one hotspot frame and then another. If their suggestion has merit, it has negative implications for the deep-mantle plume hypothesis.

In order to explain the existence of two coherent hotspot sets, an alternative formulation is proposed. The most critical departure from contemporary thought concerning hotpots is to assume that the hotspot reference frames have a shallow origin – in the upper part of the mesosphere, just below the asthenosphere. In contrast, beginning with Morgan (1971), most hotspot models presume a deep-mantle source – plumes. For the moment, the ultimate origin of the hotspots themselves is not considered. Whatever the origin of hotspots (including the plume hypothesis), it is assumed that within each of the two sets the hotspots are fixed relative to one another within relatively shallow reference frames.

Further, we observe that the Pacific set, with evidence for it existing on three plates (Pacific, Nazca, and Cocos), is almost entirely encompassed by subduction zones. The subduction boundaries are either active at present and/or were active for a significant portion of the past 100 my or so. In other words, the Atlantic–Indian Ocean set is separated from the Pacific set by subduction boundaries. In some cases, the depth of detectable subduction extends at least 750 km deep (from seismicity; the zones extend even deeper based on seismic tomography), well into the mesosphere, while in other cases, detectable or inferred subduction does not extend much deeper than the adjacent overlying lithosphere.

The Atlantic–Indian Ocean hotspot set occurs beneath at least six major plates: North and South America, Africa (which could be interpreted as at least two contemporary plates), Antarctica, Australia, and India (which can be treated as two to four distinct contemporary plates). The Eurasian plate has only Iceland as a postulated contemporary hotspot with a defined trace, although hotspots have been inferred for Svalbard and Eifel. The Iceland trace does not appear to be consistent with the reconstructions fitting the remaining Atlantic–Indian Ocean hotspot set. A complex, predominantly convergence boundary separates Eurasia from Africa and India.

11.3 Mesoplates

The apparent existence of a coherent hotspot set beneath a number of plates is interpreted heuristically herein as evidence for a coherent "mesoplate" in the upper mesosphere. That is, the coherence of a proper hotspot reference frame across largely accretionary and shear plate boundaries implies plate-like behavior within the hotspot source. The rationale for introducing mesoplates extends beyond an explanation for distinct hotspot reference frames. There are other, intriguing implications of inferring a mesoplate beneath the Pacific Ocean and another beneath the Atlantic and Indian Ocean (and conceivably, a third beneath Eurasia and the Norwegian-Greenland Sea).

Preconceptions of the nature of the mantle would provoke an immediate denial of treatment of the upper part of the mesosphere as "rigid"; its viscosity is too low to be characterized as rigid. Further, the boundaries are not easily defined, since there is no seismicity delineating them. However, as with the lithosphere, the kinematic rigidity of plates is a manifestation of the small magnitude of differential circumferential stresses relative to mechanical rigidity rather than the intrinsic "strength" of the mesosphere. If a body moves coherently without significant internal deformation, it is interpreted as "kinematically rigid". In the lithosphere, earthquakes indicate local dynamic rigidity overcome by shear-stress-induced fracture failure.

Figure 11.1 illustrates the idea of mesoplates. Mesoplates are kinematically rigid layers within the mesosphere, beneath one or more plates.

Putting dynamics aside for the moment, consider what the kinematic models of the motions of the plates relative to the hotspots indicate. From Appendix 2, using the relative plate reconstructions from the various oceans, the hotspot models of Müller et al. (1993) and Raymond et al. (2000), plus the spline interpolation methods presented here, it is possible to calculate the relative motion of the Pacific and Atlantic–Indian Ocean sets; such reconstructions are implied by the demonstration of the misfit of the two models when applied to hotspot traces of the other ocean sets (e.g., Raymond et al. 2000; Norton 2000) as shown in Figs. 11.3–11.6.

For convenience, the Atlantic–Indian Ocean hotspot set is referred to as the "Tristan" set (Fig. 11.2). The hotspots of the Pacific Ocean, beneath the Pacific, Cocos, and Nazca plates, are collectively referred to as the "Hawaiian" set. Conceivably, the Hawaiian set could also include the hotspots of the Tasman Sea, as proposed by Gaina et al. (2000).

Figures 11.3–11.6 show not only the predicted motion of the Pacific and adjacent plates relative to the Pacific and Atlantic–Indian hotspot sets; they also include predicted loci of the motion of the Pacific hotspot set relative to the Atlantic–Indian set. The predicted differential loci are calculated at various points along the eastern Pacific (including inferred hotspots) as well as the margins of the ocean.

The calculated differential motion implies distinct changes, which are of different magnitudes and directions, depending on where the calculations are

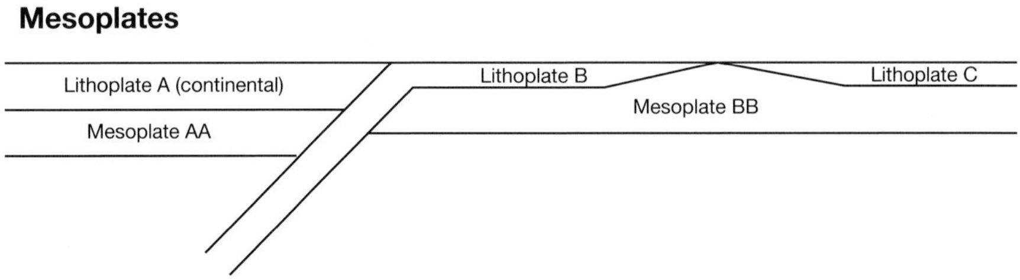

Fig. 11.1. Cross section illustrating the mesoplate concept.

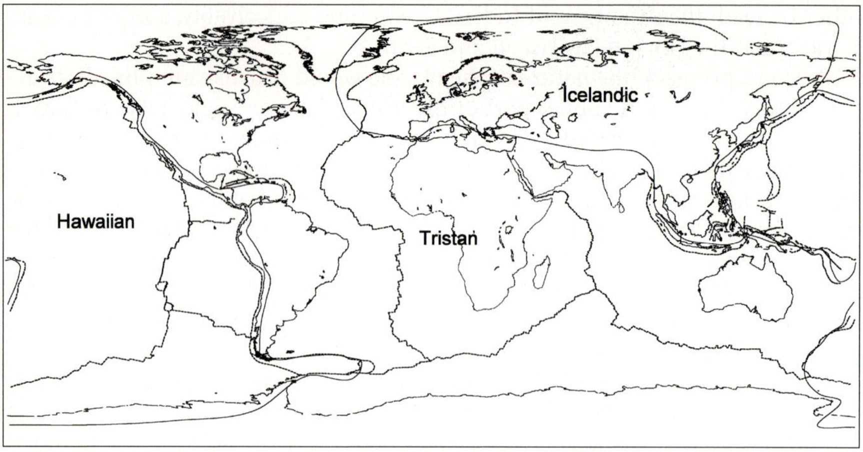

Fig. 11.2. Postulated boundaries of Hawaiian, Tristan, and Icelandic mesoplates.

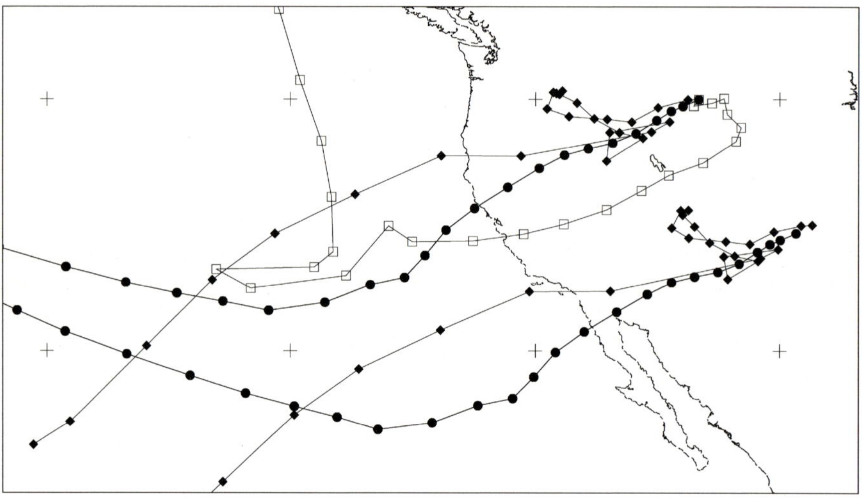

Fig. 11.3. Reconstructed loci of inferred Yellowstone hotspot relative to North America in Tristan (circles) and Hawaiian (diamonds) reference frames and of Hawaiian frame relative to Tristan frame (squares). North American motion in the Tristan frame is significantly greater in the Hawaiian frame.

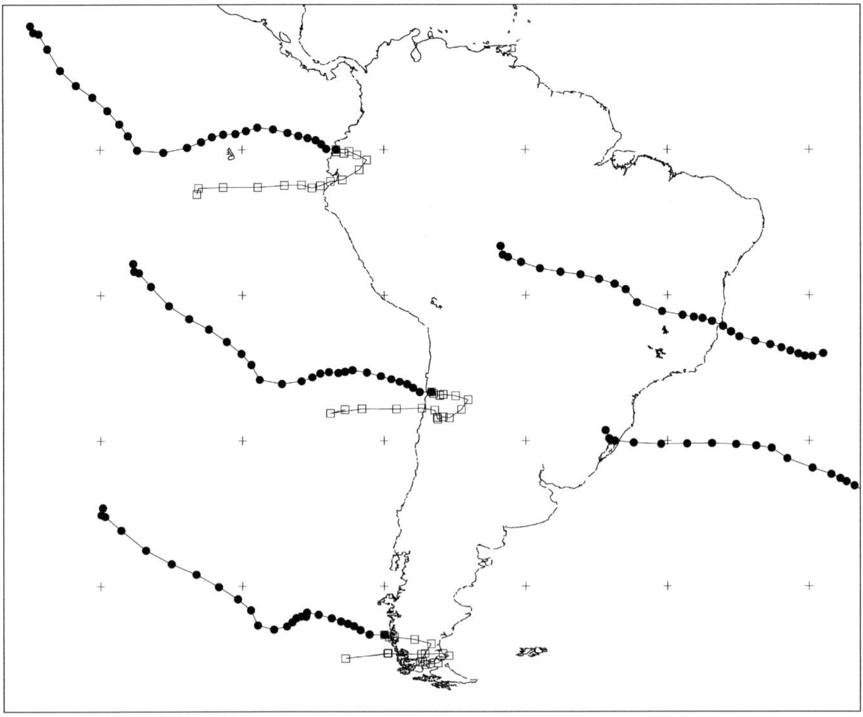

Fig. 11.4. Reconstructed loci of hotspot motion relative to South American plate in Tristan (circles) and Hawaiian (diamonds) reference frames and of Hawaiian relative to Tristan reference frame (squares). Positions at 5-my increments.

made. In the northeast Pacific, the inferred Hawaiian mesoplate has been moving slowly relative to the Atlantic–Indian ("Tristan") mesoplate for the past 25 Ma, with net displacement to the south-southeast. Prior to 25 Ma, the Hawaiian mesoplate moved at a more rapid rate to the southwest relative to the Tristan mesoplate. Interestingly, comparison of the Pacific–Tristan locus with the locus of North American motion in the Tristan frame indicates subparallelism; that is, for the period from 75 to 25 Ma, both North America and the Hawaiian mesoplate were moving to the southwest relative to the Tristan mesoplate. This generalization is applicable to the entire western margin of North America from central Alaska to Mexico.

For the southeast Pacific, the locus of Hawaiian mesoplate motion relative to the Tristan mesoplate has largely been latitudinal, with two distinct changes in the sense of motion. For the period prior to 25 Ma, the Hawaiian mesoplate was moving to the west until 45 Ma when it began moving to the southwest. At 25 Ma, the Hawaiian mesoplate began moving to the east.

Relative to the southwest margin of the Pacific, the geometry of the mesoplate motions is similar to that of the southeast margin. From 75 to 45 Ma, the Hawaiian mesoplate was moving to the east relative to the Tristan mesoplate; from 45 to 25 Ma, motions were to the southwest; since 45 Ma, more or less easterly motion has occurred.

Similar analyses can be applied to the other margins of the Pacific Ocean basin. However, it is quite reasonable to infer that a third mesoplate, beneath Eurasia and the northernmost Atlantic, may also exist, especially since the hotspot models that apply to the Atlantic–Indian hotspots do not fit Iceland. Unfortunately, since only Iceland seems to have some definition as a hotspot trace, and even it is poorly known, it would be difficult to derive motion of the Eurasia plate relative to a possible Eurasian mesoplate. However, there are some additional constraints that might apply to inferring its motion. There are two submarine ridges, just south of the Charlie-Gibbs fracture zone in the northern Atlantic, that correspond with magnetic Chron 25, on either flank of the Mid-Atlantic Ridge. In analogy with the Tuamotu and Nazca Ridges in the east-central Pacific, the ridges may reflect a period of time in which a hotspot responsible for them both was located beneath the Mid-Atlantic Ridge spreading center. Unfortunately, there are no obvious seamount chains associated with the pair of submarine ridges – chains that might represent the older and/or younger continuation of the hotspot that may have produced the ridges. When compared with the Tristan hotspot model of Müller et al. (1993), the reconstructed position of the spreading center does not imply ridge-parallel fixity relative to the Tristan mesoplate at Chron 25 (50.5 Ma). Thus the ridges might represent a Eurasian mesoplate source.

In a similar vein, the Canary Island hotspot is presently beneath Africa and corresponds, at least in part, with a small arcuate chain of islands and seamounts, When reconstructed using the Tristan hotspot parameters for 130 Ma, the resulting trace misses an offshore seamount chain and a zone of mafic intrusives in onshore Nova Scotia, dated at 140 Ma, by about 400 km. If the Canary hotspot was previously part of the Eurasian mesoplate, then the seamount chain and onshore diabases might provide additional Eurasian mesoplate kinematic constraints.

In jumping to the inference of two or three mesoplates, several additional assumptions also need to be more explicitly articulated. In particular, the boundaries of the mesoplates need to be defined (Fig. 11.2). It is asserted that the current boundaries of the Pacific Mesoplate correspond with the active subduction boundaries involving subducting oceanic plate older than 20 Ma: beneath South America, the Australian plate (Tonga–Kermadec), the easternmost western Pacific subduction zones (Philippine, Marianas, Japan, Kurile, Kamchatka) and beneath the Alaskan peninsula in the North Pacific. The subducting (Pacific or Nazca) plate separates the Pacific mesoplate from the nearby Tristan or Eurasian mesoplate (a small mesoplate could be postulated to underlie the Philippine plate since the plate boundaries are almost entirely characterized by subduction).

For the remaining margins of the Pacific, the inferred boundaries between the mesoplates are less obvious. For western North America, the boundary is inferred to underlie the Cordillera. Consideration of the Cenozoic subduction/transform history in light of the inferred plate and mesoplate kinematics places some constraints on the location of the boundary (Fig. 11.7).

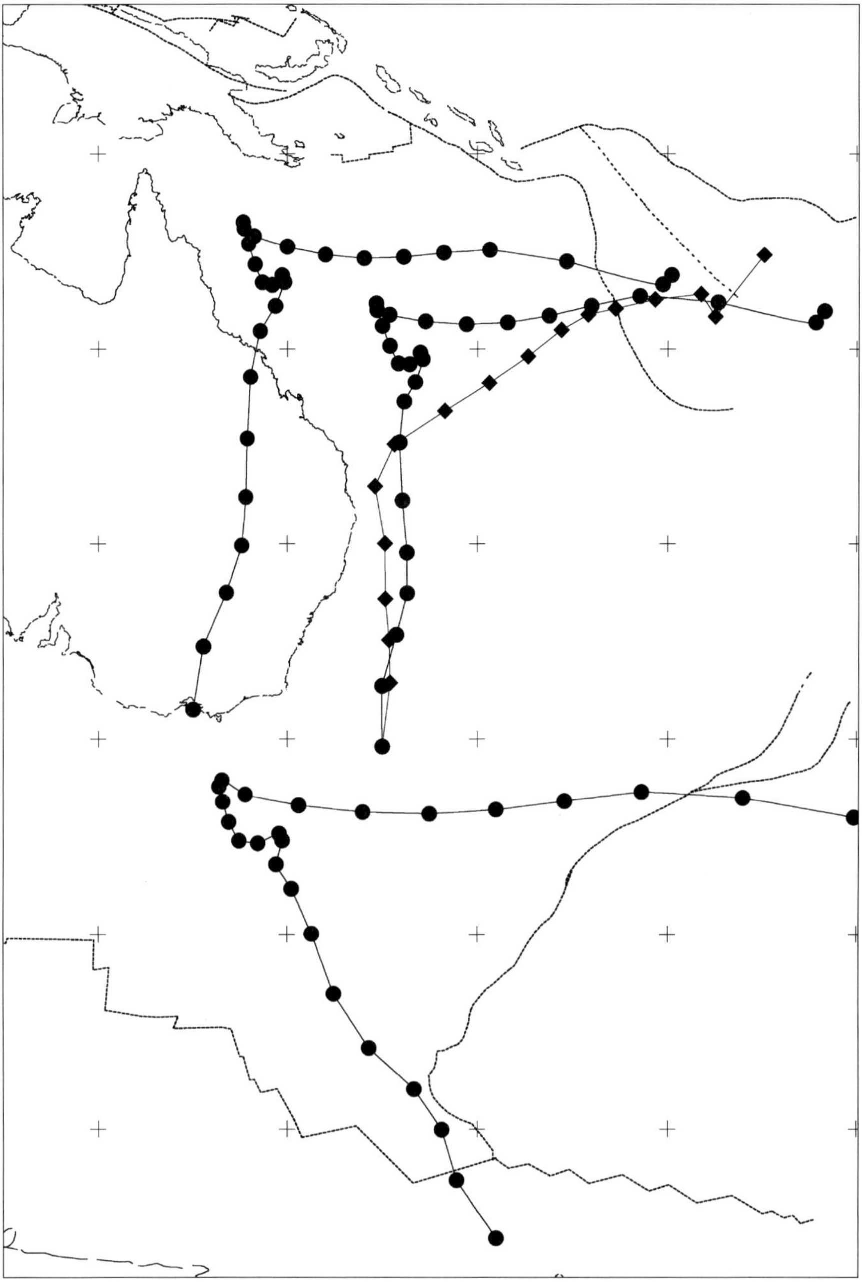

Fig. 11.5. Reconstructed loci of hotspot motion relative to Australian and Antarctic plates in Tristan (circles) and Hawaiian (diamonds – Australian plate only) reference frames. Symbols at 5-my increments.

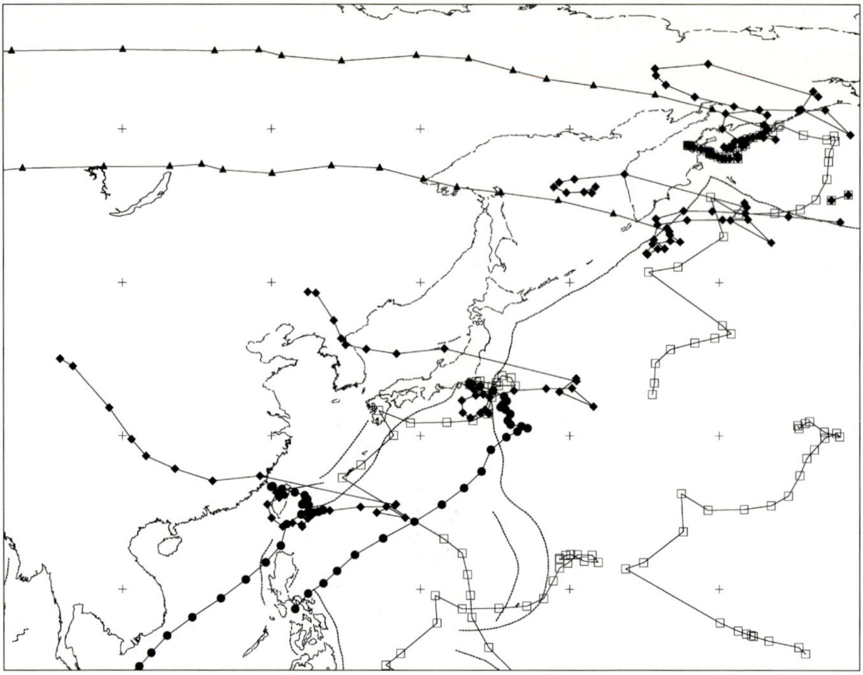

Fig. 11.6. Loci: Tristan relative to Hawaiian mesoplate (squares), Pacific relative to Eurasian plate (triangles), Eurasian relative to Tristan (diamonds), Eurasian relative to Hawaiian (circles), 0–130 Ma, western Pacific.

Beneath subduction zones involving young (less than 20 Ma) subducting lithosphere, such as Mexico and Central America and the Juan de Fuca subduction zone, the boundary could be inferred to exist well to the east of the present subduction boundaries. The rationale for this inference is the assumption that, upon subduction, the young and, therefore, hot plate becomes part of the asthenosphere and then mesoplate once it reaches subasthenospheric depths.

The subparallelism of calculated motion of the Pacific mesoplate relative to the Tristan mesoplate with North American plate motion relative to the Tristan mesoplate between 75 and 25 Ma introduces a tantalizing dynamic inference. Southwestern motion of North America relative to Pangea, roughly normal to the Farallon/Kula subduction zones, and, similarly, western and southwestern motion of South America relative to the Tristan mesoplate, normal to the Farallon subduction zone, may have been the principal driving force for displacing the Pacific mesoplate relative to the Tristan mesoplate (or, equivalently, the motion of the mesoplate facilitated the North and South American plate motions). Thus the discrepancy between the Atlantic–Indian and Pacific hotspot sets represents relative motion between the Tristan and Pacific mesoplates, driven by the American plates.

The existence of two (perhaps three) sets of hotspots – sets that are in motion relative to one another, but fixed internally over a large area of the earth – is the

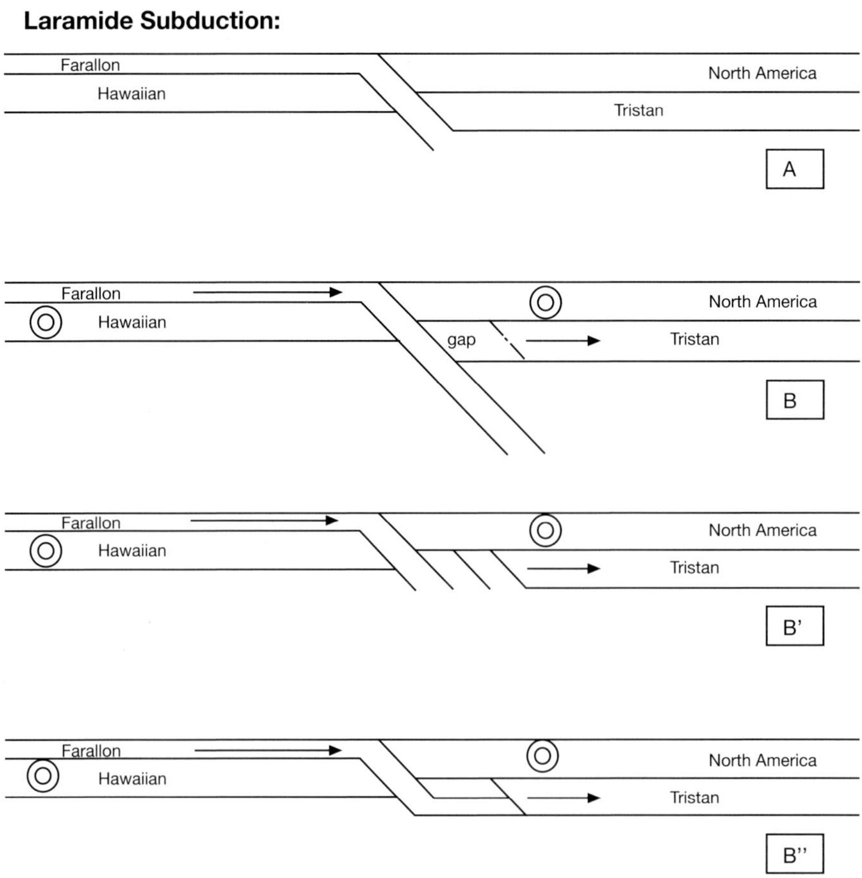

Fig. 11.7. Inferred evolution of the North American and Farallon plates and the Hawaiian mesoplate relative to the Tristan mesoplate.

primary motivation for inferring the existence of two or three mesoplates. The gross correspondence of the displacement of the American plates relative to the Tristan mesoplate with the similar displacement of the Pacific mesoplate argues for further development of the mesoplate hypothesis.

How thick are mesoplates? What is the nature of the boundary between mesoplates. What happens in the region between divergent mesoplate boundaries and what happens along boundaries where mesoplates converge?

Kinematic requirements suggest that in an area of inferred divergence, such as the western Americas between 75 and 25 Ma, part of the gap is filled by low-angle subducted lithosphere, which becomes part of the mesoplate. There is significant evidence suggesting that subduction beneath western North America from the latest Cretaceous into the Mid-Cenozoic occurred at a very low angle (e.g., Cross and Pilger 1978a, b), in analogy with contemporary low-angle subduction beneath Peru and central Chile (Pilger 1981b, 1984). The origin of low-angle subduction is probably a combination of factors: relative youth of subducting oceanic lithosphere, the presence of low-density aseismic ridges, and active overriding of a subducting plate relative to the mesosphere (Cross and Pilger 1982).

While low-angle subducted lithosphere could fill the gap, it is conceivable that the thickness of the mesoplate is greater than the thickness of the lithosphere. Upwelling from deeper in the mantle would be required to fill the remaining gap.

The possibility that an aseismic ridge or seamount chain was responsible for Laramide low-angle subduction provides a further question: where is the hotspot that might have produced the ridge or chain? The Yellowstone hotspot is the most obvious candidate. Reconstructions of North America relative to the Tristan and Pacific mesoplates are consistent with such an interpretation. Assuming that the Yellowstone has been part of the Tristan hotspot set during much of the Cenozoic, reconstructed North American plate motion implies that the hotspot trace extends to the southwest into present-day Southern California. However, the mesoplate model would imply that the hotspot was previously part of the Pacific mesoplate. The age of the transfer of the hotspot from the Pacific to the Tristan mesoplate would be uncertain, depending on where the boundary was in the Early Cenozoic. More significantly, the reconstructions imply that the hotspot was located beneath a Pacific plate in the Late Cretaceous and perhaps into the Cenozoic. Further, the reconstructed location of the hotspot is appropriate for it to have generated a trace on a Pacific (Kula or Farallon?) plate that would have been subducted in Laramide time. This speculation is tempered somewhat with the recognition that evidence for the Yellowstone hotspot trace on the North American plate is strongest for only the period of time since 16 Ma. Conceivably, the Coast Range of western Oregon, which appears to contain an accreted island-seamount chain of Early Cenozoic age, could be another remnant of the elusive Yellowstone hotspot trace.

In the case of converging mesoplates, the expected behavior is less obvious. In the example of the period between 75 and 45 Ma, the zone of Tristan and Pacific mesoplate convergence probably corresponds with the complicated zone of shear and convergence between the Australian and Pacific plates; for the Eurasian mesoplate (if it exists apart from the Tristan mesoplate), the zone of convergence would correspond with the plate boundaries between the Eurasian and Pacific plates and possibly the Philippine and Pacific plates. If the Eurasian mesoplate is moving slowly with respect to the Tristan mesoplate, reconstructed motion of the Pacific mesoplate relative to the Tristan implies largely southwestern, coast-parallel motion. Conceivably, a thick mesoplate beneath eastern Eurasia could be inhibiting northwestern motion of the Pacific mesoplate.

If the inferred mesoplate behaves in a largely passive way, its centripetal displacement is guided by subduction boundaries. Further, the mesoplate feeds spreading centers above it and experiences phase changes at its base as it is displaced downward in the vicinity of subduction zones.

To summarize, the properties of mesoplates as initially postulated are as follows.

- The definition of a mesoplate as a distinct plate-like entity is its lack of significant lateral internal deformation.
- The upper boundary of mesoplates (the lower boundary of the asthenosphere) is a phase change – the solidus.
- Plate subduction zones that extend the lithosphere from the Earth's surface into the mesosphere are the principal controls on mesoplate boundaries. The existence and relative motion of mesoplates are largely a consequence of the development of and coupled to the relative motion of subduction boundaries. The remaining (non-subduction) boundaries are presumably, though not necessarily, more diffuse.
- Spreading centers draw their supply of new oceanic crust and lithospheric mantle by vertical flow from mesoplates into the asthenosphere; that is, as plates diverge, isostatic forces induce vertical displacement of the mesoplate, through the solidus into the gap between the diverging plates. Spreading centers do not necessarily (if at all) correspond with mesoplate boundaries.
- Zones of divergence between adjacent mesoplates require upwelling of deeper mesosphere, which is added to one of the new mesoplates.
- Zones of convergence between adjacent mesoplates result in downwelling of mesoplate material into the deeper mesosphere; the material that is displaced downward is detached from the mesoplate. (This property is perhaps the least well constrained.)
- Should a subduction zone separating two mesoplates cease to exist, it is possible that the two mesoplates might well be united into a single larger mesoplate.
- A hotspot adjacent to a mesoplate boundary might be transferred to an adjacent mesoplate by a change in mesoplate kinematics and/or boundary configurations, as suggested for the Tasman seamounts (Gaina et al. 2000), Canary Islands, and Yellowstone.
- The lower boundary of a mesoplate is poorly constrained, although the kinematics of subducted plate becoming part of the mesoplate could provide constraints on mesoplate thickness by application of volume and mass conservation considerations.

These properties of mesoplates form a speculative set, especially since the inference of their existence is at present entirely indirect and circumstantial. Rheology of mesoplates, as distinct from the asthenosphere and deeper parts of the mesosphere, may be worth investigating. For example, could mesoplates have a distinct quasi-elastic flexural rigidity, in analogy with lithospheric plates? The way in which mesoplates have been defined might be interpreted to rule out measurable flexural rigidity, but, as with the lithosphere, there could be different time scales of such a property. More broadly, the question arises as to whether mesoplates might have any consistent physical properties from place to place. Kinematic rigidity may occur within the somewhat heterogeneous materials that might comprise mesoplates and not explicitly represent homogeneous rheologies.

CHAPTER 12

Mesoplates and the Origin of Hotspots: A Top-Down Model

12.1 Aside

Since Morgan published the first hotspot motion model for the Pacific in 1972, the formal conversation among researchers evolved into a debate as to whether hotspots do indeed define a single, fixed, internal reference frame. However, discrepancies between Atlantic–Indian Ocean and Pacific Ocean sets became the primary focus of attention. In just the past few years, some workers have begun to write in terms of two or three separate reference frames. If there are two or three distinct hotspot reference frames – mesoplates – this presents interesting problems for the deep mantle plume hypothesis. Perhaps it is time to consider the implications of a few distinct hotspot frames for the origin of hotspots.

12.2 Review

In Chap. 11, the case for "mesoplates" was advanced, along with the corollary assertion that the relative position of hotspots, whatever their origin, is essentially fixed within its mesoplate. In this chapter, it is argued that many, if not all, hotspots originate within mesoplates and the overlying asthenosphere. The elements of this argument are largely kinematic and empirical; the age pattern of hotspot traces relative to the age and/or structure of the lithoplates of which they are a part provide significant insight into the origin of the traces.

A key component of the mesoplate hypothesis is recognition that stable continent interiors record contemporary and paleostress fields that correspond with inferred plate motions in the hotspot reference frame (i.e., the model of Müller et al. 1993). As the stress fields can only represent processes internal to a plate or interaction of a plate with adjacent plates and/or the underlying asthenosphere, this correspondence implies a shallow origin of the hotspot reference frame.

Stable continent forms only a small areal fraction of lithoplates. Yet there are similar features of oceanic plates that also imply a relatively shallow reference frame. In particular, there are the bathymetric and gravimetric lineations observed over the east-central Pacific. Haxby and Weissel (1986) recognized that many of the lineations parallel the direction of Pacific plate motion in the hotspot reference frame (the "Hawaiian" mesoplate). The wavelength of the features, combined with flexural modeling (McAdoo and Sandwell 1989), imply a shallow origin, "frozen-in" to the plates. Either plates were deformed subsequent to their formation by sea-floor spreading, or magmatic patterns at the spreading center introduced a record of plate motions (e.g., Schouten et al. 1987). In either case, a deep-mantle origin can be ruled out as an explanation for the lineations.

Again, correspondence of the lineations with hotspot trace orientations, combined with shallow origin of the lineations, implies shallow "fixity" of the hotspots.

12.3 Small Island-Seamount Chains of the Pacific

Where does this leave the *origin* of the hotspots, then? The numerous smaller island-seamount chains of the subequatorial portion of the Pacific plate (i.e., Pitcairn–Gambier, Marquesas, Austral–Cook, Society Islands: Figures 6.12–6.19) present a challenge for the hotspot model, particularly insofar as there is evidence for "reactivation" of some chains, or, perhaps, the overprint of multiple hotspot encounters (Figs. 6.16, 6.18). The seamount chains of the Gulf of Alaska show similar patterns (Figs. 6.22–6.25).

The South Pacific chains, with the exception of the Easter–Tuamotu trace, are all apparently short-lived. There is no obvious evidence for significant older extensions of the Society, Austral–Cook, Marquesas, or Pitcairn–Gambier traces.

The proximity of the subduction zone beneath the Alaskan peninsula precludes assessment of the older extensions of the Gulf of Alaska seamount chains. Cross and Pilger (1982) did suggest that the Pratt-Welker chain could extend into the subduction zone, as an explanation for relatively low-angle subduction and the absence of contemporary volcanism along its inferred extension, following the hypothesis of Pilger (1981b) for the relation of hotspot traces and low-angle subduction.

A particularly intriguing relation of chains to the structure of the older lithoplate on which they rest can be observed for each chain, both in the subequatorial Pacific and the Gulf of Alaska. The older portion of each chain exists to the south of a paleofracture zone, to the north of which is significantly older lithoplate (Figs. 12.1, 12.2). Conversely, the older extensions of small island-seamount chains do not appear to exist to the south of fracture zones that separate older lithosphere to the north from younger lithosphere to the south in the eastern Pacific.

12.4 A Model for the Small Island-Seamount Chains of the Pacific

Thermal modeling and seismic evidence imply that oceanic plate thickens with age. The model of Parker and Oldenburg (1973) relates thickness (Z) to the square root of age (t, in my): $Z=9.4\ t^{1/2}$ km. The age patterns and the hotspot model imply motion of the subequatorial portion of the Pacific plate over the

Fig. 12.1. Sample locations for subequatorial island-seamount chains, calculated hotspot loci, and seafloor isochrons (chron numbers, age in *parentheses*). Note that each chain exists on lithoplate that is younger than lithoplate to the north, across a paleotransform.

underlying mesoplate in a northwesterly direction since about 45 Ma (the Hawaiian–Emperor bend).

Thus, as the portion of the Pacific plate with a fracture zone (separating older and thicker plate on the north from younger and thinner plate on the south) moves to the northwest, underlying asthenosphere and deeper mesoplate, over which the fracture zone passes, rise isostatically. Depressurization of fertile asthenosphere could result in enhanced melting within the rising asthenosphere and partial melting of underlying mesosphere, replenishing the asthenosphere. Figure 12.3 illustrates this concept.

Most of the smaller island-seamount chains of the Pacific plate are south of a fracture zone that separates older plate on the north from younger plate on the south. It is still possible that additional factors contribute to the propensity of the rising asthenosphere and underlying mesoplate to melt, especially the possibility of regional petrologic and/or thermal differences between the subequatorial Pacific and the central Pacific asthenosphere/mesoplate to the north. Further, once a hotspot is produced by this mechanism, it might well persist even if older and thicker plate might subsequently pass over it.

The Line Islands include sample ages that require a number of hotspots in order to fit them (Figs. 6.10, 6.11). While the structure of the lithoplate on which the Line Islands are built is uncertain, more recent analyses (e.g., Norton 2000) suggest that fracture zones crossing the chain have offsets of the same sense as

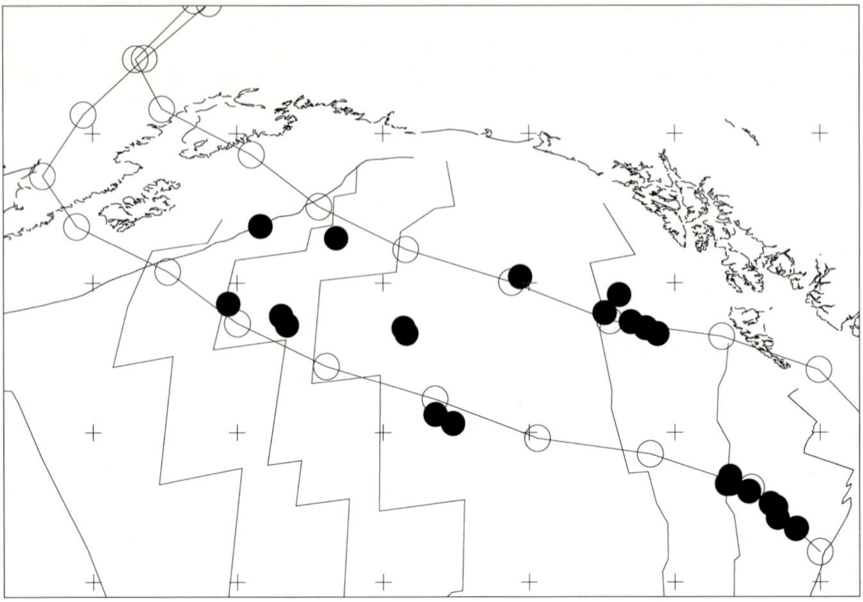

Fig. 12.2. Sample locations for Gulf of Alaska seamount chains, calculated loci, and seafloor isochrons. As in Fig. 12.1, note that the oldest known portion of each chain exists on lithoplate that is younger than lithoplate to the north, across a paleotransform.

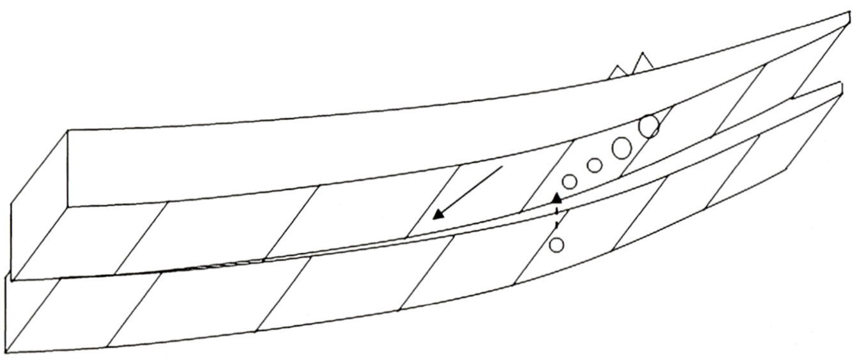

Fig. 12.3. A "view" of the underside of two segments of oceanic lithoplate separated by a fracture zone. The more distant segment is offset to the right. Thus older and thicker lithosphere on the far side of the fracture zone is juxtaposed against younger, thinner lithosphere. *Arrow* Direction of motion of the lithoplate relative to the underlying asthenosphere and mesosphere; *circles* possible locus of enhanced melting of asthenosphere as a result of depressurization as thinner plate passes over the zone. A number of such loci could come into existence, resulting in modifications to lithoplate structure (as evidenced by bathymetric and gravity lineations in the subequatorial Pacific; Haxby and Weissel 1986) or in small island-seamount chains, such as those shown in Figs. 12.1 and 12.2.

those further south, with older plate on the north, younger on the south. Thus the Line Island traces, which appear to have the appropriate orientation for given age ranges according to the hotspot model (e.g., Raymond et al. 2000), may have originated in a manner similar to that inferred for the subequatorial and Gulf of Alaska traces. Passage of a fracture zone offset (as the Pacific plate moved to the north-northwest) over successive zones of fertile asthenosphere/mesosphere could produce the observed overprinted episodes of magmatism.

12.5 Large Hotspot Traces – A Different Model Is Required

What of the origin of the large hotspot traces? The generalizations that apply to the smaller chains are not true of the Hawaiian–Emperor chain. The Louisville trace appears to have grown onto progressively younger and thinner plate, subparallel with fracture zones for at least its younger extent. The Tuamotu–Easter chain appears to have been accompanied by progressive shifting of the ridge axis to correspond with the trace between roughly 50 and 30 Ma, suggesting that its underlying hotspot influenced seafloor spreading and consequent plate structure, rather than having the hotspot influenced by plate structure as proposed for the smaller chains.

In the other oceans, there are no obvious hotspot traces that show the pattern observed for the smaller chains of the Pacific. In many cases, plate motions in the hotspot frame (relative to the Tristan mesoplate) are subparallel with fracture zones, rather than oblique as observed in the Pacific. One exception is the older part of the South Atlantic. The Tristan da Cunha trace trends obliquely to fracture zones prior to 40 Ma (O'Connor and Duncan 1990). However, like the Easter–Tuamotu trace, there is evidence for ridge jumping, perhaps controlled by an underlying hotspot.

The patterns of magmatism in onshore eastern South America indicate widespread activity prior to and subsequent to the opening of the South Atlantic Ocean (Figs. 6.36, 6.39), with lesser activity in onshore southwestern Africa. The existing hotspot model of Müller et al. (1993) implies that the Tristan da Cunha hotspot is the continuation of the Parana (and Endenka in Africa) activity, following Morgan (1981, 1983). The plume-head-and-tail model of Richards et al. (1989), with its deep plume origin, finds application here. However, is there an alternative model for producing the volcanism and the long-lived hotspot from shallower depths?

As a gross approximation, let it be assumed that progressive replenishment of oceanic plate source material at the mid-oceanic ridge, as the spreading process depletes it, is entirely vertical, in the local proper motion frame of reference, extending to a presently undefined depth within the mantle. This undefined level is referred to as the depth of lateral replenishment and is analogous to the isostatic compensation depth in Pratt-Airy isostasy. If the ridge is itself in motion in the proper frame of reference, then any zone beneath the ridge is a locus of vertical replenishment for a only a discrete period of time. However, if a segment of the spreading ridge is stationary in the proper frame for a longer period of time, then that portion of the asthenosphere and mesoplate beneath the ridge is a locus of vertical flow for an equivalently longer period of time. Thus, some portions of the deeper mantle are tapped at a greater rate than others, as controlled by the spreading rate, width of the zone of vertical flow, the rate of replenishment, and the duration of the period of proper stationarity.

As Schouten et al. (1986) have inferred, bathymetric lineations, which seem to record proper motion of the ridge, imply vertical replenishment of the spreading-center source region. Might ridge-centered hotspots represent extreme cases of replenishment, which introduced anomalous compositions from much greater depths than typical for ordinary ridges? In other words, hotspots/plumes are initiated or enhanced at a time of proper ridge stationarity and, possibly, rapid spreading, resulting in extreme depletion of shallow mesoplate and consequent tapping of deeper, more "primitive" and fertile mesoplate.

Thus, the Easter hotspot, responsible for the Tuamotu and Nazca Ridges, may have originated when the Pacific–Farallon ridge became stationary for several millions of years (Fig. 12.4). The predicted locus of the hotspot trace on the Pacific plate corresponds with the Tuamotu ridge between roughly 45 and 30 Ma, but the older portion of the locus extends into older Pacific lithoplate that has no evidence of a ridge or seamounts of the appropriate age (Figs. 6.6, 6.7, 12.1); adjacent portions of the Line Islands are too far away and available dated samples are too old. Thus, the hotspot appears to have come into existence at about 45 Ma.

But, what about non-ridge centered oceanic hotspots, or those which have continental manifestations? It is suggested that hotspots such as Hawaii and the hotspot responsible for Louisville ridge might represent formerly spreading ridge-centered hotspots that originated from a prolonged period of ridge segment stationarity in the proper frame. It is suggested here that the oceanic plateaus of the northwestern Pacific (e.g. Hess and Shatsky Rises) represent proper stationarity of the various paleospreading centers along which they are inferred to have formed. The character of the plateaus, especially their substantial volume, might suggest that the spreading rate during their formation was also very great. If the spreading center were stationary relative to source regions (asthenosphere and mesoplate) located vertically beneath, relatively higher spreading rates would result in tapping of deep mantle. The resulting hotspots associated with the oceanic plateaus could be represented by the fertile mesoplate that is inferred to be responsible for the subequatorial hotspot traces of the east-central Pacific (e.g., Henderson et al. 1984), whose origin is discussed above. Unfortunately, older portions of both the Hawaiian–Emperor and Louisville traces have been subducted, so that it is not possible to test the stationary-ridge hypothesis for the origin of their hotspots.

Insofar as originally (?) continental hotspots are concerned, especially those inferred to be associated with major continental plateau basalts, consider the possible kinematics of rifting with which the Norwegian–Greenland Sea is associated. Unfortunately, it is not possible to determine whether the mesosphere beneath western Eurasia and northeastern North America is part of the Tristan or Hawaiian mesoplate, or a separate (Iceland) mesoplate. Assuming that the mesosphere of the area is part of the Tristan mesoplate, the proper plate kinematic model implies that, between 60 and 45 Ma, significant changes in Eurasian plate motion occurred, both in direction and rate (Fig. 6.59). Initial rifting may have occurred in such a way as to maintain stationarity of portions of the rift in the proper frame. Thus, the initial accommodation of rifting by rise of mantle material would have been more pronounced in the region of relative stationarity, tapping deeper and, presumably, more fertile source regions.

Between 70 and 55 Ma, the proper kinematic model implies not only slow proper motion of Greenland–Eurasia relative to the Tristan mesoplate (Fig.

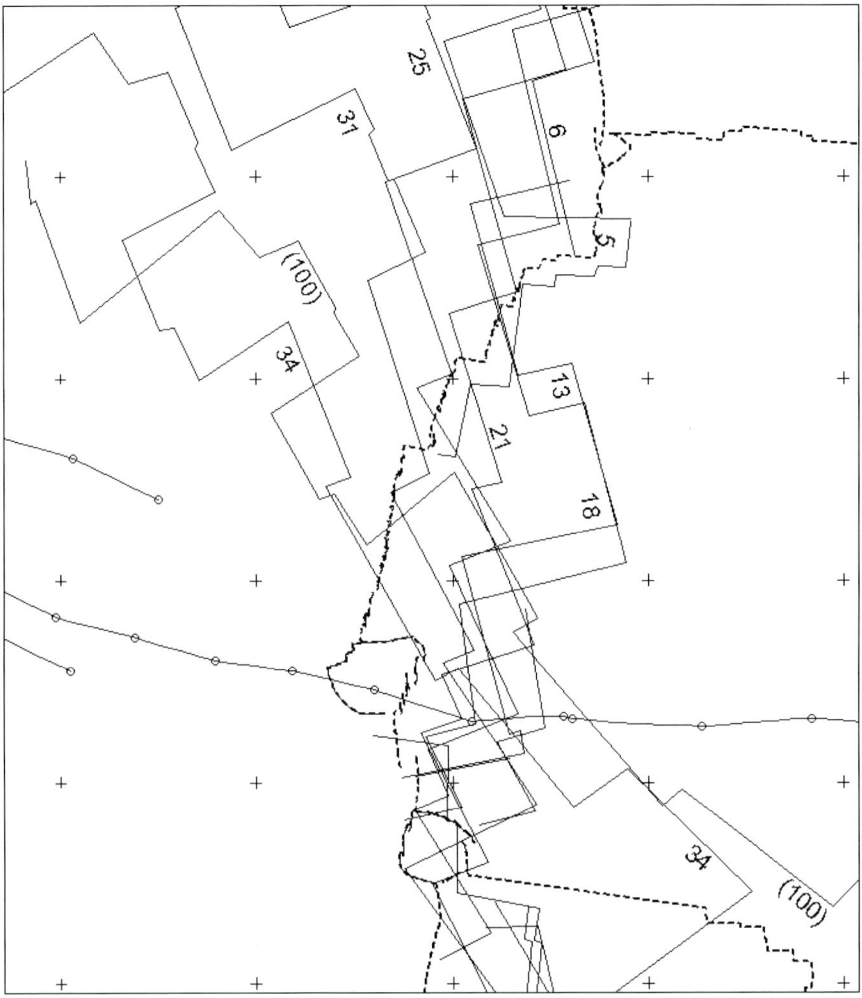

Fig. 12.4. Magnetic isochrons from the Pacific plate rotated to their location of origin in the east-central Pacific Ocean, according to the model of Raymond et al. (2000), as modified in this volume, together with loci of hotspot motion for the Easter, Marquesas, and Pitcairn hotspots relative to the Pacific and Nazca plates. Note the clustering of isochrons between anomalies 25 and 13 in the vicinity of the Easter hotspot. The Easter hotspot is inferred to have generated the Tuamotu and Nazca ridges when the spreading ridge was centered over the hotspot.

6.59), it also implies motion subparallel with the newly developing rift, especially in the region of abundant magmatism along the East Greenland margin. It is therefore suggested that the abundant magmatism associated with the opening of the Norwegian–Greenland Sea is not a result of a preexistent plume, but, rather, is the result of rifting that tapped deep-mantle regions due to stationarity of that segment of the developing rift/ridge that manifested the anomalous igneous activity.

Similarly, the present-day Iceland hotspot may have originated by analogous processes in Baffin Bay or the Arctic Ocean, perhaps represented by the Alpha Ridge. The proximity of Iceland to the British Tertiary and voluminous flood basalts associated with the early rifting of the Norwegian–Greenland Sea is a contemporary accident. The hotspot produced by the rifting would now be located south of Britain if it were still active; absence of obvious contemporary manifestation of volcanism in that area would suggest that the hotspot is now inactive and that the anomalous mantle brought closer to the surface by rift stationarity has now been exhausted of fertile material.

The inferences regarding Greenland–Eurasia, above, depend on the assumption that the mesosphere of the region was (and is) part of the Tristan mesoplate. If a distinct Iceland mesoplate exists and had significant movement relative to the Tristan mesoplate, the interpretation is obviously not supported by the kinematics of Greenland and Eurasia relative to the Tristan mesoplate.

What about the mechanisms of rifting and plate fragmentation? Consider the present-day East African Rift system. The model of Müller et al. (1993) (as do each of the previous Africa-hotspot kinematic models) implies proper motion of the African plate subparallel with the contemporary Rift for much of the Cenozoic (Fig. 6.66). Late-Cenozoic paleostress and contemporary stresses appear to be governed by the vagaries of the advanced rifting process and by the relatively high topography of the rift zone, due to the elevated isotherms at depth. It is suggested that initial rifting was largely controlled by the prevailing rift- and proper motion-parallel maximum horizontal principal compressive (MHPC) stresses. Volcanism along the rift is difficult to correlate with the rifting process, beyond the gross spatial correspondence of volcanism with the broad uplift upon which the rift is located.

At the northern-most end of the Rift system is the Afar Triangle, a region of abundant "plume-like" volcanism, much of which is subaerial. The hotspot-plate model (Fig. 6.66) suggests that, for the past 10 my, proper motion of the Afar region has been to the east-northeast, over the region of Late-Cenozoic seafloor spreading in the Gulf of Aden. Spreading rates in the Gulf of Aden are moderately slow, and, in the proper motion frame, the western end of the ridge, which projects into the Afar Triangle, is latitudinally stable; the ridge is moving properly parallel with itself. Thus, assuming vertical mantle flow as compensation for rifting and spreading, the latitudinal stationarity of the western end of the Gulf of Aden spreading center would result in progressively deeper tapping of more primitive and fertile mantle sources, producing Afar "hotspot" magmatism. In a similar manner, Mid to Late Cenozoic proper motion of the northeastern margin of Africa was close to rift-parallel, implying intraplate stress control on the rifting process.

Might not other zones of anomalous volcanism represent either primary proper ridge stationarity or motion parallel with the ridge, or the persistence of a melting anomaly formed in such a way, long after its proximity to the ridge has passed? Consider the other zones of anomalous volcanism in North Africa. It is suggested that the anomalous volcanism of Tibesti and similar regions may represent melting anomalies which originated beneath a Tethyan spreading center and which were subsequently overridden by the African plate. This suggestion implies that such melting anomalies are not of recent origin and the present-day topographic anomalies of North Africa have been moving progressively southward (relative to Africa). Available geological maps of North Africa suggest Late-Cenozoic volcanics due indeed exist to the north of Tibesti and the other centers.

In an earlier article (Pilger 1982), it was suggested that anomalous volcanism in Eastern Australia partially reflected changing stress fields within the Australian–Indian plate. A southward progression in age of apparently alkaline volcanic centers in the Australian Highlands also appears to approximately correspond with termination of a persistent tholeiitic volcanic episode and evidence for a change in intraplate stress field (from MHPC parallel with the Highlands to normal). The East Australian volcanics had been postulated to have a plume origin (e.g., Wellman and McDougall 1974; Vogt et al. 1977). Sutherland (1983) argued for a hotspot-like origin of the central volcanics from melting anomalies formed beneath an Early Cenozoic spreading center in the Coral Sea. However, explanations for the tholeiitic pulse have remained uncertain.

Interestingly, the modeled proper motion of Australia in the Late Cretaceous was parallel with the synchronously formed Coral Sea and Southeastern Indian Ocean margins of the continent. That is, proper motion of the rift margins was rift-parallel, inducing greater vertical flow of mantle as rifting and early seafloor spreading occurred. Subsequently, Australia began moving more northerly, overriding the former Coral Sea subrift/spreading center mantle. As Sutherland (1983) proposed, the Central volcanics of Eastern Australia may mark the location of the Coral Sea spreading center. The proper reconstructions of Australia also imply that the eastern margin of the continent overrode mantle, which would have been beneath the Tasman Sea spreading center, which apparently initiated in the Late Cretaceous (Weissel and Hayes 1977). Alternatively, as noted in the discussion of tholeiitic volcanism in the Highlands, above, progressive propagation of seafloor spreading from south to north in the Tasman Sea may have been responsible for the "pre-hotspot" volcanic activity. Further, the progressive cessation of volcanism marked by the central volcanics could represent the "last gasp" of magmatism from the extended zone beneath the Highlands, produced by the Tasman rift/spreading propagation.

The Balleny hotspot, in analogy with the East Australia anomalies, may have formed in association with rifting between Australia and Antarctica. As with the inferred Coral Sea hotspot(s) now beneath offshore Victoria, proper motion of the developing Australian–Antarctic rift was rift-parallel in the latest Cretaceous and Early Cenozoic (Fig. 6. 59). With its formation, northward proper motion of the Australian–Antarctic ridge resulted in the Balleny hotspot being located solely beneath Antarctica subsequent to its formation.

The mechanism for the formation of the East Australian, Tasman Sea, and Balleny hotspots can, therefore, be inferred. Early rifting along the future margins of the Coral Sea and Southeast Indian ocean margins was oriented parallel with Australian–Antarctic proper motion (as the predicted stress field model would imply; Fig. 6.59). Because of rift-parallel motion, deeper portion of the mantle were progressively tapped, producing "plumes" of various dimensions. With changing proper motion of Australia, the plate passed over the newly generated mantle anomalies, producing the familiar terminal East Australian central volcanic traces and Tasman Sea seamount/guyot chains. Similarly, at least one such melting anomaly was produced by a comparable mechanism beneath the southern Australian margin (the Balleny trace).

The suggested correspondence of a change in paleostress field as igneous activity progressively subsided (Pilger 1982) remains a related issue, since the mechanism for reorientation is uncertain. Perhaps the mantle anomaly represents the postulated early Coral Sea rift/spreading source, as well as a fundamental boundary between suboceanic and subcontinental mantle with different effects on intraplate stress fields (e.g., less effective traction of the Australian continental lithosphere against previously suboceanic mesosphere).

What of the great continental flood basalt provinces postulated by many to indicate the initial impingement of a rising deep-mantle plume upon the lithosphere? The continental margin of the Norwegian–Greenland Sea has been included by some with continental flood basalts (e.g., White and McKenzie 1989) and can be interpreted in terms of proper ridge/rift stationarity. Similarly, the Deccan Traps occur adjacent to a continental margin that is subparallel, in part, to the locus of Indian plate motion at the Cretaceous–Cenozoic boundary (Fig. 6.31). Part of the Seychelles may also reflect temporary ridge stationarity.

The proper motion model of Müller et al. (1993) extends only as early as 130 Ma, close in age to the Parana flood basalts of Brazil. There are suggestions that early rifting may have been parallel to plate motions in this region, augmented by the possible affect of ridge propagation from the south (Fig. 6.37). Like the East Australian highlands, the locus of deep extension may have been offset along a detachment from the zone of shallow rifting, focusing melting beneath and to the west of the future Brazilian Atlantic continental margin. The significantly larger volume of Parana volcanism, compared with the Australian highlands, could represent a greater amount of extension over a much shorter period of time; conceivably, extension might have even extended into the subcontinental mesoplate itself.

The Columbia River basalts provide an interesting case study themselves. Most of the volcanism that is recognized as part of the Columbia episode beginning at roughly 18 Ma (Figs. 6.54–6.55). The pattern of Columbia and younger volcanism of the Snake River Plain–Yellowstone trace suggests that the hotspot responsible for it was (and is) located within the Pacific mesoplate. The predicted reconstructed trace on the North American plate implies significant changes in plate motion relative to the mesoplate, especially a reversal in motion between 30 and 20 Ma. Prior to 25 Ma, proper plate motion was to the east, relative to the Pacific Mesoplate; then, around 20 Ma a reversal occurred, with the intervening motion implying near-stationarity, with a slight northeasterly direction of motion, between 25 and 20 Ma. What is uncertain is when the Columbia River–Yellowstone portion of North America encroached upon the Pacific mesoplate (or, alternatively, when the mesosphere beneath the western United States became part of the Pacific mesoplate). Conversely, the Yellowstone hotspot may have come into existence with the onset of rifting beneath the future Columbia River plateau, producing a hotspot within the underlying mesoplate.

12.6 Summary

A cascade of arguments combine to support the hypothesis that hotspots are not only fixed within mesoplates, they originate within mesoplates and the overlying asthenosphere. These include:

- Correspondence of plate motions in the hotspot reference frame with contemporary and paleostress orientations in stable continental interiors.
- Correspondence of plate motions in the hotspot reference frame with lineations in gravimetric and bathymetric fields.
- Existence of small, short-lived island-seamount chains on relatively younger lithoplate leeward of fracture zones separating older and thicker lithoplate from the younger lithoplate.
- Apparent inception of the Easter hotspot with the onset of stationarity of the Pacific–Farallon spreading center, about 45 Ma, implying that the stationarity and fast spreading resulted in tapping of deeper mesoplate.
- Correspondence of plateau basalts with rift margins (Parana and Deccan) or other extensional environments (Columbia River).
- Apparent onset of progressive rift extension prior to volcanism in East Africa.
- Kinematic correspondence of East Australian Highlands terminal volcanic pattern and Tasman seamount hotspot with apparent Coral Sea spreading-center stationarity.
- Correspondence of inferred, reconstructed Balleny hotspot with the Cretaceous rift between Australia and Antarctica.

CHAPTER 13

Fracture Reservoirs and Intraplate Stresses

13.1 Aside

Plate Tectonics has been almost completely developed by academic and government geoscientists. However, the largest proportion of geologists and geophysicists is employed in the oil and gas industry. Has Plate Tectonics played a significant role in hydrocarbon exploration over the past 30 years or so? For the most part the answer has to be negative. The origin of continental margins and interior basins in the context of rift tectonics has certainly been informed by plate tectonic modeling, including thermal evolution which probably controls hydrocarbon maturation. The geologic time scale has been refined with incorporation of magnetic reversals from oceanic plates. Perhaps most importantly, sequence stratigraphy has had thermal evolution of sedimentary basin as an implicit or even explicit component, as it has been applied to sedimentary sequence and facies characterization as part of the potential reservoir evaluation process. Nevertheless, it is apparent that Plate Tectonics has, at best, only played a peripheral role in advancing the exploration process.

Conversely, information acquired during oil and gas exploration provides a key component of this volume. With recognition that boreholes typically are deformed into elliptical shapes after drilling, and further recognition that the ellipticity tends to have a regionally consistent orientation, it became apparent that sedimentary basins exhibit regional stress fields. More significantly, the orientation of the stress fields inferred from boreholes corresponds with inferred motions of plates in the hotspot reference frame (as described in Chap. 7). In this volume, demonstration of the correspondence of stress fields with proper plate motion in the geologic past as well as the present provides an opportunity for Plate Tectonics to return the favor to hydrocarbon explorationists by providing a predictive ability for contemporary and paleostress fields.

13.2 Fracture Reservoirs

The application of low-angle ("horizontal") drilling technology to the development of and exploration for fractured reservoirs in the last 20 years provides the motivation for the improvement in fracture and stress field predictive abilities, in order to more effectively utilize this new technology. It is, therefore, highly desirable that improved understanding of the origin of fracturing, and its relationship to basin subsidence, diagenesis, trap formation, and hydrocarbon maturation and migration, be obtained. Further, not only the existence of fractures, but also the extent to which they are open (or can be opened or even induced artificially) and are filled with fluids are both of critical importance. At present, the success of low-angle drilling programs has largely depended on empirical experience supplemented by some structural analysis. That is, existing fields that were discovered by conventional, near-vertical drilling are being extended by the new technology; the azimuth in which wells are drilled is usually normal to the predicted trend of open fractures delineated by existing production, fracture logs, and/or other geological and geophysical constraints.

Where exploration is occurring, horizontal wells are typically oriented normal to the strike of basinal strata or with known fault/fold systems assumed to have controlled fracturing. Many conventional fractured reservoirs in the U.S. Gulf Coast, for example, can be shown to be associated with faults and/or fault-controlled flexures. Improved prediction of the orientation of possible fracture systems and of the present-day stress field, whether controlled by faulting, folding, or other mechanisms, can improve the ability of horizontal-drilling exploration, by optimizing the chances for intersecting open fractures.

13.3 Horizontal-Drilling Environments

It may not be widely known even among explorationists, but there are basically three very different approaches to applied horizontal drilling for fracture reservoirs. In the first case, typified by the Austin Chalk play in South Texas, the reservoir targets consist of preexisting, open natural fractures in the chalk and the associated Buda limestone. Thus, an essential development technology involves drilling horizontal wells at roughly right angles to the inferred fracture trends, so as to intersect as many fractures per borehole unit length as possible (Fig. 13.1). Prior to the application of horizontal drilling, vertical boreholes had a greatly reduced chance of encountering fractures simply because most fractures have a nearly vertical dip and fracture spacing is commonly significantly greater than typical borehole diameters. Even with horizontal drilling, it is critical that the best estimates of fracture orientation, as well as fracture spacing, be available in order to encounter the greatest density of fractures in which hydrocarbons might occur.

The second and third types of application of horizontal drilling involve production from fractures artificially induced in the potential reservoir rock. In these cases, fractures are produced by artificially increasing the fluid pressure through perforations in the borehole casing adjacent to the desired formation, beyond the elastic limit of the rock. The induced fractures will propagate parallel with the maximum horizontal principal compressive (MHPC, or σ)stress direction. In the second type, drilling is normal to the strike of the anticipated fractures (thus normal to the MHPC stress direction; Fig. 13.2). In the third type, optimal production occurs when the expected induced fractures (from the orientation of the MHPC stress) are oriented parallel with the borehole (Fig. 13.3).

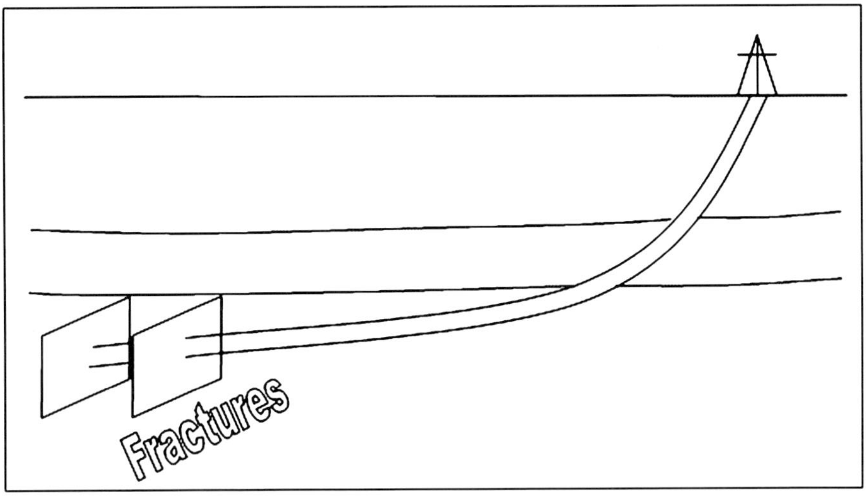

Fig. 13.1. Horizontal well intersecting fractures.

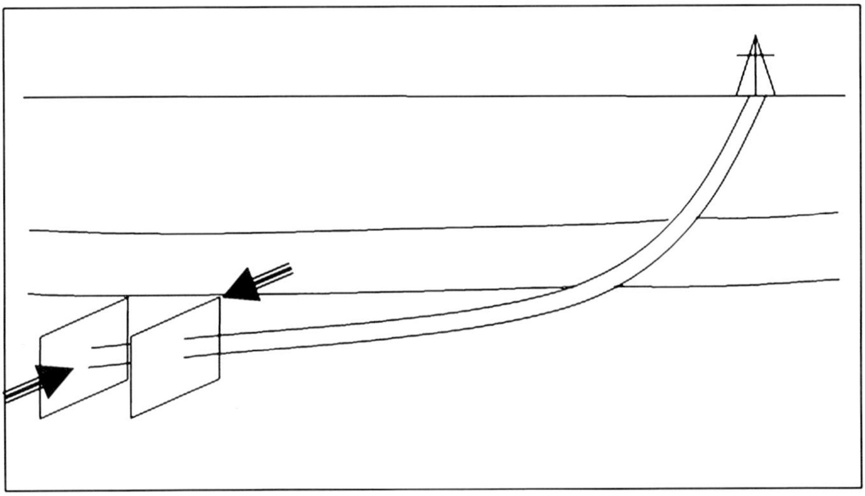

Fig. 13.2. Same as Fig. 13.1. Arrows Orientation of principal horizontal compressive stresses parallel with fractures.

The rationale for drilling parallel with or normal to the MHPC stress direction depends on the existence and, if present, the orientation of preexisting natural fractures.

In the first two cases (for example, the Austin Chalk of the U.S. Gulf Coast and Bakken Shale of the Williston Basin), horizontal drilling is usually normal to the paleo-MHPC stress direction. In the second case, drilling is normal to the contemporary MHPC stress field and normal to the paleostress (fracture) direction

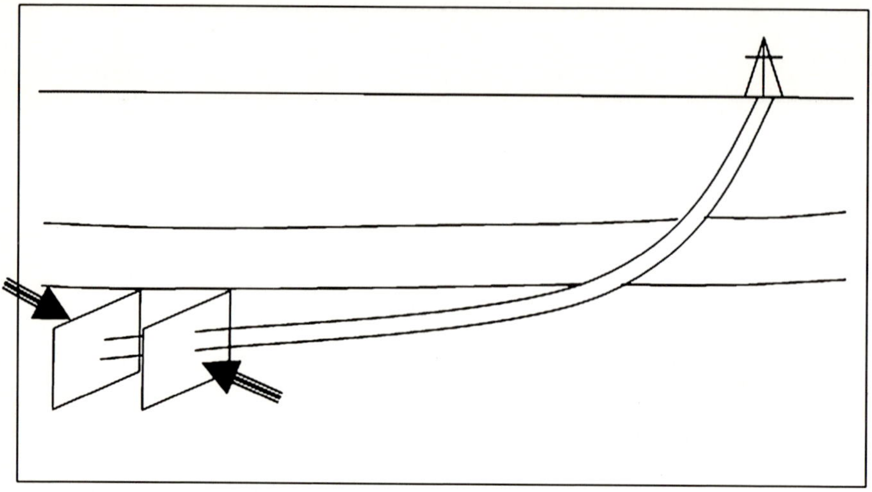

Fig. 13.3. Same as Fig. 13.1. Arrows Orientation of principal horizontal compressive stresses normal to fractures.

(that is, contemporary and paleostress directions are parallel). In the third case (Cretaceous Chalk of the Danish North Sea), drilling is parallel with the contemporary MHPC stress field and normal to the paleostress direction. In order for the fractures in cases one and two to be open, it is necessary that either the contemporary MHPC stress magnitude be small relative to the natural fluid pressure, or that its orientation be at least subparallel with the paleo-MHPC stress orientation (recorded by the fractures). Maximum production should result from drilling normal to paleofractures (and, of course, the contemporary MHPC stresses).

In the third case, the necessity of hydrofracturing parallel with the borehole is a result of the absence of significant natural fractures parallel with contemporary MHPC stresses. Since the MHPC stresses are oriented at a high angle to existing fractures, the production of artificial fractures parallel with MHPC stresses should result in a network of new fractures linking fluid-filled paleofractures. The existing stress field will further serve to enhance production by squeezing fluids from the natural fractures into the induced fractures.

The three cases of applied horizontal drilling for production from fractured reservoirs imply the need for as much knowledge concerning the existence and orientation of existing fractures and the orientation of the contemporary stress field as possible. Abundant information implies that, over the history of a sedimentary basin, it is quite possible that stress field orientations and magnitudes have varied significantly. However, the existence of a stress field, even of a moderately high magnitude, does not imply that fracturing of rocks in the basin will necessarily occur during the duration of that field orientation without other factors also being involved.

13.4 Mechanisms for Fracturing

There are several working hypotheses that might be applied to fracture orientation prediction. These hypotheses are underlain by basic assumptions derived from published theory, laboratory results, and field observations. Significant fracturing of a rock is indicative of its relatively brittle response to induced stress.

Commonly, ductility contrasts with more easily deformed rocks can result in fracture of a stronger unit, under the application of a high differential stress field. The contrast in ductility may be a result of contrasts in diagenetic histories as well as lithology. It is not uncommon for carbonates to become lithified more rapidly than shales and mudstones. As a result of rapid deposition and progressive diagenesis, it is conceivable that the shale/mudstones can become overpressured beneath a carbonate seal. Consequent brittle failure of the carbonate occurs when the progressively increasing ambient fluid pressure becomes too great. Diagenetic reactions that release previously bound fluids may increase overpressuring. The higher fluid content of the shale, combined with the greater ductility of clays relative to carbonates, contributes to the brittle behavior of the carbonate.

The orientation of the resulting fractures will be controlled by the orientation of the stress field operable at the time of failure. Extensional fractures should be oriented normal to σ_3, the least principal stress, with σ_1 and σ_2 within the plane of fracture. The orientation of the principal stresses will be determined by a combination of regional and local tectonic factors. In an actively subsiding basin (such as the U.S. Gulf Coast), these factors probably include:

1. Down-to-the-basin growth faults and associated folds. Gravitational instability of the rapidly prograding shelf of the northern coast of the Gulf of Mexico commonly results in periodic basinward faulting. Some of the faults appear to sole into overpressured shale masses or salt. In some cases, apparent overthrusting of a complex toe of the slumping mass may accompany fault movement far into the deep basin. Fracturing of brittle rocks within the fault blocks may occur by splaying out of portions of the fault, rollover folding, due to displacement of the block along the curved surface of the sole fault, or antithetic faulting. In the presence of extensional stress fields, fracture strike orientations should parallel the strike of adjacent faults and axes of rollover folds. Bending of previously planar brittle strata within folds may result in extensional fractures whose spacing is closest where curvature is greatest.
2. Salt kinematics. Salt flowage, diapirism, and sill emplacement are important factors in the structural evolution of the northern Gulf Basin. Gravitational instability of the deep salt relative to overlying dense, compacted and cemented sediments can result in deformation as salt is displaced laterally and/or vertically, with consequent folding and faulting of overlying strata. Again, the presence of relatively brittle units (mostly sandstones) can result in their fracture rather than ductile deformation. Drape folds above salt structures would be expected to produce fracture strike orientations parallel

to the axes of folding. Compressive structures (faults and folds) in the Gulf Basin appear to be relatively uncommon except immediately adjacent to diapirs or along the continental rise, where overthrust and fold belts (the Perdido and Mississippi fan fold belts) are present. Fracture orientations associated with compressive structures can also be predicted, depending on the complexity of each individual structure.

3. Regional tectonic stress field. On a large scale, the presence of a regional stress field, perhaps modified by local tectonic and/or diagenetic effects (e.g., salt kinematics, growth faulting, differential compaction/cementation), may control the orientation of fractures, even if the stress field is not large enough to be responsible for fracturing. For example, early lithification of a carbonate stratum overlying and sealing more slowly compacting argillaceous sediment could, as fluid pressures increase within the shales, result in natural hydrofracturing of the carbonate. The orientation of the fractures could then be controlled by the regional stress field, even if the actual deformation were induced by gravitational loading and diagenesis. In such a case, σ_1 would be oriented vertically, with σ_2 (the MHPC stress) oriented parallel with fracture strike.
   Regional stress fields may originate from several processes. Adjacent to plate or microplate boundaries, plate interactions produce systematic stress fields whose orientations depend on the relative convergence, divergence, or shear across the boundary, together with gravitational loading in association with overthrusting or subsequent sedimentary basin formation, and preexistent inhomogeneities. On a large scale, plate motions may involve significant traction between the plate and underlying asthenosphere or mesosphere, resulting in compressive stresses oriented in the direction of plate motion relative to the mesosphere (e.g., Solomon and Sleep 1975); it is this mechanism that is inferred to be responsible for the intraplate stress fields described in Chaps. 7 and 8.
   Within continental North America, the direction of contemporary plate motion in the proper reference frames is from northeast to southwest (Chap. 7). As noted in this volume, a large number of measurements of the principal horizontal stress orientations within the stable cratonic interior of North America are also oriented northeast-southwest (Zoback and Zoback 1980; see also Fig. 7.1). Similar measurements in western Europe suggest a nearly north-south orientation of MHPC stresses from in situ measurements (e.g., Zoback et al. 1989). However, only the model of Minster et al. (1974) implies north-northwest to south-southwest movement of the western continental part of the Eurasian plate; the model of Gripp and Gordon (1991) implies more west-to-east motion (Chap. 7). In Chap. 8, the close correspondence between measured strikes of dated igneous dikes (and similar geological indicators of the orientation of MHPC stress) and the direction of plate motion relative to the hotspot frame of reference is demonstrated for examples from North America and, to a lesser extent, Europe, where such data are mostly available, for the past 130 my.
4. Preexisting inhomogeneities. Major lithologic boundaries, such as regional thrust-faults, paleosubduction zones, former extensional detachment faults,

> tabular igneous bodies, or even major regional facies changes, may localize and reorient stress fields produced by other processes. Reactivation, even inversion, of preexisting structures is a common phenomenon at virtually all scales of tectonism. For example, the regional stress field in much of cratonic North America is consistent with a principal horizontal stress orientation of northeast-southwest, from Quebec to Texas and from northern Alberta to Arkansas. However, in the northern Appalachians, the state of stress is somewhat more complicated (Fig. 7.2). There appear to be somewhat consistent departures from the regional stress field along the western flank of the Appalachians and along the East Coast continental margin, extending to offshore Labrador. In the original study by Zoback and Zoback (1980), these departures seemed to be even more pronounced; subsequent study suggests that the departures are not as great as first inferred (Zoback et al. 1989).

The location of the deflections corresponds with known faults of major displacement, in many cases. Subjective evaluation of the stress field suggests that the Appalachian–East Coast trends consist of lateral inhomogeneities that modify the regional stress field in a consistent manner. Deglaciation may also play a role in the stress fields of the circum-Hudson Bay region (e.g., Stein et al. 1979), while Late-Cenozoic extension of the western United States and associated thermal uplift produce a complicated stress field that can be interpreted in terms of microplates and distributed deformation of the western margin of North America.

13.5 South Texas: Fractured Austin Chalk Fields

One of the areally largest fracture-reservoir fields, Pearsall in South Texas, is anomalous in many respects, compared with such fields as Giddings, farther north (Fig. 13.4). In Giddings, fractured reservoirs appeared to be closely related to faults and fault-controlled rollover folding. In contrast, significant production in the southern part of Pearsall Field appears to be from fracture systems that seem to lack association with either recognizable faults or folds. Pearsall anticline (forming the northern third of the field) is an asymmetric structure, with a steeper north-dipping flank and a much more gently-dipping south flank. The area of abundant fracture production in Pearsall field extends well south of the anticline into an area of relatively constant dip in which significant faults are also apparently absent. Thus, the fracturing of the Austin Chalk in Pearsall field cannot be convincingly related to either mapped faults or folds. Existing mapping of and available seismic sections across the field show neither faults nor folds in much of the area of known fracture production.

Available information on fracture orientations in Pearsall field is sparse, due to the high level of contemporary development and exploration activity and the desire for proprietary advantage. Available information on fracture orientations is summarized in Fig. 13.5, while contemporary stress field orientations (from borehole ellipticity) is summarized in Fig. 13.6. One might assume that the orientation of recent low-angle boreholes in Pearsall field may represent the

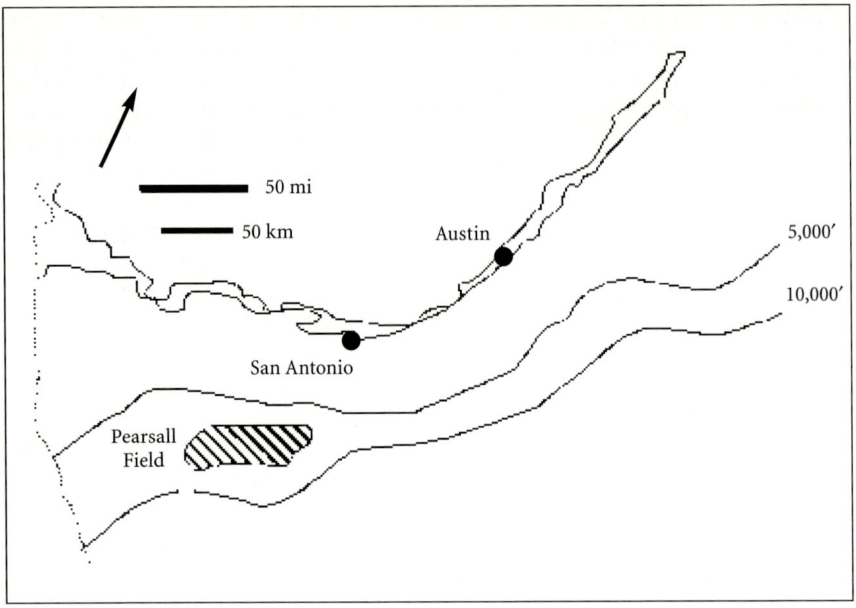

Fig. 13.4. Austin Chalk outcrop and subsea contours, South Texas, United States, with Pearsall field.s

operating companies' best estimates of the normal direction to fracture systems; obviously, however, this kind of interpretation is fraught with pitfalls. Nevertheless, many borehole azimuths are very close to normal to the fracture strikes as described in the literature.

In the southwest part of Pearsall field, most reported fracture strikes are close to northeast-southwest, ranging from slightly oblique to highly oblique to the regional strike of the Austin Chalk. In one well in the center of the southwest quadrant of Frio County, fractures strike nearly east-west, while the base of Austin Chalk is slightly more easterly than the fracture strikes. In Gonzales County, dominant fracture strike azimuths from one well are slightly north of east, with a secondary, nearly northeast-southwest trend; both trends are also apparent in nearby fault systems with which the fractures are probably associated.

In summary, for southern Pearsall Field, very few faults or folds have been recognized that may have fracturing associated with them. Further, fracture strike azimuths may or may not be parallel with the regional strike of the base and top of the Austin Chalk, nor are the observed strikes necessarily parallel with the strike of Austin isopachs. (Differential uplift/subsidence of the Maverick and South Texas basins began in the Late Cretaceous, such that present-day structure contours are not indicative of paleostrike of the saddle area between the two basins; isopachs should be more indicative of paleoslope and paleostrike.) It seems likely that fractured reservoirs in the Gonzales and Giddings areas formed in close association with down-to-the-basin growth faults, antithetic

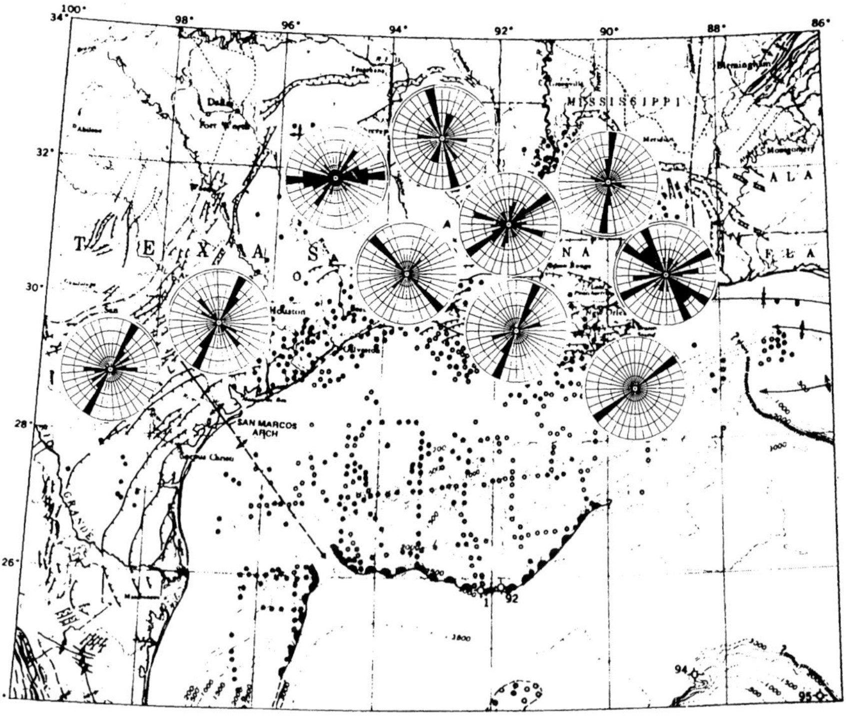

Fig. 13.5. Maximum horizontal principal compressive (MHPC) paleostress orientations from fracture log and oriented core analyses, Gulf Coast, United States, on tectonic map base. (After Martin 1978).

faults, and rollover folding. The absence of such structures in much of Pearsall Field precludes formation of Pearsall fractures by recognized faulting and folding. Conceivably, a pervasive stress field due to downslope gravitational potential (to the south and southeast) could have governed the fractures in Pearsall area, but the poor correspondence between fracture azimuths and stratigraphic trends weakens this possibility. Moreover, the orientation of even some of the faults in adjacent areas, such as those in Gonzales and Karnes Counties, are not readily explained by simple down-to-the-basin displacement. Some of the faults trend more east-west than might be expected from regional strike of the Gulf Coast.

A few tens of miles to the southwest of Pearsall Field is Chittim anticline, a structure apparently formed as a result of Laramide tectonism (e.g., Ewing 1987). The orientation of Chittim anticline and larger, known Laramide structures to the southwest, across the Rio Grande in Mexico, is approximately northwest-southeast. It is therefore approximately normal to the orientation of some of the reported fracture trends in Pearsall field. Thus, it is possible that the deformational event responsible for the Laramide structures to the southwest could have produced the stress field that was dominant during the fracturing of

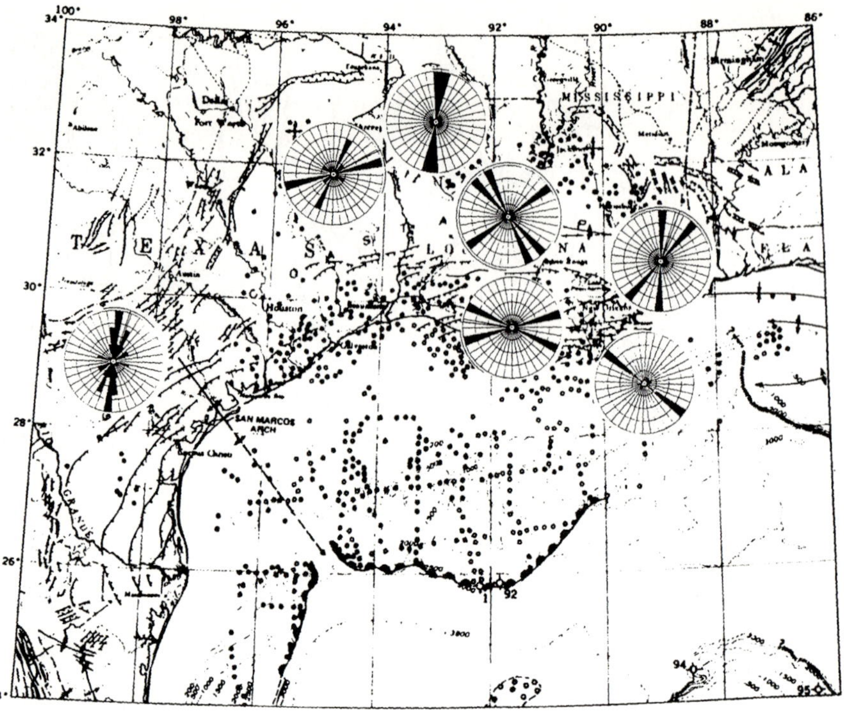

Fig. 13.6. Contemporary MHPC stress orientations from wellbore breakouts, based on multi-arm dip-meter surveys, Gulf Coast, United States, on tectonic map base. (After Martin 1979).

the Austin Chalk in the Pearsall area (Ewing 1987). Parallelism with the central-south Texas Gulf Coast inhibits distinction between a Laramide tectonic or gravitional, downslope origin for the stress field in this area, however. In any case, the Laramide tectonic event or the gravitational mechanism cannot readily explain more east-west trending fractures in Pearsall Field and the faults of similar orientation farther to the north. There is, however, evidence for early Laramide, east-west oriented compression much farther to the northwest, in the Colorado-Wyoming Rocky Mountains (Lorenz 1985; Cross 1986), and early Laramide dikes in Arizona and New Mexico imply east-west oriented MHPC stresses as well (Figs. 8.14, 8.15; Rehrig and Heidrick 1976).

Apparent regional trends in the presumed Paleozoic basement of the Texas Gulf Coast range from northwest-southeast in the central Texas coastal area to more east-west to the west of San Antonio. It is possible that the Late-Paleozoic deformational fabric in the pre-Mesozoic basement, produced by closure of the proto-Atlantic (Iaepetus) Ocean and consequent Ouachita tectonism, could be affecting the stress field in the overlying Mesozoic and Cenozoic cover. The deflection of the subsurface Ouachita belt around the Llano Uplift appears to have controlled subsequent post-Paleozoic faulting which has been active since at least the Late Cretaceous, if not as early as the initial rifting which produced the

Gulf of Mexico basin. The Balcones, Mexia–Talco, and correlative fault zones are closely parallel with inferred subsurface Ouachita trends. The Balcones fault zone appears to affect Paleozoic units, while the Mexia–Talco zone appears to sole-out into Jurassic salt, close to the updip limit of the salt. Thus, regional stress fields may be modified by the anisotropy of the basement, affecting the local stress fields responsible for small-scale faulting and fracturing, in analogy with the present-day stress field of the northern Appalachians, as discussed above. It is difficult to make more predictions for stress fields modified by basement structures, except to suggest that only preexisting structures of particular orientations with respect to a prevailing stress field are likely to be reactivated. That is, control by Ouachita faults and folds of later stress fields probably occurs only when the faults are oriented parallel with the MHPC stress. Evidence for involvement of Paleozoic basement in faulting (e.g., the Balcones zone) suggests that regional, tectonic mechanisms may still be important.

13.6
Possible Controls on Austin Chalk Fracturing – South Texas

The predominant factors controlling the stress field in the Austin Chalk when fracturing occurred probably consisted of gravitationally induced, down-to-the-basin extension, facilitated by detachments rooted in Jurassic salt and subtly controlled by the crustal structure inherited from both the Ouachita orogeny and the subsequent rifting and opening of the Gulf of Mexico. Presumably, enhanced flexure, due to loading of the continental margin by the deposition of the prograding Wilcox–Midway in the Early Cenozoic, was the primary mechanisms for producing the gravitational instability. In thermal mechanic models of the evolution of sedimentary basins, slopes of the flanks of the basins are the greatest early in the history of the basin. If the Texas Gulf Coast basin originated as a result of Late Jurassic rifting (e.g., Pilger 1978b, 1981a), by latest Cretaceous time (65 Ma), approximately 105 my after the end of rifting, near thermal equilibrium should have been attained, with consequent minimal slopes (if sedimentation and subsidence were in balance; e.g., Nunn et al. 1984). The massive influx of Wilcox–Midway clastics, prograding onto the thinned transitional crust beneath the northwestern Gulf, would then be the principal mechanism for unbalancing thermal/mechanical equilibrium. In the Giddings area, principal faulting and associated fracturing are interpreted to be Eocene in age (Kulch 1990), corresponding with the flexural-loading mechanism.

Evidence for pre-Cenozoic faulting and fracturing is apparent in the Gonzales County area, therefore prior to Wilcox–Midway progradation, based on observed Cretaceous growth on faults that cut the Austin Chalk and that die out in the lower Tertiary. The faulting apparently involves detachments rooted within Jurassic salt, slightly down-dip from the most landward fault. There is no evidence for significant thickening of the Late Cretaceous section to the southwest of the area, so the onset of faulting in this area at this time is somewhat puzzling.

A potential explanation for the anomalous, east-west oriented Late Cretaceous structures in South Texas may come from globally derived models for regional stress fields. Recent kinematic models of the Atlantic and Indian Ocean plates relative to hotspots provide semi-independent means of predicting regional stress fields, assuming that traction between the plate and the mesosphere induces compressive stresses in the direction of proper plate motion, a notion supported by present-day observations (Zoback and Zoback 1980; Zoback et al. 1989). As noted earlier in this volume, the contemporary MHPC stresses within North America parallel estimated contemporary plate motion, while the hotspot model of Müller et al. (1993), as interpolated using the methods presented in this volume, is compatible with paleostress indicators (dated dikes, faults, veins, vertical stylolites) over the past 130 my. Part of the supporting evidence for the stress indicator/plate motion correlation is the orientation of dikes, fractures, and faults in South and East Texas (Fig. 13.5).

The approximate parallelism of the Balcones and Luling fault trends with the Texas Gulf Coast and the subsurface Ouachita belt (except where the coast and Ouachita belt diverge in South Texas) is certainly one of the most impressive of observations that may bear on the origin of the faults and fractures in the Upper Cretaceous and Tertiary rocks of South and Central Texas. It is difficult to readily distinguish between the control of fracturing due to changing regional stress fields or to the onset of massive sedimentation and consequent flexure of the margins of the Gulf Basin at the end of the Cretaceous. It may well be that the two processes (regional compressive stress controlled by plate traction and localization of sedimentation) are closely related.

Since the 1950s, at least, the diachroneity of sedimentation in the Tertiary of the Gulf Coast has been well known. The principal locus of Early-Tertiary sedimentation was along the South Texas Coast; it then migrated progressively north and eastward to where present-day clastic sedimentation is maximum off the southeastern Louisiana coast (Fig. 13.7; Hardin 1957; Winker 1982). It is interesting to note that the centers of sedimentation parallel the hypothetical flowlines from the hotspot model (Figs. 8.1–8.16). Cloetingh and others (1990) have proposed that the orientation of compressive stresses acting on an elongate sedimentary basin, such as might be deposited along a continental margin, has an effect on the amount of subsidence the basin may be able to accommodate. Compression normal to the long axis of the basin (e.g., the Gulf Coast) will inhibit subsidence, while compression parallel with the basin will enhance subsidence. From the North-America-hotspot locus shown in Fig. 10.6, it is apparent that the locus of maximum Tertiary sedimentation occurs along the portion of the coast that is parallel with the direction of North American plate motion. Thus, in a qualitative sense, Cloetingh et al.'s prediction is borne out.

An additional factor that may have influenced the Wilcox–Midway sediment pulse was possible regional uplift of northern Mexico and the southwestern United States in the Early Cenozoic. Another consequence of the plate motion change implied by the hotspot model is that North American began overriding the previous magmatic arc source region. Note that the hotspot model implies that, prior to about 80 Ma, North America was moving to the north-northwest relative to the mantle (in present coordinates), such that the southwestern margin of North America was moving almost parallel with itself and, therefore, the source regions of the Cretaceous volcanic arc terranes of the Cordillera (the Sierra Nevada and Peninsular Ranges). If the source regions became fixed rela-

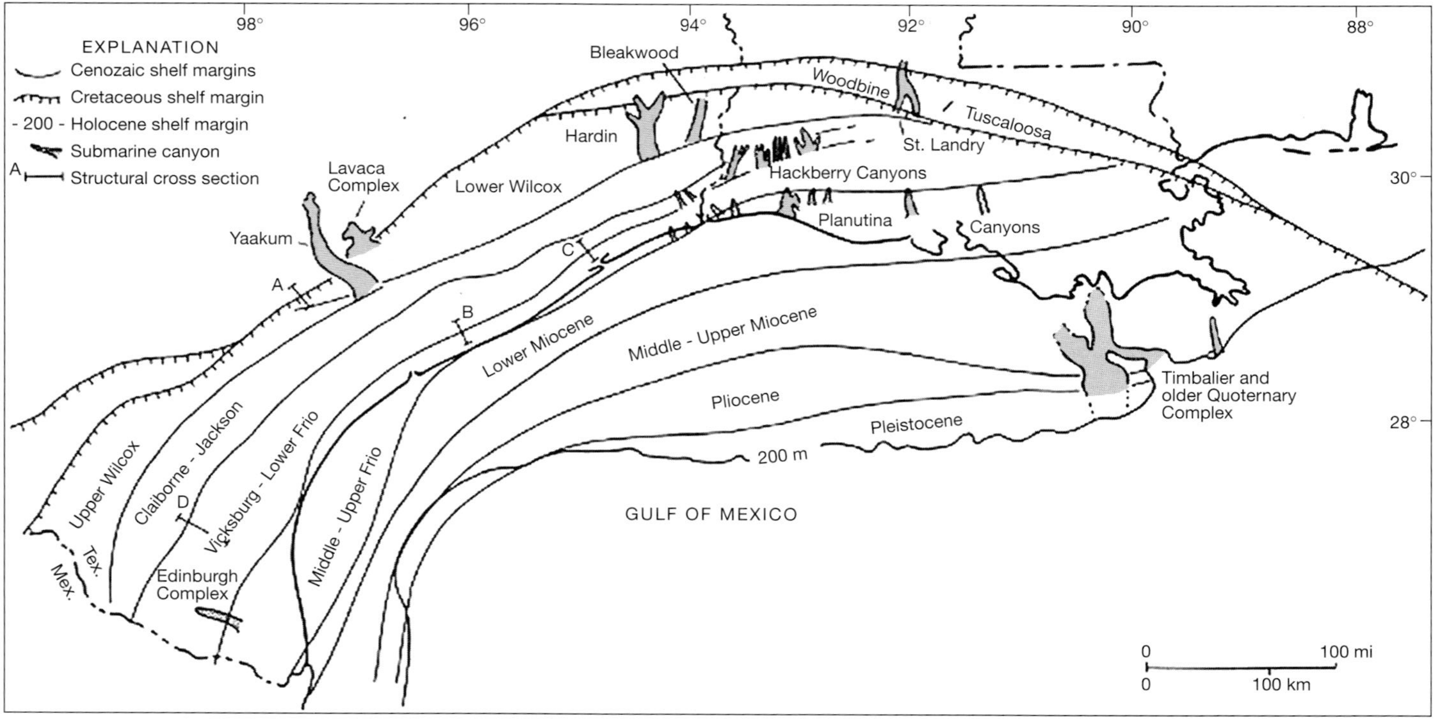

Fig. 13.7. Depositional trends of the Gulf Coast, Upper Cretaceous through Cenozoic. (After Winkler 1982).

tive to the mantle, Mexico would have moved to the northwest over them. The changes in plate motion after 80 Ma would have resulted in more easterly progressive overriding of the source regions, resulting in uplift of the southwestern United States, thereby resulting in increased erosion and supply of sediment to the adjacent Texas and northern Mexico Gulf Coast. It the uplift were felt as far away as South Texas, it, too, would have provided an additional regional tilt toward the Gulf, enhancing down-slope stresses.

13.7 Likely Factors Responsible for Fractures in Austin Chalk, South Texas

Insofar as the state of stress of South Texas in the Late Cretaceous and Early Tertiary is concerned, a combination of regional stresses due to North American plate motion, enhanced by flexural loading by Late Cretaceous and Early Tertiary clastic sedimentation, very likely controlled the orientation of fractures and faults formed at that time. Factors such as salt swells and igneous intrusions may have modified the local stress field. The change in inferred plate motion at about 65 Ma implies that the MHPC stress was oriented approximately east-west; after that time, the orientation became more nearly northeast-southwest. Thus, early (Late Cretaceous) faults and fractures are more likely to be oriented east-west than later (Early Cenozoic) structures. Interestingly, farther north, that portion of the Mount Pleasant fault zone in East Texas and western Louisiana that apparently cuts Paleozoic basement beneath the southern Sabine Uplift near the Sabine River (AAPG 1989) was most active in the Late Cretaceous; post-Cretaceous rocks are unaffected; However, farther to the west, where the fault zone trends southwest-northeast, faulting was active well into the Tertiary and may still be active. Fracture trends observed in oriented core and fracture logs from East Texas and the Palo Duro Basin (Brown et al. 1980; Collins et al. 1987; Laubach 1989) show orientations that are also compatible with the stress model.

If fracturing is a result of diagenetic and burial hydrofracturing in the presence of an oriented stress field, then the fracture orientation will depend on the apparent time of hydrofracture. Progressive younging of the Austin Chalk from South Texas into northeast Texas may imply that east-west-oriented fractures in the chalk are less likely to be present to the north, because of the more southwesterly plate motion after 65 Ma, unless other factors are involved. Similarly, the regional stress field predicted from the hotspot model provides a basis for predicting fracture orientation for other Late Cretaceous and younger rock units susceptible to fracture.

In order for fractures to be open, either the fluid pressure in the fractured rock is great enough to bear part of the overburden stress or the ambient regional stress field is strong enough and oriented appropriately with respect to the paleofractures in order for the fractures to be open. In the simplest situation, assuming the magnitude of the regional stress field is great enough, vertical paleofractures need to be oriented within 45° of the MHPC stress in order for the fractures to be open. Thus, in the case of the Austin Chalk of South Texas, MHPC stresses are generally parallel to subparallel with the present coastline (Zoback and Zoback 1980; Zoback et al. 1989) and are parallel to subparallel with paleofracture strikes (see also Figs. 13.6, 13.7). Conceivably, MHPC stresses could have been maintained in essentially the same orientation, if not the same magnitude, over the entire period from the Late Cretaceous or Early Tertiary, when the fractures first formed, to the present, if gravitational loading by the Cenozoic clastic wedges has been the principal control on the ambient stress field. Coincidentally, the present-day azimuth of North American plate motion relative to the mesosphere is also subparallel with the central Gulf Coast of Texas and the strike of the Austin Chalk in South Texas. The plate motion models do not imply that this orientation has been constant over the past 65 Ma, however (Figs. 8.1–8.16). During the Mid- to Late Tertiary, more east-west motion is implied by the plate motion model (and supported by paleostress orientations in Colorado and Utah). In any case, the success of Austin Chalk fracture production can probably be attributed to the parallelism of paleo- and contemporary MHPC stresses, even if the stress field in the intervening time between fracture formation and the present may have varied.

In the northcentral to northeastern Gulf Coast, surface joint measurements in the Late Cretaceous Selma Chalk (Bittner et al. 1988) show two predominant orientations (Fig. 8.6): westnorthwest-eastsoutheast (parallel with the strike of the Selma) and northeast-southwest (parallel with present North-American-mesosphere plate motion). Conceivably, the contemporary MHPC stress field could be nonparallel with the dominant paleo-MHPC stress field in this area, unless the contemporary stress field is dominated by gravitational loading. Thus, the first fracture set may be older than the second, as a result of a reorientation of the ambient stress field. The few available contemporary stress measurements imply that a transition in fields is present in the northeastern Gulf Coast region. Acquisition of additional stress-field measurements from the area is most desirable, as part of the assessment of Selma fracture reservoir potential. If the northeast-southwest trend is controlled by North American plate motion, it may represent fractures that are present only in the near-surface. Commonly, the orientation of a joint set observed in near-surface rocks corresponds with the regional stress field (e.g., Plumb et al. 1984) and apparently represents a shallow unloading phenomenon, as fractures of similar orientation are not necessarily present at greater depths (e.g., more than 1,000–1,500 m in the Appalachian Basin; Evans and Mroz 1990).

In the Bakken Shale of the Williston Basin, present measured MHPC stresses tend to be more east-west oriented than predicted by the plate-mesosphere model. To the northwest, in Alberta, and to the east and southeast, in the central Midwestern United States, measurements are more compatible with the plate-mesosphere model (Fig. 8.1). Thus, the northern and eastern edge of the Williston Basin may represent another stress field transition. The inferred stress boundary between those measurements which correspond with the plate-mesosphere model and those which do not approximately coincides with the area of distinctly increasing elevations that are inferred to represent Late-Cenozoic uplift of the western United States (e.g., Suppe et al. 1995). If this uplift is a result of progressive overriding by North America of hotter mantle associated with Basin Range extension, quite possibly drag of the plate against the mesosphere may be significantly reduced, and other factors may be dominating the stress field.

In the case of the Bakken Shale, production is significantly enhanced by hydrofracturing, suggesting that the magnitude of the regional stress field is also relatively low; that is, the regional stress field is inadequate to maintain open paleofractures, even though the strike of the natural fractures is parallel with the measured MHPC stress. The regional stress field, nevertheless, is of great enough magnitude and anisotropy to control the orientation of the fractures induced by artificial increase in fluid pressure. Parallelism of paleo- and induced fractures in the Bakken means that optimal orientation of horizontal drilling will be normal to fractures.

In the Cretaceous chalk of the North Sea's Dan Field (Fig. 13.8), the regional MHPC stress orientation is close to north-south, while there are apparently few existing fractures with this orientation (Fig. 13.8; Jorgensen 1991). While boreholes in the Austin Chalk and Bakken Shale are best oriented normal to the contemporary MHPC stress field, boreholes in the Dan Field are best oriented normal to the field. Hydrofractures are consequently induced parallel with the borehole. Presumably, the induced fractures link with existing paleofractures of

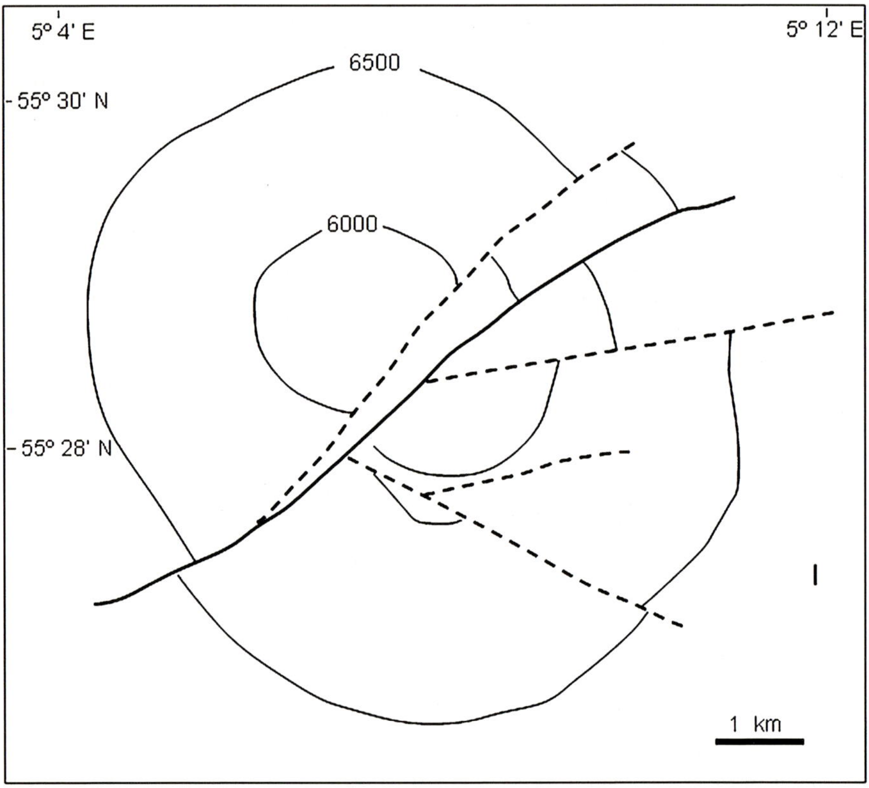

Fig. 13.8. Dan Field, Danish North Sea: contours on top of Cretaceous chalk.

various orientations; the prevailing compressive stresses will then close the highly pressured existing fractures (which trend at a high angle to the present stress field) and pore spaces as they drain into the newly induced fractures.

13.8 An Exploration Strategy for Fracture Reservoirs

There are several key components of a strategy for fracture reservoir prospecting beyond the usual elements applicable to non-fracture reservoirs (source rock, hydrocarbon maturity, seal, closure, timing). Most intense fracturing probably forms in association with faults or folds; overpressuring of shales (which may sandwich the potential reservoir) due to retarded compaction or hydrocarbon maturation may enhance the tendency for fracturing with less net deformation. Thus, establishment of the orientation of major structures (faults and folds) will probably provide important predictive ability for the orientation of associated fractures (which will usually parallel the strike of the fault or the trend of fold axes or be normal to the trend of fold axes). However, fractures are commonly observed in "tight" sandstones and limestones, as well as calcareous shales, even in the absence of obvious folding and faulting. Fractures can form during any change in the ambient stress field due to either subsidence or uplift. The orientation of the fractures is controlled by the MHPC stress field at the time of fracturing. In the case of the Austin Chalk, fracturing occurred during subsidence, with principal fractures orientation controlled by faulting and folding, flexuring of the continental margin and the orientation of the MHPC stress field. It is also necessary to determine the orientation of the contemporary stress field which may indicate which, if any, fracture set is most likely to be open. The fracture set most closely parallel with (and within 30° of) the MHPC stress is most likely to be open. Thus horizontal drilling is best oriented normal to the MHPC stress orientation if the orientation is parallel with a major fracture set.

If no fracture set exists that is closely parallel with the MHPC stress, then it may be desirable to drill horizontal wells parallel with MHPC stresses and induce fracturing that will link up existing fractures whose drainage will be enhanced because of their oblique orientation to the MHPC stress. In addition to the Danish North Sea, it may prove that optimal exploitation of parts of the Austin Chalk in South Texas and the Selma Chalk in the northeastern Gulf Coast may involve these tactics (if and only if the MHPC stress is indeed oriented at a high angle to the dominant fracture orientation).

In summary, it is extremely desirable that more than the nature and orientation of potential fractures be taken into account in exploration for fracture reservoirs. It is also essential that the present-day stress field be characterized and incorporated into analysis of the potential reservoir.

CHAPTER 14

Summary and Conclusion

14.1 Aside

The typical approach to reading a scientific article is to skim the abstract, and, if it is of interest, skip to the last section or two and skim the discussion and conclusion. Since a monograph commonly lacks an abstract, this last chapter may well be where some readers begin.

This volume, unlike a single article, involves several unique contributions that, nevertheless, comprise one long argument. There has been over the last 10 years a temptation to divide this volume up into a series of papers, each building upon the preceding. At one time, a very long paper was submitted for publication, but this approach did not succeed, in part because the paper was still incomplete (although the referees might have had a different take).

The theses of this monograph are so interrelated that I couldn't see how to easily segment them. Yes, the application of these ideas to hydrocarbon exploration could almost stand alone. To the basic researcher, an application chapter is probably viewed as inappropriate. But, from the perspective of the explorationist, the full story and rationale for characterizing regional stress fields as an empirical consequence of plate tectonics must be told. Besides, for many years, exploration geology/geophysics has paid my salary, although it has not explicitly paid for this research.

So, I believe that every chapter of this book is of a piece. I cannot make the case for geokinematics and mesoplates or the origin of hotspots, intraplate stresses, bathymetric lineations, or plate structures by themselves. And, I can't apply geokinematics to hydrocarbons without mesoplates and the rest.

14.2 Summary

Diverse lines of evidence suggest that surface and near-surface phenomena, especially volcanic ("hotspot") traces, intraplate stress fields, and bathymetric and oceanic gravity field lineations record motion of lithospheric plates with respect to a limited number of internal reference frames. These references frames are anchored in "mesoplates", of which there are two and possibly three at present. Mesoplates are inferred largely on the basis of evidence for two sets of hotspots whose traces are consistent for either of two reference frames. The recognized hotspot sets are (1) the inferred traces in the Pacific Ocean, on the Pacific and Nazca plates, including major traces produced by hotspots beneath Hawaii, Easter Island, the southeast end of the Louisville Ridge, the Galapagos Islands, and Juan Fernandez and lesser traces in the South Pacific and Gulf of Alaska, and (2) traces on the plates surrounding and beneath the Atlantic and Indian Oceans. A third set, beneath Eurasia and part of North America, may be represented by the Iceland trace. The ability to construct self-consistent kinematic models that fit each of the two well-defined hotspot trace sets provides the rationale for interpreting the hotspots as being imbedded in "mesoplates". Mesoplates are a construction based on the empirical ability to characterize the hotspots within a set as relatively fixed and, therefore, defining a kinematically rigid reference frame. The depth of mesoplates is not explicitly defined on the basis of their inferred existence as reference frames.

It is the consistency of intraplate stress fields, especially within the North American plate, with hotspot-derived kinematic models that is the principal evidence for the relatively shallow origin of hotspot reference frames. The existence of an instantaneous, fixed hotspot reference frame was one of the arguments advanced in favor of the plume hypothesis (Morgan 1971, 1972a). However, it is difficult to see how consistency of stress field orientations with hotspot frames could be observed if the two reference frames (stresses and hotspots) had significantly different depths of origin. Therefore, it is inferred that the relative fixity of two or three sets of hotspots within two or three reference frames and correspondence with contemporary and paleostress fields are indications of the presence of shallow mesoplates.

In addition, oceanic bathymetric and gravity lineations appear to record the motion of spreading ridges with respect to the same hotspot reference frames. As with intraplate stress fields, it is difficult to see how such lineations could record motion relative to a deep reference frame. The existence of marginal basins above subduction zones is also consistent with hotspot-reference-frame kinematic models, implying relatively shallow controls on the maintenance of the geometry of subduction zones.

To summarize, there are two well-defined hotspot reference frames that are consistent not only with hotspot traces, but also intracontinental paleostress fields, oceanic lineations, and the formation of marginal basins in association with subduction zones. This leaves the origin of the hotspots within a mesoplate as the next question to be addressed. Conceivably, following Morgan (1971, 1972a), hotspots could represent deep-mantle plumes that originate near the base of the mantle; their relative fixity could be a shallow phenomenon. The dilemma with this interpretation is that there are certain anomalous volcanic traces that apparently record a component of motion of plates in a hotspot reference frame, but do not reproduce the detailed locus of hotspot motion. The two best examples are from the East African Rift and the East Australian Highlands. In East Australia, as discussed above, a long period of basaltic volcanism came to an end with emplacement of alkaline central volcanic fields in a latitudinally time-transgressive manner. The central volcanics do not align in a single locus, however, while the latitudinal pattern of their emplacement is nevertheless consistent with that predicted by the hotspot model. Some workers have suggested more than one hotspot as the origin of separately aligned traces. However, the explanation for the earlier volcanism has been largely absent.

In the case of the East African rift, inception of volcanism from north to south parallels the latitudinal motion of the African plate in the hotspot reference frame. However, the observed pattern of dated volcanics does not reproduce the anticipated detailed locus of plate motion in the hotspot reference frame. East African volcanism is assumed to be associated with crustal rifting along regions of recent extension that largely appear to lack significant volcanism. There is a suggestion that rifting predates volcanism, and, therefore, may also be time transgressive. However, evidence for time-transgressive rifting is still lacking. Conversely, there could be a mechanism by which volcanism, in both regions, represents motion of the lithosphere with respect to a mesoplate, with the locus of volcanism controlled by shallow, extensional tectonism rather than deep-mantle processes.

Further, the origin of hotspots, in general, may reflect similar processes as inferred for the East African Rift-related and East Australian Highlands volcanism. That is, lithospheric tectonics could perturb the shallow mesoplate, either due to extensional faulting that penetrates the mesoplate or relative stationarity of the ridge in the mesospheric reference frame, inducing focused isostatic uplift of mesoplate in a localized region. With rise of mesoplate material through the solidus, melting is induced, producing anomalous volcanism – thus hotspot traces. Persistence of hotspot traces could further reflect limited intraplate extension normal to the hotspot trace (and, therefore, normal to the direction of plate motion relative to the mesoplate); the induced melting spot at the base of the lithosphere would tend to focus stresses and induce additional extension, consequent isostatic uplift, and renewed melting.

Further, there is the additional observation that a number of "lesser" inferred hotspot traces in the South Pacific and Gulf of Alaska appear to have originated on the south and younger side of a paleotransform fault. As the plate moves to the northwest, asthenosphere and mesoplate beneath the plate rise to shallower depths on the leeward side of the transform faults. The isostatic rise results in greater partial melting of the asthenosphere and anatexis of the upper surface of the rising mesoplate, converting the latter to asthenosphere. Thus, the Societies, Austral–Cook, Pitcairn, Pratt-Welker and Cobb island-seamount chains manifest the isostatic effect and record motion of the Pacific plate over the Hawaiian mesoplate.

The mesoplate hypothesis does not explicitly require a shallow origin for hotspots. Rather, it only requires that hotspots, whatever their origin, achieve a relative stationarity at shallow, upper mesospheric depths. However, the existence of peculiar volcanic provinces, especially the East Australian Highlands, and also the volcanics of the East African Rift and similar terranes, along with the minor hotspot traces of the Pacific, seem to require some interaction between the lithosphere, asthenosphere, and mesosphere in order to produce the observed space-time patterns.

In order to undertake some of the analysis presented in this volume, a new technique of plate reconstruction interpolation and kinematic calculation was developed. The kinematic/interpolation approach does stand on its own, but finds most ready application to the correlation of intraplate stresses and plate motions in the mesoplate reference frame.

14.3 Future Work

Data and more data are always the plea. A profound amount of work by numerous earth scientists provided the data sources that this study required. Magnetic, gravimetric, and bathymetric surveys of the world oceans, and earthquake hypocentral location and focal mechanism calculations were critical to the formulation, testing, and elaboration of plate tectonics, along with dredging and deep-ocean drilling and the analysis of resulting sediment and rock samples. Geologic mapping of the continents with biostratigraphic correlation and structural analysis provided the interpretive framework, especially in North America and Western Europe. Isotopic dating is perhaps the most critical component: constraining the geologic and geomagnetic time scale, defining the age–distance pattern of possible hotspot traces, and quantifying the "paleo" in "paleostresses".

To further test and develop the ideas presented here, there are critical data needs in a number of areas.

One aspect of data collection that still remains open is the determination of orientation of existing isotopically dated dikes. A large number of published petrologically motivated studies include isotopic dates on dikes without characterization of their orientation. In many cases, it should be possible to determine the orientations of such dikes from published detailed maps. A fraction of the data tabulated in Appendix 3 consists of dual sources (isotopic age in one; structural orientation in another).

The Mesozoic dikes of New England and adjacent Quebec have had some significant isotopic dating, but could benefit from additional analysis. As the dikes appear to have distinct orientation trends, supplemental dating could prove illuminating as few paleostress measurements are available for the Early Mesozoic of North America.

The igneous rocks of the Eifel region extending into Eastern Europe (Western European Cenozoic volcanic province) are a prime candidate for more extensive isotopic dating. Whether they form a classic Wilson-Morgan hotspot trace or perhaps something like the Eastern Australian Highlands remains uncertain. Available information is not clear about whether exposed dikes comprise any portion of the Western European Cenozoic volcanic province or whether erosion has not been deep enough. Clearly more paleostress information is desirable, particularly since most available paleostresses from Western Europe are only stratigraphically constrained, without the precision of isotopic dating.

There are parts of the world oceans that could benefit from high-resolution magnetic and bathymetric surveys that would provide valuable further tests of the mesoplate/hotspot hypothesis. The east-central Pacific is particularly challenging in this regard. There are few anomaly identifications younger than Chron 7. The present existence of two microplates in the region are hints of the probably complex tectonics of the past 25 Ma. Further, the relation of the postulate Easter hotspot traces to the fabric of the Nazca and Pacific plates is uncertain for post-Chron 7 plates. Additional characterization of plate age and structure for pre-Chron 7 would also be desirable.

The fabric of the crust surrounding and beneath the Line Islands is also uncertain. It is suggested that the Line Islands, with its multiple volcanic episodes, may reflect the same process inferred to have produced the short island-seamount chains to the south and east (Societies, Austral–Cook, Marquesas, and Pitcairn). Published interpretations of the seafloor fabric of the Line Islands regions imply paleotransform offsets of the right sense (younger to the south), but the interpretations have very few constraints due to the low amplitude of magnetic anomalies close to the equator.

Another region that could benefit from more detailed magnetic and bathymetric surveys is the central South Atlantic in the vicinity of the Walvis Ridge and the Rio Grande Rise, inferred to have been produced by the Trista da Cunha hotspot. Existing data imply jumping ridges and the possibility of microplates in the Early Cenozoic.

One item that was left out of this volume was the evidence for the geometry of existing and paleosubduction zones that tomographic seismology is providing. There are hints of subduction zone deformation and reorganization that appear to correspond with major changes in plate motions relative to mesoplates and of mesoplates relative to one another. There is clear opportunity for further exploration of this correspondence by, for example, kinematic modeling constrained to subduction boundaries as defined by tomography. There is not yet enough high-resolution tomography to provide the desired constraints. Further, as some of the most intriguing results are beneath Eurasia, there is the absence of adequate constraints on a sub-Eurasian (Icelandic) mesoplate to provide the desired kinematic information. Nevertheless, observed deformation of the subduction boundaries, where the information is available, appears to reflect lithoplate-lithoplate, lithoplate-mesoplate, and mesoplate-mesoplate motions.

In addition to more and more data, there is also that for which geokinematics can only be a prelude: geodynamics. Those who undertake geodynamic modeling need to consider the large empirical arguments advanced in this volume. Convection models need to produce hotspots and hotspot frames; they need to produce intraplate stresses that match observed contemporary and paleostresses. Maybe computers are still not big enough; maybe we need clusters of Linux clusters, but, until these fascinating correspondences of hotspots and stresses and lineations are convincingly modeled, the geodynamic models are incomplete.

14.4 A Final Speculation

The origin of the "big" hotspots and plateau basalts remains a major issue, even in the context of the mesoplate hypothesis. It is suggested that the plateau basalts are the initial manifestations of rapid rifting of lithoplate and consequent changes in the asthenosphere and mesoplate without involvement of a deep-mantle plume. Rapid thinning of the lithosphere results in isostatic uplift of asthenosphere, enhancing partial melting and the inception of volcanism.

Doubts could be expressed as to whether lithospheric thinning and enhanced asthenospheric melting are enough to produce the volumes of basalt observed. Could another factor (other than a plume) be involved?

It is suggested that the rifting event affects the mantle beneath the lithoplate in a more profound way. Perhaps the major faults that form during rapid rifting propagate into the mesoplate as well. The resulting "fracturing" of the mesoplate could produce a deeper rift, enhancing uplift and anatexis. Further, this rift could be preserved, becoming a long-lived hotspot.

14.5 Prospects

The simplicity that is geokinematics as advanced in this volume – Plate Tectonics extended to a deeper level – is itself a challenging result. If further work provides additional support for geokinematics, earth scientists confront an issue that many of the other sciences encounter. Out of the assumed complexity of a physical system – the Earth, a simple formulation of lithoplate and mesoplate tectonics has emerged. Information theory and thermodyamics, as they converge on entropy, and multifractals, for which entropy is also a measure, can possibly provide the desired answer. The simplicity of plate tectonics is the manifestation of an attractor that maximizes entropy relative to the gravitational field and internal energy production of the earth.

Earth scientists are no where near "complete" understanding of plate tectonics. The mesoplate hypothesis – geokinematics – can be a short-lived base station for the next assault on the challenging mountain that is understanding the Earth.

References

Atwater T (1970) Implications of plate tectonics for the Cenozoic tectonic evolution of western North America. Geol Soc Am Bull 81:3513–3536

Atwater T, Molnar P (1973) Relative motion of the Pacific and North American plates deduced from sea-floor spreading in the Atlantic, Indian and South Pacific Oceans. In: Kovach RL, Nur A (eds) Proceedings of the Conference on Tectonic Problems of the San Andreas Fault. Geological Sciences Stanford University XIII:136–148

Baksi A (1999) Reevaluation of plate motion models based on hotspot tracks in the Atlantic and Indian Oceans. J Geol 107:13–26

Bergerat F (1987) Stress fields in the European platform at the time of Africa-Eurasia collision. Tectonics 6:99–132

Best MB (1988) Early Miocene change in direction of least principal stress, southwestern United States: conflicting inferences from dikes and metamorphic core-detachment fault terranes. Tectonics 7:249–259

Bittner E, King DT Jr, Hoston I (1988) Fracturing in the Upper Cretaceous Selma Group chalky marls, inner coastal plain of Alabama: stratigraphic (facies) control of joint development and regional joint-strike orientations. Gulf Coast Assoc Geol Soc Trans 38: 277–282

Brown RO, Forgotson JM, Forgotson JM Jr (1980) Predicting the orientation of hydraulic created fractures in the Cotton Valley Formation of East Texas. Society of Petroleum Engineers of AIME Transactions 9269:1–12

Bullard EC, Everett JE, Smith AG (1965) Fit of continents around Atlantic. In: Blackett PMS, Bullard EC, Runcorn SK (eds) A symposium on continental drift. R Soc Lond Philos Trans Ser A 258:41–75

Butler RF, Coney PJ (1981) A revised magnetic polarity time scale for the Paleocene and early Eocene and implications for Pacific plate motion. Geophys Res Lett 8:301–304

Cande S, LaBrecque J, Haxby W (1988) Plate kinematics of the South Atlantic: Chron C34 to present. J Geophys Res 93:13479–13492

Cande SC, Kent DV (1995) Revised calibration of the geomagnetic time scale for the Late Cretaceous and Cenozoic. J Geophys Res 100:6093–6095

Cande SC, Raymond CA, Stock J, Haxby W (1995) Geophysics of the Pitman Fracture Zone and Pacific-Antarctic Plate motions during the Cenozoic. Science 270:947–953

Cande SC, Stock J, Müller RD, Isihara T (2000) Cenozoic motion between East and West Antarctica. Nature 404:140–150

Chase CG (1972) The n-plate problem of plate tectonics. Geophys J R Astron Soc 29:117–122

Chase CG (1978) Plate kinematics: the Americas, east Africa, and the rest of the world. Earth Planet Sci Lett 37:355–368

Chu D, Gordon RG (1999) Evidence for motion between Nubia and Somalia along the Southwest Indian Ridge. Nature 398:64–67

Clague DA, Jarrard RD (1973) Tertiary Pacific plate motion deduced from the Hawaiian-Emperor chain. Geol Soc Am Bull 84:1135–1154

Clague DA, Dalrymple GB (1987) The Hawaiian-Emperor volcanic chain, part I. Geologic evolution. US Geol Surv Profess Pap 1350:5–73

Clague DA, Dalyrymple GB (1989) Tectonics, geochronology, and origin of the Hawaiian-Emperor chain. In: Winterer EL, Hussong DM, Decke RW (eds) The geology of North America, N, The Eastern Pacific and Hawaii. Geological Society of America, Boulder, Colorado, USA, pp 188–217

Cloetingh S, Gradstein F, Kooi H, Grant A, Kaminski M (1990) Plate reorganization: a cause of rapid late Neogene subsidence and sedimentation around the North Atlantic? J Geol Soc Lond 147:495–506

Collins EW, Luneau BA (1986) Fracture analyses of the Palo Duro basin area, Texas panhandle and eastern New Mexico. University of Texas Bureau of Economic Geology, Geol Circ 8606, 39 pp

Cox A (1973) Plate tectonics and geomagnetic reversals – readings with introductions. Freeman, San Francisco, 702 pp

Cross TA (1986) Tectonic controls of foreland basin subsidence and Laramide-style deformation, western United States. Foreland basins: Int Assoc Sedimentol Spec Publ 8:15–39

Cross TA, Pilger RH Jr (1978a) Constraints on absolute motion and plate interaction inferred from Cenozoic igneous activity in the western United States. Am J Sci 278:865–902

Cross TA, Pilger RH Jr (1978b) Tectonic controls of Late Cretaceous sedimentation, western interior, USA. Nature 270:653–657

Crough ST, Morgan WJ, Hargraves RB (1980) Kimberlites: their relation to mantle hot spots. Earth Planet Sci Letts 50:260–274

DeMets C, Gordon RG, Argus DF, Stein S (1990) Current plate motions. Geophys J Int 101:425–478

DeMets C, Gordon R, Argus D, Stein S (1994) Effect of recent revisions to the geomagnetic reversal time scale on estimates of current plate motions. Geophys Res Lett 21:2191–2194

Du Toit AL (1937) Our wandering continents. Oliver and Boyd, Edinburgh, 366 pp

Duncan RA (1981) Hotspots in the southern oceans – an absolute frame of reference for motion of the Gondwana continents. Tectonophysics 74:29–42

Duncan RA (1984) Age progressive volcanism in the New England Seamounts and the opening of the central Atlantic Ocean. J Geophys Res 89:9980–9990

Duncan RA (1991) Age distribution of volcanism along aseismic ridges in the eastern Indian Ocean. In: Weissel J, Peirce J, Taylor E, Alt J, et al, Proceedings of the Ocean Drilling Project. Scientific Results 121. College Station, TX, pp 507–517

Duncan RA, Clague DA (1985) Pacific plate motion recorded by linear volcanic chains. In: Nairn AE M, Stehli FG, Uyeda S (eds) The Ocean Basins and Margins, 7A. The Pacific Ocean. Plenum Press, New York, pp 89–121

Duncan RA, Hargraves RB (1990) 40Ar/39Ar geochronology of basement rocks from the Mascarene Plateau, the Chagos Bank, and the Maldives Ridge. In: Duncan RA, Backman J, Peterson LC, et al (eds.) Proceedings of the Ocean Drilling Project. Scientific Results 121. College Station, Tx, 115:43–51

Duncan RA, Richards MA (1991) Hotspots, mantle plumes, flood basalts, and true polar wander. Rev Geophys 29:31–50

Eby GN (1985) Age relations, chemistry, and petrogenesis of mafic alkaline dikes from the Monteregian Hills and younger White Mountain igneous provinces. Can J Earth Sci 22:1103–1111

Emerick CM, Duncan RA (1982) Age progressive volcanism in the Comores Archipelago, western Indian Ocean and implications for Somali plate tectonics. Earth Planet Sci Lett 60:415–428

Ewing TE (1987) The Frio River Line in South Texas – transition from Cordilleran to northern Gulf tectonic regimes. Gulf Coast Assoc Geol Soc Trans 37:87–94

Fleitout L, Dalloubeix C, Moriceau C (1989) Small-wavelength geoid and topography anomalies in the South Atlantic: a clue to new hot-spot tracks and lithosphere deformation. Geophys Res Lett 16:637–640

Forsythe D, Uyeda S (1975) On the relative importance of the driving forces of plate motion. Geophys J R Astron Soc 43:163–200

Francheteau J (1970) Paleomagnetism and Plate Tectonics. PhD dissertation, Univ California, San Diego, Scripps Institution of Oceanography, La Jolla, California, 345 pp

Gaina C, Müller RD, Cande SC (2000) Absolute plate motion, mantle flow, and volcanism at the boundary between the Pacific and Indian Ocean mantle domains since 90 Ma. In: Richards MA, Gordon RG, van der Hilst RD (eds) The history and dynamics of global plate motions. Geophysical Monograph 121, American Geophysical Union, Washington, DC, pp. 189–210

Gripp AE, Gordon RG (1991) Current plate velocities relative to the hotspots incorporating the NUVEL-1 global plate motion model. Geophys Res Lett 17:1109–1112

Harada Y, Hamano Y (2000) Recent progress on the plate motions relative to hotspots. In: Richards MA, Gordon RG, van der Hilst RD (eds) The History and Dynamics of Global Plate Motions. Geophysical Monograph 121. American Geophysical Union, Washington, DC, pp 327–338

Hardin GC Jr (1957) Notes on Ceonzoic sedimentation in the Gulf Coast Geosyncline, USA. In: Rainwater EH, Zingula RP (eds) Geology of the Gulf Coast and Central Texas. Houston Geological Society, Houston, pp 1–15

Haxby WF, Weissel JF (1986) Evidence for small-scale mantle convection from Seasat altimetry data. J Geophys Res 91:3507–3520

Heidrick TL, Titley SR (1982) Fracture and dike patterns in Laramide plutons and their structural and tectonic implications. In: Titley SR (ed) Advances in Geology of the Porphyry Copper Deposits, Southwestern North America. Univ Arizona Press, Tucson, pp 73–91

Heirtzler JR, Dickson GO, Herron EM, Pitman WC III, Le Pichon, X (1968) Marine magnetic anomalies, geomagnetic field reversals, and motions of the ocean floor and continents. J Geophys Res 73:2119–2136

Henderson LJ, Gordon RG, Engebretson DC (1984) Mesozoic aseismic ridges on the Farallon plate and southward migration of shallow subduction during the Laramide orogeny. Tectonics 3:121–132

Hess HH (1962) History of ocean basins. In: Engel AE J, James HL, Leonard BF (eds) Petrologic Studies: a Volume in Honor of AF Buddington. Geol Soc Am pp 599–520

Herz N (1977) Timing of spreading in the South Atlantic: information from Brazilian alkalic rocks, Geol Soc Amer Bull 88:101–112

Hey R, Johnson GL, Lowrie A (1977) Recent plate motions in the Galapagos area. Geol Soc Am Bull 88:1385–1403

IAGA (2001) Global paleomagentic database, http://www.ngu.no/dragon/Palmag/paleomag.htm

Jarrard RD, Clague DA (1977) Implications of Pacific island and seamount ages for the origin of volcanic chains. Rev Geophys Space Phys 15:57–76

Isacks B, Oliver J, Sykes LR (1968) Seismology and the new global tectonics. J Geophys Res 73:5855–5899

Jorgensen LN (1991) Dan Field – Danish North Sea. Am Assoc Petrol Geol Bull 75:1991

Jurdy DM, Gordon RG (1984) Global plate motions relative to the hot spots 64 to 56 Ma. J Geophys Res 89:9927–9936

Klitgord KD, Schouten H (1986) Plate kinematics of the central Atlantic. In: Vogt PR, Tucholke BE (eds) The Geology of North America, vol M. The Western North Atlantic Region, Geological Society of America, Boulder, Colorado, pp 351–378

Kulch N (1990) Seismic and horizontal drilling unlock Austin Chalk. World Oil, September, pp 47–54

LePichon X (1968) Sea-floor spreading and continental drift. J Geophys Res 73:3661–3697

LePichon X, Francheteau J, Bonnin J (1973) Plate tectonics. Developments in geotectonics 6. Elsevier, Amsterdam, 300 pp

Letouzey J (1986) Cenozoic paleostress pattern in the Alpine Foreland and structural interpretation in a platform basin. Tectonophysics 132:215–231

Livermore R, Bruguier N, Cunningham A, Domasch U, Eagles G, Fretzdorff S, German C, Maldonado A, Morris P, JR39b team (1999) Deep-towed Sonar and Seismic Survey on the East Scotia Ridge. InterRidge News 8:34–37

Lorenz JC (1985) Tectonic and stress histories of the Piceance Creek Basin and the MWX site, from 75 million years ago to the present. Sandia Report, SAND84-2603, UC-92, Sandia National Laboratories, Albuquerque, New Mexico, 50 pp

McAdoo DC, Sandwell DT (1989) On the source of cross-grain lineations in the central Pacific gravity field. J Geophys Res 94:9341–9352

McDougall I, Duncan RA (1988) Age progressive volcanism in the Tasmantid Seamounts. Earth Planet Sci Lett 89:207–220

McHone JG (1978) Distribution, orientations, and ages of mafic dikes in central New England. Geol Soc Am Bull 89:1645–1655

McKenzie DP, Morgan WJ (1969) Evolution of triple junctions. Nature 224:125–133

McKenzie DP, Parker R (1967) The North Pacific: example of tectonics on a sphere. Nature 216:1276–1280

McKenzie DP, Sclater J (1971) The evolution of the Indian Ocean since the Late Cretaceous. Geophys J R Astron Soc 24:437–528

McKenzie DP, Davies D, Molnar P (1970) Plate tectonics of the Red Sea and East Africa. Nature 226:243–248

McNutt M, Caress DW, Reynolds J, Jordahl KA, Duncan RW (1997) Failure of plume theory to explain midplate volcanism in the southern Austral Islands. Nature 389:481

Menard HW, Atwater T (1968) Changes in direction of sea-floor spreading. Nature 219:463–467

Minster JB, Jordan TH (1978) Present day plate motions. J Geophys Res 83:5331–5354

Minster JB, Jordan TH, Molnar P, Haines E (1974) Numerical modeling of instantaneous plate tectonics. Geophys J R Astron Soc 36:541–576

Moberly R (1972) Origin of lithosphere behind island arcs, with reference to the western Pacific. Geol Soc Am Mem 132:35–55

Molnar P, Atwater T (1973) Relative motion of hot spots in the mantle. Nature 246:288–291

Molnar P, Francheteau J (1975) The relative motion of 'hot spots' in the Atlantic and Indian oceans during the Cenozoic. Geophys J R Astron Soc 43:763–774

Molnar P, Stock J (1987) Relative motions of hotspots in the Pacific, Atlantic and Indian Oceans since Late Cretaceous time. Nature 327:587–591

Moore WJ, McKee EH (1983) Phanerozoic magmatism and mineralization the Tooele 1 Degree by 2 degree quadrangle, Utah. Geol Soc Am Mem 157:183–190

Morgan WJ (1968) Rises, trenches, great faults, and crustal blocks. J Geophys Res 73:1959–1982

Morgan WJ (1971) Convection plumes in the lower mantle. Nature 230:42–43

Morgan WJ (1972a) Plate motions and deep mantle convection. In: Shagam R, Hargraves RB, Morgan WJ, Van Houten FB, Burk CA, Holland HD, Hollister, LC (eds) Studies in Earth and Space Sciences (Hess Volume). Geol Soc Am Mem 132:7–22

Morgan WJ (1972b) Deep mantle convection and mantle plumes. Am Assoc Petrol Geol Bull 56:203–213

Morgan WJ (1978) Rodriguez, Darwin, Amsterdam, ..., a second type of hotspot island. J Geophys Res 83:5355–5360

Morgan WJ (1981) Hotspot traces and the opening of the Atlantic and Indian Oceans. In: Emiliani D (ed) The Sea, vol 7. The Oceanic Lithosphere. Wiley-Interscience, New York, pp 443–475

Morgan WJ (1983) Hotspot tracks and the early rifting of the Atlantic. Tectonophysics 94:123–139

Morley LW, Larochelle A (1964) Paleomagnetism as a means of dating geological events. R Soc Can Spec Publ 8:39–51

Müller RD, Royer J-Y, Lawver LA (1993) Revised plate motions relative to the hotspots from combined Atlantic and Indian Ocean hotspot tracks. Geology 21:275–278

Müller RD, Cande SC, Royer J-Y, Roest WR, Maschenkov S (1999) New constraints on the Late Cretaceous/Tertiary plate tectonic evolution of the Caribbean. In: Mann P (ed) Caribbean Basins. Sedimentary Basins of the World 4. Elsevier Science, Amsterdam, pp 39–55

Nakamura K (1977) Volcanoes as possible indicators of tectonic stress orientation – principle and proposal. J Volcanol Geotherm Res 2:1–16

Nakamura K, Uyeda S (1980) Stress gradient in arc-back arc regions and plate subduction. J Geophys Res 85:6419–6428
Norton IO (1995) Plate motions in the North Pacific: the 43 Ma non-event. Tectonics 14:1080–1094
Norton IO (2000) Global hotspot reference frames and plate motion. In: Richards MA, Gordon RG, van der Hilst RD (eds) The History and Dynamics of Global Plate Motions. Geophysical Monograph 121. American Geophysical Union, Washington, DC, pp 339–357
Nunn JA, Scardina AD, Pilger RH Jr (1984) Thermal evolution of the North-Central Gulf Coast. Tectonics 3:723–740
O'Connor JM, Duncan RA (1990) Evolution of the Walvis Ridge and Rio Grande Rise hotspot system: implications for African and South American plate motions over plumes. J Geophys Res 95:17,475–17,502
O'Connor JM, Stoffers P, McWilliams MO (1995) Time-space mapping of Easter Chain volcanism. Earth Planet Sci Lett 136:197–212
O'Connor JM, Stoffers P, Wijbrans JR (1998) Migration rate of volcanism along the Foundation Chain, SE Pacific. Earth Planet Sci Lett 164:41–59
Pardo-Casas F, Molnar, P (1987) Relative motion for the Nazca (Farallon) and South America plates since Late Cretaceous time. Tectonics 6:233–248
Pilger RH Jr (1978a) A method for finite plate reconstructions, with applications to Pacific-Nazca plate evolution. Geophys Res Lett 5:469–472
Pilger RH Jr (1978b) A closed Gulf of Mexico, pre-Atlantic Ocean plate reconstruction and the early rift history of the Gulf and North Atlantic. Gulf Coast Association of Geological Societies Transactions 28:385–393
Pilger RH Jr (1981a) The opening of the Gulf of Mexico: implication for the tectonic evolution of the northern Gulf Coast. Gulf Coast Assoc Geol Soc Trans 31:377–381
Pilger RH Jr (1981b) Plate reconstructions, aseismic ridges, and low-angle subduction beneath the Andes. Geol Soc Am Bull 92:448–456
Pilger RH Jr, Henyey TL (1979) Pacific-North American plate interaction and Neogene volcanism in coastal California. Tectonophysics 57:189–209
Pilger RH Jr (1982) The origin of hotspot traces: evidence from Eastern Australia. J Geophys Res 87:1825–1834
Pilger RH Jr (1983) Kinematics of the South American subduction zone from global plate reconstructions, Geodynamics of the Eastern Pacific Region, Caribbean and Scotia Arcs. Am Geophys Union Geodynam Ser 9:113–126
Pilger RH Jr (1984) Cenozoic plate kinematics, subduction, and magmatism: South American Andes. J Geol Soc Lond 141:793–802
Pilger RH Jr, Handschumacher DW (1981) The fixed hotspot hypothesis and origin of the Easter-Sala y Gomez trace. Geol Soc Am Bull 92:437–446
Pitman WE III, Talwani M (1972) Sea-floor spreading in the North Atlantic. Geol Soc Am Bull 83:619–646
Plumb R, Engelder T, Sbar M (1984) Near-surface in situ stress 2. A comparison with stress directions inferred from earthquakes, joints, and topography near Blue Lake, New York. J Geophys Res 89:9333–9349
Press WH, Teukolsky SA, Vetterling WT, Flanner BP (1992) Numerical recipes in C, 2nd edn. Cambridge University Press, Cambridge 994 p
Price JG, Henry CD (1984) Stress orientations during Oligocene volcanism in Trans-Pecos Texas: timing the transition from Laramide compression to Basin and range tension. Geology 12:238–241
Raymond CA, Stock JM, Cande (2000) Fast paleogene motion of the Pacific hotspots from revised global plate circuit constraints. In: Richards MA, Gordon RG, van der Hilst RD (eds) The History and Dynamics of Global Plate Motions. Geophysical Monograph 121. American Geophysical Union, Washington, DC, pp 359–375
Rehrig WA, Heidrick TL (1976) Regional tectonic stress during the Laramide and Late Tertiary intrusive periods, Basin and Range Province, Arizona. Arizona Geol Soc Dig 10: 205–223

Richards MA, Duncan RA, Courtillot VE (1989) Flood basalts and hotspot tracks: plume heads and tails. Science 246:103–107

Richardson RM, Solomon SC, Sleep NH (1979) Tectonic stress in the plates. Rev Geophys Space Phys 17:981–1019

Richter FM (1973) Dynamical models for seafloor spreading. Rev Geophys Space Phys 11:223–287

Rosa JWC, Molnar P (1988) Uncertainties in reconstructions of the Pacific, Farallon, Vancouver, and Kula plates and constraints on the rigidity of the Pacific and Farallon (and Vancouver) plates between 72 and 35 Ma. J Geophys Res 93:2997–3008

Royer J-Y, Gordon R (1997) The motion and boundary between the Capricorn and Australian plates. Science 277:1268–1274

Schouten H, Dick HJ B, Klitgord KC (1987) Migration of mid-ocean-ridge volcanic segments. Nature 326:835–839

Schult FR, Gordon RG (1984) Root mean square velocities of the continents with respect to the hot spots since the Early Jurassic. J Geophys Res 89:1789–1800

Shaw PR, Cande SC (1990) High-resolution inversion for South Atlantic plate kinematics using joint altimeter and magnetic anomaly data. J Geophys Res 95:2625–2644

Sleep NH (1990) Hotspots and mantle plumes: some phenomenology. J Geophys Res 95:6715–6730

Sleep NH (1992) Hotspot volcanism and mantle plumes. Annu Rev Earth Planet Sci 20:19–43

Sleep NH, Richard MA, Hager BH (1988) Onset of mantle plumes in the presence of preexisting convection. J Geophys Res 93:7672–7689

Smith EGC (1981) Calculation of poles of instantaneous rotation from poles of finite rotation. Geophys J R Astron Soc 65:223–227

Solomon SC, Sleep NH (1975) Some simple physical models for absolute plate motions. J Geophys Res 79:2557–2567

Srivastava SP, Roest (1989) Seafloor spreading history II-IV. In: Bell JS (coordinator) East Coast Basin Atlas Series: Labrador Sea Atlantic Geoscience Centre. Geologic Survey of Canada, Map sheets L17-2–L17-6

Stein S, Sleep N, Geller R, Wang S, Kroeger G (1979) Earthquakes along the passive margin of eastern Canada, Geophys Res Lett 5:537–540

Steinberger B, O'Connell RJ (2000) Effects of mantle flow on hotspot motion. In: Richards MA, Gordon RG, van der Hilst RD (eds) The History and Dynamics of Global Plate Motions. Geophysical Monograph 121. American Geophysical Union, Washington, DC, pp 377–398

Suppe J, Powell C, Berry R (1975) Regional topography, seismicity, Quaternary volcanism, and the present day tectonics of the western United States. Am J Sci 275A:397–436

Sutherland FL (1981) Migration in relation to possible tectonic and regional controls in eastern Australian volcanism. J Volcanol Geotherm Res 9:181–213

Sykes LR (1967) Mechanism of earthquakes and nature of faulting on the mid-ocean ridges. J Geophys Res 72:2131–2153

Sykes LR (1978) Intraplate seismicity, reactivation of preexisting zones of weakness, alkaline magmatism, and other tectonism postdating continental fragmentation. Rev Geophys Space Phys 16:621–688

Takahashi M, Saito K (1997) Miocene intra-arc bending at arc-arc collision zone, central Japan. Island Arc 6(2):226–238

Turcotte DL, Oxburgh ER (1978) Intra-plate volcanism. Philos Trans R Soc Lond Ser A 288:561–579

Vine FJ, Matthews DH (1963) Magnetic anomalies over oceanic ridges. Nature 199:947–949

Wegener A (1912) Die Entstehung der Kontinente. Geol Rundsch 3:276–292

Weissel JK, Hayes DE (1972) Magnetic anomalies in the Southeast Indian Ocean. In: Hayes DE (ed) Antarctic oceanology II. The Antarctic-New Zealand Sector American Geophysical Union. Antarctic Res Ser 19:165–196

Weissel JK, Hayes DE (1977) Evolution of the Tasman Sea reappraised. Earth Planet Sci Lett 36:77–84

Weisstein EW (1998) "Euler's displacement theorem". The CRC Concise Encyclopedia of Mathematics. CRC Press, Boca Raton, Florida, p 574
Wellman P, McDougall I (1974) Cainozoic igneous activity in eastern Australia. Tectonophysics 23:49–65
White R, McKenzie D (1989) Magmatism at rift zones: the generation of volcanic continental margins and flood basalts. J Geophys Res 94:7685–7729
Wilson JT (1963a) A possible origin of the Hawaiian Islands. Can J Phys 41:863–860
Wilson JT (1963b) Evidence from islands on the spreading of ocean floors. Nature 197:536–538
Wilson JT (1965a) Evidence from ocean islands suggesting movement in the Earth. Philos Trans R Soc Lond Ser A 258:145–165
Wilson JT (1965b) A new class of faults and their bearing on continental drift. Nature 207:343–347
Winker CD (1982) Cenozoic shelf margins, northwestern Gulf of Mexico. Geol Soc Am Bull 83:575–600
Winterer EL, Sandwell DT (1987) Evidence from en-echelon cross-grain ridges for tensional cracks in the Pacific plate. Nature 329:534–537
Zellmer KE, Taylor B (2001) A three-plate kinematic model for Lau Basin opening. Geochemistry, Geophysics, Geosystems 2, paper number 2000GC000106, 26 pp
Zoback ML, Zoback MD (1980) State of stress in the conterminous United States. J Geophys Res 85:6113–6156
Zoback ML, Zoback MD, Adams J, et al (1989) Global patterns of tectonic stress. Nature 341:291–298

Supplementary References: Additional Sources for Appendix 5

Albers JP, Stewart JH (1972) Geology and mineral deposits of Esmeralda County, Nevada. Nevada Bureau Mines Geol Bull 78, 88 pp
Aldrich MJ Jr, Chapin CE, Laughlin AW (1986) Stress history and tectonic development of the Rio Grande Rift, New Mexico. J Geophys Res 91:6199–6211
Anderson JJ (1971) Geology of the southwestern high plateaus of Utah: Bear Valley Formation, an Oligocene-Miocene volcanic arenite. Geol Soc Am Bull 82:1179–1203
Anderson JJ (1986) Geologic map of the Circleville quadrangle, Beaver, Piute, and Garfield Counties, Utah. Utah Geolog Mineral Surv Map 80
Anderson JJ, Rowley PD (1986) Geologic map of the Circleville quadrangle, Beaver, Piute, Iron and Garfield Counties, Utah. Utah Geol Mineral Surv Map 82
Anderson JJ, Livari TA, Rowley PD (1988) Geologic map of the Little Creek Peak quadrangle, Garfield and Iron Counties, Utah. Utah Geol Mineral Surv
Anderson RE, Longwell CR, Marvin RF (1972) Significance of K-Ar ages of Tertiary rocks from the Lake Mead region, Nevada-Arizona. Geol Soc Am Bull 83:273–288
Archibald DA, Clark AH, Farrar, E, Zaw UK (1978) Potassium-argon ages of intrusion and scheelite mineralization, Cantung, Tungsten, Northwest Territories. Can J Earth Sci 15:1205–1207
Armstrong RL (1970a) Geochronology of Tertiary igneous rocks, eastern Basin and Range province, western Utah, eastern Nevada, and vicinity, United States. Geochim Cosmochim Acta 34:203–232
Armstrong RL (1970b) K-Ar dating using neutron activation for Ar analysis: comparison with isotopic dilution analysis. Geochim Cosmochim Acta 34:233–236
Armstrong RL, Harakal JE, Hollister VF (1982) Eocene mineralization at Mount Tolman (Keller), Washington and Silver Dyke, Montana. Isochron/West 33:9–10
Aubele JC, Crumpler LS, Shafiquallah M (1986) K-Ar ages of Late Cenozoic rocks of the central and eastern parts of the Springerville volcanic field, east-central Arizona. Isochron/West 46:3–5
Bailey RA, Dalrymple GB, Lanphere MA (1976) Volcanism, structure, and geochronology of Long Valley Caldera, Mono County, California. J Geophys Res 81:725–744

Baldridge WS, Damon PE, Shafiqullah, M, Bridwell RJ (1980) Evolution of the central Rio Grande rift, New Mexico – new potassium-argon ages. Earth Planet Sci Lett 51:309–321

Baldwin OD, Adams JAS (1971) K-40/Ar-40 ages of the alkalic igneous rocks of the Balcones fault trend of Texas. Texas J Sci 31:223–231

Barnes VE (1977) Del Rio Sheet, Geologic Atlas of Texas. Univ Texas at Austin Bureau of Econ Geol

Baudry N, Kroenke L (1991) Intermediate wavelength (400–600 km), South Pacific geoidal undulations: their relationship to linear volcanic chains. Earth Planet Sci Lett 102:430–443

Beaty DW, Naeser CW, Lynch WC (1987) The origin and significance of the strata-bound, carbonate-hosted gold deposits at Tennessee Pass, Colorado. Econ Geol 82:2158–2178

Best MG (1987) Geologic map and sections of the area between Hamlin Valley and the Escalante Desert, Iron County, Utah, United States Geological Survey Map I-1774

Best MG, Mehnert HH, Keith JD, Naeser CW (1987a) Miocene magmatism and tectonism in and near the southern Wah Wah Mountains, southwestern Utah. US Geol Surv Profess Pap 1433B:29–47

Best MG, Grant SK, Cleary, J, Hutsinpiller, Saunders DM (1987b) Geologic map of the Indian Peak Range, Beaver County, Utah. United States Geological Survey Map I-1795

Best MG, Morris HT, Knopf RW, Keith JD (1987c) Geologic map of the Southern Pine area, Beaver and Iron Counties, Utah. United States Geological Survey Map I-1794

Best MG, Lemmon DM, Morris HT (1988) Geologic map of the Milford quadrangle, east half of the Frisco Quadrangle, Beaver County, Utah. United States Geological Survey Map I-1904

Bikerman M (1967) Isotopic studies in the Roskruge Mountains, Pima County, Arizona. Geol Soc Am Bull 78:1029–1036

Blusson SL (1968) Geology and tungsten deposits near the headwaters of Flat River Yukon Territory and Southern District of Mackenzie, Canada. Geol Surv Can Pap 67(22):77 pp

Bohannon RG (1979) Strike-slip faults of the Lake Mead region of southern Nevada. In: Armentrout JM, Cole MR, TerBest HJr (eds) Cenozoic Paleogeography of the Western United States. Pacific Coast Section Society of Economic Paleontologists and Mineralogists 3:129–140

Bonham HF Jr, Garside LS (1979) Geology of the Tonopah, Lone Mountain, Klondike, and Northern Mud Lake Quadrangles, Nevada. Nevada Bureau Mines Geol Rep 92, 142 pp

Bookstrom AA, Naeser CW, Shannon JR (1987) Isotopic age determinations, unaltered and hydrothermally altered igneous rocks, north-central Colorado mineral belt. Isochron/West 49:13–20

Brimhall GH Jr (1977) Early fracture-controlled disseminated mineralization at Butte, Montana. Econ Geol 72:37–59

Brookins DG (1970) The kimberlites of Riley County, Kansas. Kansas Geol Surv Bull 200:32 pp

Bruhn RL, Yusas MR, Huertas F (1982) Mechanics of low-angle normal faulting: an example from Roosevelt Hot Springs geothermal area, Utah. Tectonophysics 86:343–361

Burke KBS, Hamilton JB, Gupta VK (1973) The Caraquet Dike: its tectonic significance. Can J Earth Sci 10:1760–1768

Carlson JE, Laird DW, Peterson JA, Schilling JH, Silberman ML, Stewart JH (1975) Preliminary map showing distribution and isotopic ages of Mesozoic and Cenozoic intrusive rocks in Nevada. United States Geological Survey Open-File Report OFR75-499

Carr WJ, Dickey DD, Quinlivan WD (1980) Geologic map of the Vidal NW, Vidal Junction, and parts of the Savahia Peak SW and Savahia Peak quadrangles, San Bernardino County, California. United States Geological Survey Map I-1126

Clemens DM, Wood,S H (1993) Late Cenozoic volcanic stratigraphy and geochronology of the Mount Bennett Hills, Central Snake River Plain, Idaho. Isochron/West 60:3–14

Coats RR, Marvin RF, Stern TW (1965) Reconnaissance of mineral ages of plutons in Elko County, Nevada, and vicinity. US Geol Surv Profess Pap 525D:D11–D15

Collins EW (1987) Characterization of fractures in limestones, northern segment of the Edwards aquifer and Balcones fault zone, central Texas. Gulf Coast Assoc Geol Soc Trans 37:43–54

Collins EW, Luneau BA (1986) Fracture analyses of the Palo Duro Basin area, Texas Panhandle and eastern New Mexico. Univ Texas Bureau Econ Geol Geol Circ 86-6, 39 p
Compton RR, Todd VR, Zartman RE, Naeser CW (1977) Oligocene and Miocene metamorphism, folding, and low-angle faulting in northwestern Utah. Geol Soc Am Bull 88:1237–1250
Cooke BJ, Godwin CI (1984) Geology mineral equilibria, and isotopic studies of the McDame tungsten skarn prospect, north-central British Columbia. Econ Geol 79:826–847
Corbett (1968) Tertiary volcanism of the Specimen-Lulu-Iron Mountain area, north-central Colorado. Colorado School of Mines Quarterly 63:11–37
Cornwall HR (1972) Geology and Mineral deposits of Southern Nye County, Nevada. Nevada Bureau Mines Geol Bull 77, 49 pp
Creasey SC (1980) Chronology of intrusion and deposition of porphyry copper ores, Globe-Miami District, Arizona. Econ Geol 75:830–844
Criss RE, Champion DE, McIntyre DH (1985) Oxygen isotope, aeromagnetic, and gravity anomalies associated with hydrothermally altered zones in the Yankee Fork Mining District, Custer County, Idaho Econ Geol 80:1977–1296
Crowe BM, Crowell JE, Krummenacher D (1979) Regional stratigraphy, K-Ar ages and tectonic implications of Cenozoic volcanic rocks, southeastern California. Am J Sci 279:186–216
Cunningham CG, Stevens TA (1979) Geologic map of the Deer Trail Mountain-Alunite Ridge mining area, west–central Utah. United States Geological Survey Map I–1230
Cyr JB, Pease RB, Schroeter TG (1984) Geology and mineralization at Equity Silver Mine. Econ Geol 79:947–968
Dasch EJ, Armstrong RL, Clabaugh SE (1969) Age of Rim Rock dike swarm, Trans-Pecos, Texas. Geol Soc Am Bull 80:1819–1824
Dethier DP, Aldrich MJ Jr, Shafiquallah M (1986) New K-Ar ages for Miocene volcanic rocks from the northeastern Jemez Mountain and Tejana Mesa, New Mexico. Isochron/West 47:12–14
Detrick RS, Crough ST(1978) Island subsidence, hotspots, and lithospheric thinning. J Geophys Res 83:1236–1244
Dickinson RG, Leopold EB, Marvin RF (1968) Late Cretaceous uplift and volcanism on the north flank of the San Juan Mountains, Colorado. Colorado School Mines Quart 63:125–132
Dilles JH (1987) Petrology of the Yerington Batholith, Nevada: evidence for evolution of porphyry copper ore fluids. Econ Geol 82:1750–1789
Eberly LD, Stanley TB Jr (1978) Mesozoic stratigraphy and geologic history of southwestern Arizona. Geol Soc Am Bull 89:921–940
Edwards G, McLaughlin WA (1972) Shell list no. 1 – Ka-Ar and Rb-Sr age determinations of California, Nevada, and Utah rocks and minerals. Isochron/West 3:1–7
Ekren EB, Byers FM Jr (1976) Ash-flow fissure in west-central Nevada. Geology 4:247–251
Ekren EB, Byers FM Jr (1985) Geologic map of WinWan Flat, Kinkaid NW, Kinkaid, and Indian Head Peak Quadrangles, Mineral County, Nevada, US Geol Survey Map I-1518
Ekren EB, Bath GD, Dixon, GL, Healey DL, Quinlivan WD (1974) Tertiary history of Little Fish Lake Valley, Nye County Nevada, and implications as the origin of the Great Basin, J Res US Geol Sur 2:105–118
Ekren EB, Bucknam RC, Carr, WJ, Dixon GL, Quinlivan, WD (1976) East-trending structural lineaments in central Nevada. US Geol Survey Professional Paper 986, 16 pp
Ekren EB, Orkild PB, Sargent, KA, Dixon GL (1977) Geologic map of Tertiary rocks, Lincoln County, Nevada, US Geological Survey Map I-104
Ekren EB, Rogers CL, Anderson RE, Orkild PP (1968) Age of basin and range normal faults in Nevada Test Site and Nellis Air Force Range, Nevada, Geological Society of America Memoir 110:247-250
Elston WE, Damon PE, Coney PJ, Rhodes RC, Smith EI, Bikerman M (1973) Tertiary volcanic rocks, Mogollon-Datil province, New Mexico, and surrounding region: K-Ar dates, patterns of eruption, and periods of mineralization. Geol Soc Am Bull 84:2259 2274

Epis RC, Waters RA, Scott GR (1979) Geologic map of the Black Mountain Quadrangle, Fremont and Pike Counties, Nevada. United States Geological Survey Map I–1195

Ewing TE (1986) Balcones volcanoes in South Texas, exploration methods and examples. In: Stapp WL (ed) Contributions to the Geology of South Texas. South Texas Geological Society, San Antonio, pp 368–379

Feraud G, Alric V, Rornari M, Bertrand H, Haller M (1999) 40Ar/39Ar dating of Jurassic volcanic province of Patagonia: migrating magmatism related to Gondwana break-up and subduction. Earth Planet Sci Lett 172:83–96

Ferguson HG, Muller SW (1949) Structural geology of the Hawthorne and Tonopah quadrangles, Nevada. US Geol Surv Profess Paper 216

Ferguson HG, Muller SW, Cathcart SH (1954) Geology of the Mina Quadrangle, Nevada. United States Geological Survey Map GQ–45

Finnell TL (1976) Geologic map of the Twin Sister Quadrangle, Grant County, New Mexico. United States Geological Survey Miscellaneous Field Studies Map, MF–779

Gans PB, Mahood GA, Schermer E (1989) Synextensional magmatism in the Basin and Range Province: a case study from the eastern Great Basin. Geol Soc Am Spec Pap 233:53 pp

Garside LJ, Silberman ML (1978) New K-Ar ages of volcanic and plutonic rocks from the Camp Douglas Quadrangle, Mineral County, Nevada. Isochron/West 22:29–32

Garside LJ, Bonham HF Jr, Ashley RP, Silberman ML, McKee EH (1980) Radiometric ages of volcanic and plutonic rocks and hydrothermal mineralization in Nevada – Determinations run under the USGS–NBMG Cooperative Program. Isochron/West 30:11–18

Geldmacher J, Hoernle K (2000) The 72 Ma geochemical evolution of the Madeira hotspot (eastern North Atlantic); recycling of Paleozoic (<= 500 Ma) oceanic lithosphere. Earth Planet Sci Lett 183:73–92

Gilluly J, Gates O (1965) Tectonic and geologic history of the Northern Shoshone Range, Nevada. US Geol Surv Profess Pap 465:153 pp

Gilluly J, Masursky H (1965) Geology of the Cortez Quadrangle, Nevada. US Geol Surv Bull 1175:17 pp

Goddard EN (1986) Geologic map, Judith Mountains, Ferguson County, Montana. United States Geological Survey Map I–1729

Grant SK, Best MG (1979) Geologic map of the Lund Quadrangle, Iron County, Utah. United States Geological Survey Open-file Report 79-1655

Grose TLT, McKee EH (1986) Potassium-Argon ages of Late Miocene to Late Quaternary volcanic rocks in the Susanville-Eagle Lake area, Lassen County, California. Isochron/West 45:5–11

Hammond PE (1979) A tectonic model for the evolution of the Cascade Range. In: Armentrout JM, Cole MR, TerBest HJr (eds) Cenozoic Paleogeography of Western United States, Pacific Coast Section. Soc Econ Paleontol Mineral 3:219–238

Hauser BB, Richter DH, Shafiquallah M (1985) Geologic map of the Saford Quadrangle, Chihuahau County, Arizona. United States Geological Survey Map I–1617

Heberlein DR, Godwin CI (1984) Hypogene alteration at the Berg porphyry copper-molybdenum property, north-central British Columbia. Econ Geol 79:902–918

Henry CD, Price JG (1986) Early Basin and Range development in Trans-Pecos Texas and adjacent Chihuahua: magmatism and orientation, timing, and style of extension. J Geophys Res 91:6213–6224

Henry CD, McDowell FW, Price JG, Smyth RC (1986) Compilation of Potassium-Argon ages of Tertiary igneous rocks, Trans-Pecos Texas. University of Texas at Austin, Bureau of Econ Geol. Geol Circ 86-2:34 pp

Henry CD, Price JG, Miser DE (1989) Geology and Tertiary igneous activity of the Hen Egg Mountain and Christmas Mountains Quadrangles, Big Bend region, Trans-Pecos Texas. Univ Texas at Austin. Bureau Econ Geol Rep Invest 183:105 pp

Hon K, Ludwig KR, Simmons KR, Slack JF, Grauch RI (1985) U-Pb isochron age and Pb isotope systematics of the Golden Fleece Vein – implications for the relationship of mineralization to the Lake City Caldera, western San Juan Mountains, Colorado. Econ Geol 80:410–417

Honjo N, McElwee KR, Duncan RA, Leeman WP (1986) K-Ar ages of volcanic rocks from the Magic Reservoir eruption center, Snake River plain, Idaho. Isochron/West 46:15–19

Hoover JD, Ensenat SE, Barnes CB, Dyer R (1988) Early Trans-Pecos magmatism: petrology and geochemistry of Eocene intrusive rocks in the El Paso area. New Mexico Geological Society Guidebook, 39th field conference, Southwestern New Mexico

Hose RK, Blake MC Jr, Smith RM (1976) Geology of White Pine County, Nevada. Nevada Bureau Mines Geol Bull 85(1):1–35

Huntsman JR (1984) Caribou Mountain: a porphyry copper deposit in southeastern Idaho. Econ Geol 79:748–754

James LP, McKee EH (1985) Silver-lead-zinc ores related to possible Laramide plutonium near Alta, Salt Lake County, Utah. Econ Geol 85:501–505

John DA (1989) Evolution of hydrothermal fluids in the Park Premier Stock, central Wasatch Mountains, Utah. Econ Geol 84:879–902

Johnson MG (1989) Geology and mineral deposits of Pershing County, Nevada. Nevada Bureau Mines Geol Bull 89, 121 pp

Johnson MG, Todd VR (1978) A summary of radiometric age determinations of igneous rocks from southeastern Arizona. Isochron/West 8:1–20

Keith JD, Shanks WC III, Archibald DA, Farrar E (1986) Volcanic and intrusive history of the Pine Grove porphyry molybdenum system, southwestern Utah. Econ Geol 81:553–577

Keith SB, Barrett LF (1976) Tectonics of the central Dragoon Mountains; a new look. In: Wilt JC, Jenney JP(eds) Tectonic digest: Arizona Geol Soc Dig 10:169–204

Kleinhampl FJ, Ziony JI (1985) Geology of northern Nye County, Nevada. Nevada Bureau Mines Geol Bull 92:142 pp

Krueger HW, Schilling JH (1971) Geochron/Nevada Bureau of Mines K/Ar age determinations – list 1. Isochron/West 71(1):9–14

Lanier G, John EC, Swense AJ, Reid JBard CE, Caddey SW, Wilson JC (1978) General geology of the Bingham Mine, Bingham Canyon, Utah. Econ Geol 73:1228–1241

Laughlin AW, Aldrich MJ Jr, Ander ME, Heiken GH, Vaniman DT (1982a) Tectonic setting and history of Late-Cenozoic volcanism in west-central New Mexico. New Mexico Geological Society Guidebook, 33rd Field Conference, Albuquerque Country II, pp 279–284

Laughlin AW, Kress VC, Aldrich MJ Jr (1982b) K-Ar ages of dike rocks, Big Bend National Park, Texas. Isochron/West 35:17–18

Laubach SE (1989) Fracture analysis of the Travis Peak Formation, western flank of the Sabine Arch, East Texas. Univ Texas Bureau Econ Geol Rep Invest 185:55 pp

Lee DE, van Loenen RE (1971) Hybrid granitoid rocks of the southern Snake Range, Nevada. US Geol Surv Profess Pap 668, 48 pp

Lee DE, Marvin RF, Stern TW, Peterman ZE (1970) Modification of potassium-argon ages by Tertiary thrusting in the Snake Range, White Pine County Nevada, and nearby Utah. United States Geological Survey Professional Paper 700–D. pp D196–D206

Lindsey DA (1981) Geologic map and discussion of selected mineral resources of the North and South Moccasin Mountains, Fergus County, Montana. United States Geological Survey Map I–1362

Lipman PW (1983) The Miocene Questa caldera, northern New Mexico: Relation to batholith emplacement and associated molybdenum mineralization. Proceedings of the Denver Region Exploration Geologists Society Symposium, Denver, pp 133–148

Luedke RG, Smith RL (1982) showing distribution, composition, and age of late Cenozoic volcanic centers in Oregon and Washington. United States Geological Survey Miscellaneous Investigations Series Map MI-1091-D

Luedke RG, Smith RL (1978a) Map showing distribution and age of late Cenozoic volcanic centers in Arizona and New Mexico. United States Geological Survey Miscellaneous Investigations Series Map MI-1091-A

Luedke RG, Smith RL (1978b) Map showing distribution and age of late Cenozoic volcanic centers in Colorado, Utah, and southwestern Wyoming. United States Geological Survey Miscellaneous Investigations Series Map MI-1091-B

Lund K, Snee LW, Evans KV (1986) Age and genesis of precious metals deposits, Buffalo Hump District, central Idaho: implications for depth of emplacement of quartz veins. Econ Geol 81:990–996

Mackenzie WB, Bookstrom AA (1976) Geology of the Majuba Hill area, Pershing County, Nevada. Nevada Bureau Mines Geol Bull 86, 23 pp

Maldonado F (1985) Geological map of the Jackass Flats area, Nye County, Nevada. United States Geological Survey Miscellaneous Investigations Map I–1519

Mansker WL, Richards BD, Cole GP (1987) A note on newly discovered kimberlites in Riley County Kansas. Geol Soc Am Spec Pap 215:197–204

Marvin RF, Dobson SW (1979) Radiometric age compilation B. United States Geological Survey, Isochron/West 26:3–32

Marvin RF, Hearn BC Jr, Mehnert HH, Naeser CW, Zartman RE, Lindsey DA (1980) Late Cretaceous-Paleocene-Eocene igneous activity in north central Montana. Isochron/West 29:5–26

Marvin RF, Mehnert HH, McKee EH (1973a) A summary of radiometric ages of Tertiary volcanic rocks in Nevada and eastern California, part III: Southeastern Nevada. Isochron/West 6:1–30

Marvin RF, Stern TW, Creasey SC, Mehnert HH (1973b) Radiometric ages of igneous rocks from Pima, Santa Cruz, and Cochise Counties, southeastern Arizona. US Geol Surv Bull 1379, 67 pp

Marvin RF, Weir KL, Mehnert HH, Merritt VM (1974a) K-Ar ages of selected Tertiary igneous rocks in southwestern Montana. Isochron/West 10:17–20

Marvin RF, Young EJ, Mehnert HH, Naeser CW, Zartman RE, Lindsey DA (1974b) Summary of radiometric age determinations on Mesozoic and Cenozoic igneous rocks and uranium and base metal deposits in Colorado. Isochron/West 11:1–41

Marvin RF, Naeser CW, Mehnert HH (1978) Tabulation of radiometric ages – including unpublished K-Ar and fission-track ages for rocks in southeastern Arizona and southwestern New Mexico. New Mexico Geol Soc Guidebook 29:243–252

Mauger RL, Damon PE, Giletti B (1965) Isotopic dating of Arizona ore deposits. AIME Trans 232:81–87

McKee EH (1971) Tertiary igneous chronology of the Great Basin of the western United States – implications for tectonic models. Geol Soc Am Bull 82:3497–3502

McKee EH (1986) Geologic map of the Roberts Wilderness Study Area, Eureka County, Nevada. United States Geological Survey MF-1844

McKee EH, Koski RA (1981) K-Ar ages of igneous rocks and vein minerals from the Christmas Mine area, Arizona. Isochron/West 32:7–11

McKee EH, Silberman ML (1970) Geochronology of Tertiary igneous rocks in central Nevada. Geol Soc Am Bull 81:2317–2328

McKee EH, Silberman ML, Marvin RE, Obradovich JD (1971) A summary of radiometric ages of Tertiary volcanic rocks in Nevada and eastern California, part I: Central Nevada. Isochron/West 1:21–42

McKenzie WB, Bookstrom AA (1976) Geology of the Majuba Hill area, Pershing County, Nevada. Nevada Bureau Mines Geol Bull 86, 23 pp

Megrue GH, Norton E, Strangway DW (1972) Tectonic history of the Ethiopian Rift as deduced by K-Ar ages and paleomagnetic measurements of basaltic dikes. J Geophys Res 77: 5744–5754

Mehnert HH (1973) Radiometric ages of igneous rocks, Pima, Santa Cruz, and Cochise counties, southeast Arizona. US Geol Surv Bull 1379:27 pp

Mehnert HH, Lipman PW, Steven TA (1973) Age of the Lake City Caldera and related Sunshine Peak Tuff, western San Juan Mountains, Colorado. Isochron/West 6:31–33

Mehnert HH, Slack, Cebula GT (1980) K-Ar age of Alunite alteration at Red Mountain, Lake City area, western San Juan Mountains, Colorado. Isochron/West 28:9–11

Mohr P, Zanettin B (1988) The Ethiopian Flood Basalt Province. In: Macdougall JD (ed) Continental Flood Basalts. Kluwer Acad Publ, Dortrecht, pp 63–110

Morris EM (1987) The Cretaceous Arkansas alkalic province: a summary of petrology and geochemistry. Geol Soc Am Spec Pap 215:217–23

Morton JL, Silberman ML, Bonham HF Jr, Garside LJ, Noble DC (1977) K-Ar ages of volcanic rocks, plutonic rocks and ore deposits in Nevada and eastern California – determinations run under the USGS-NBMG cooperative program. Isochron/West 20:19–29

Mudge MR, Earhart RL (1983) Bedrock geologic map of part of the northern Disturbed Belt, Lewis and Clark, Teton, Pondera, Glacier, Flathead, Cascade, and Powell Counties, Nevada. United States Geological Survey Map I–1375

Naeser CW, Cunningham CG, Marvin RF, Obradovich JD (1979) Pliocene intrusive rocks and mineralization near Rico, Colorado. Econ Geol 75:122–133

Noble DC (1972) Some observations on the Cenozoic volcano-tectonic evolution of the Great Basin, western United States. Earth Planet Sci Lett 17:142–150

Noble DC, McKee EH, Larson LT (1987) Late Miocene hydrothermal activity at the Willard and Scossa Mining Districts, Pershing County, northwestern Nevada. Isochron/West 48:8–10

Nolan TB, Merriam CW, Blake MC Jr (1974) Geological map of Pinto Summit Quadrangle, Nevada. United States Geological Survey Map I-793

Pamyean EH (1989) Geologic map of the Lynnburg Quadrangle, west-central Utah. United States Geological Survey Map I–1830

Papezik VS, Hodych JP (1980) Early Mesozoic diabase dikes of the Avalon Peninsula, Newfoundland: petrochemistry, mineralogy, and origin. Can J Earth Sci 17:1417–1430

Profett JM Jr (1977) Cenozoic geology of the Yerington district, Nevada, implications for the nature and origin of basin and rage faulting. Geol Soc Am Bull 88:247–276

Radhakrishna T, Maluski H, Mitchell JC, Joseph M (1999) 40Ar/39Ar and K/Ar geochronology of the dykes from the south Indian granulite terrain.Tectonophysics 304:109–129

Ratte JC (1981) Geologic map of the Mogollon quadrangle, Catron County, New Mexico. United States Geological Survey Map, GQ–1557

Rehrig WA, Shafiquallah M, Damon PE (1980) Geochronology, geology, and listric normal faulting of the Vulture Mountains, Maricopa County, Arizona. Arizona Geol Dig 22:89–110

Renne PR, Deckart K, Ernesto M, Feraud G, Piccirillo EM (1996) Age of the Ponta Grossa dike swarm (Brazil), and implications to Parana flood volcanism. Earth Planet Sci Lett 144:199–211

Reynolds SJ (1985) Geology of the South Mountains, central Arizona. Arizona Bureau Geol Mineral Technol Bull 195:61 pp

Reynolds SJ, Rehrig WA (1980) Mid-Tertiary plutonism and mylonitization, South Mountains, central Arizona. Geol Soc Am Mem 153:159–175

Reynolds SJ, Florence FP, Welty JW, Roddy MS, Currier DA, Anderson AV, Keith SB (1986) Compilation of radiometric age determinations in Arizona. Arizona Bureau Geol Mineral Technol Bull 197:258 pp

Robinson RW, Norman DI (1984) Mineralogy and fluid inclusion study of the Southern Amethyst vein system, Creede Mining District, Colorado. Econ Geol 79:439–446

Rogers CL, Anderson RE, Ekren EB, O'Connor JT (1967) Quartzite Mountain quadrangle. United States Geological Survey Map, GQ-672

Ross DC (1961) Geology and mineral deposits of Mineral County, Nevada. Nevada Bureau Mines Bull 58:98 pp

Rowley PD, Steven TA, Kaplan AM, Monroe (1981a) NE quadrangle, Utah. United States Geological Survey Map MF-1330

Rowley PD, Steven TA, Mehnert HH (1981b) Origin and structural implications of upper Miocene rhyolite in Kingston Canyon, Piute County, Utah. Geol Soc Am Bull 92:590–602

Schafer PA (1935) Geology and ore deposits of the Neihart mining district, Cascade County, Montana. Montana Bureau Mines Geol Mem 13:62 pp

Schilling JH (1965) Isotopic age determinations of Nevada rocks. Nevada Bureau of Mines Report 10, 78 pp

Shafiqullah M, Lynch DJ, Damon PE, Pierce HW (1976) Geology, geochronology, and geochemistry of the Picacho Peak area, Pinal County, Arizona. Arizona Geol Soc Dig 10:305–324

Shawe DR, Marvin RF, Andriessen PAM, Mehnert HH, Merritt VM (1986) Ages of igneous and hydrothermal events in the round Mountain and Manhattan gold districts, Nye County, Nevada. Econ Geol 81:388–407

Shawe DR (1977) Preliminary generalized geologic map of the Round Mountain quadrangle, Nye County, Nevada. United States Geological Survey Miscellaneous Field Studies Map MF-833

Silberman ML, Dockter (1977) Age of emplacement and mineralization of the Majuba Hill intrusive complex. Pershing County, Nevada, Isochron/West 18:5–6

Silberman ML, McKee EH (1971) Ages of granitic plutons in north-central Nevada. Isochron/West 1:15–32

Silberman ML, McKee EH (1972) A summary of radiometric age determinations on Tertiary volcanic rocks from Nevada and eastern California, part II, western Nevada. Isochron/West 4:7–28

Silberman ML, McKee EH (1974) Ages of Tertiary volcanic rocks and hydrothermal precious metal deposits in central and western Nevada. Nevada Bureau Mines Geol Rep 19: 67–72

Silberman ML, Berger PR, Koskins RA (1974) K/Ar age relations of granodiorite emplacement and tungsten and gold mineralization near the Getchell Mine, Humboldt County, Nevada. Econ Geol 69:646–656

Silberman ML, Bonham HF, Garside LJ, Osborne DH (1975) New K-Ar ages of volcanic and plutonic rocks and ore deposits in western Nevada. Isochron/West 13:13–21

Sillitoe RH, Grauberger GL, Elliott JE (1985) A diatreme-hosted gold deposit at Montana Tunnels, Montana. Econ Geol 80:1707–1721

Sinton JM, Detrick RS (1992) Mid-ocean ridge magma chambers. J Geophys Res 97:197–216

Smith JG, Ketner (1978) Carlin Pinon Range area, Elko County, Nevada. United States Geological Survey Map I-1028

Smith JG, McKee EH Tatlock DB, Marvin RF (1971) Mesozoic granitic rocks in northwestern Nevada – a link between the Sierra Nevada and Idaho batholiths. Geol Soc Am Bull 82:2933–2944

Speed RC, Cogbill AH (1979a) Cenozoic volcanism of the Candelaria region, Nevada (summary). Geol Soc Am Bull 90:143–145

Speed RC, Cogbill AH (1979b) Deep fault trough of Oligocene age (summary). Geol Soc Am Bull 90:145–148

Speed RC, Cogbill AH (1979c) Candelaria fault and the left-slip faults of the Candelaria region, Nevada. Geol Soc Am Bull 90:149–163

Speed R, Armstrong RL (1971) K-Ar ages of some minerals from extrusive and intrusive rocks of western Nevada. Isochron/West 1:1–8

Speed R, McKee EH (1976) Age and origin of the Darrough Felsite, southern Toiyabe Range, Nevada. US Geol Surv J Res 4:75–81

Stager HK, Tingley JV (1988) Tungsten Deposits in Nevada. Nevada Bureau Mines Geol Bull 105:256 pp

Steven TA (1989) Geologic map of the Crystal Peak Caldera, west-central Utah, United States Geological Survey Map I-2002

Stewart JH, McKee EH (1977) Geology and mineral deposits of Lander County, Nevada. Nevada Bureau Mines Geol Bull 88:106 pp

Stewart JH, Moore WJ, Zietz I (1977) East-west patterns of Cenozoic igneous rocks, aeromagnetic anomalies, and mineral deposits, Nevada and Utah. Geol Soc Am Bull 87:67–77

Stewart K, Turner S, Kelley S, Hawkesworth C, Kirstein L, Mantovani M (1996) 3-D, 40Ar-39Ar geochronology in the Parana continental flood basalt province. Earth Planet Sci Lett 143:95–109

Stormer JC Jr (1972) Ages and nature of volcanic activity on the southern High Plains, New Mexico and Colorado. Geol Soc Am Bull 83:2443–2448

Suarez G, Molnar P (1980) Paleomagnetic data and pelagic sedimentary facies and the motion of the Pacific plate relative to the spin axis since the Late Cretaceous. J Geophys Res 85:5257–5280

Swadley WC, Carr WJ (1987) Geologic map of the Quaternary and Tertiary deposits of the Big Dune Quadrangle, Nye County, Nevada, and Ijo County, California. United States Geological Survey Map I-1767

Swanson DA, Wright TL, Helz RT (1975) Linear vent systems and estimated rates of magma production and eruption for the Yakima basalt on the Columbia River Plateau. Am J Sci 275:877–905

Swisher CC, Prothro DR (1990) Single crystal 40Ar/39Ar dating of the Eocene-Oligocene transition in North America. Science 249:760–762

Tabor R, Engles, JC, Staatz MH (1968) Quartz diorite-quartz monzonite and granite pluton of the Pasayten River area, Washington. US Geol Surv Profess Pap 600C:45–62

Taubeneck WH (1969) Dikes of Columbia River basalt in northeastern Oregon, western Idaho, and southeastern Washington. Second Columbia River Basalt symposium proceedings, 21–23 March 1969, Cheney. Eastern Washington State College Press, Washington, DC, pp 73–96

Tebbens SF, Cande SC (1997) Southeast Pacific tectonic evolution from early Oligocene to Present. J Geophys Res 102:12061–12084

Tingley JV (1975) K-Ar dates on granodiorite and related scheelite-bearing quartz veins at Tungsten, Pershing County, Nevada. Isochron/West 12:3–4

Titley SR, Thompson RC, Haynes FM, Manske SL, Robison LC, White JL (1986) Evolution of fractures and alteration in the Sierrita-Esperanza hydrothermal system, Pima County Arizona. Econ Geol 81:343–370

Vandycke S, Bergerat F, Dupuis C (1991) Meso-Cenozoic faulting and inferred palaeostresses in the Mons Basin, Belgium. Tectonophysics 192:261–271

Vikre PG (1981) Silver mineralization in the Rochester district, Pershing County, Nevada. Econ Geol 76:580–609

Vikre PG (1985) Precious-metal vein systems in the National District, Humboldt County, Nevada. Econ Geol 80:360–393

Vikre PG, McKee EH (1985) Zoning and chronology of hydrothermal events in the Humboldt Range, Pershing County, Nevada. Isochron/West 44:17–24

Vikre PG, McKee EH, Silverman MH (1988) Chronology of Miocene hydrothermal and igneous events in the western Virginia Range, Washoe, Storey, and Lyon Counties, Nevada. Econ Geol 83:864–874

Waldman MA, McCandless TE, Dummett HT (1987) Geology and petrography of the Twin Knobs # 1 lamproite, Pike County, Arkansas. Geol Soc Am Spec Pap 215:205–216

Wells JD, Elliot JE, Obradovitch JD (1971) Age of igneous rocks associated with ore deposits, Cortez-Bullhorn area, Nevada. US Geol Surv Profess Pap 750C:C127–C135

Whipple JW, Mudge MR, Earhart RL (1987) Geologic map of the Rogers Pass area, Lewis and Clark County, Montana. US Geol Surv Bull 1441:91 pp

Willden R, Thomas HH, Stern TW (1967) Oligocene or younger thrust faulting in the Ruby Mountains, northeastern Nevada. Geol Soc Am Bull 78:1345–1357

Witkind IJ, Marvin RF (1989) Significance of new potassium-argon ages from the Goldens Ranch and Moroni formations, Sanpete-Sevier Valley area, central Utah. Geol Soc Am Bull 101:534–548

Wrucke (1974) Geologic map of the Gold Acres-Tenabo area, Shoshone Range, Lander County, Nevada, United States Geological Survey Miscellaneous Field Studies Map MF-647, 1 sheet

Yates RG, Engels JC (1968) Potassium-Argon ages of some igneous rocks in northern Stevens County, Washington. US Geol Surv Profess Pap 600D:242–247

Zartman RE (1977) Geochronology of some alkalic rock provinces in eastern and central United States. Annu Rev Earth Planet Sci 5:257–286

Zartman RE, Howard JM (1987) Uranium-lead age of large zircon crystals from the Potash Sulfur Springs igneous complex, Garland County, Arkansas. Geol Soc Am Spec Pap 215:235–240

Zartman RE, Bush CA, Abston C (1995) A, National Geochronological Data Base, National Geochronological and Natural Radioelement Data Bases. US Geological Survey Digital Data Series DDS-14 (CD-ROM)

Zech RS (1988) Geologic map of the Fossil Ridge area, Gunnison County, Colorado. United States Geological Survey Map I-1883

Zoback ML (1992) First- and second-order patterns of stress in the lithosphere: The World Stress Map Project. J Geophys Res 97:11,703–11,728

Zoback ML, Thompson GA (1978) Basin and Range rifting in northern Nevada: clues from amid-Miocene rift and its subsequent offsets. Geology 6:111–116

Zoback ML, Anderson RE, Thompson GA (1981) Cainozoic evolution of the state of stress and style of tectonism of the Basin and Range province of the western United States. Philos Trans R Soc Lond Ser A 300:407–434

Appendix

Appendix 1: Derivation of Instantaneous Pole and Angle of Plate Motion from an Analytic Finite Rotation Functions

The relation to be derived is:

$$\omega_T=\lambda'_T\Omega_T+\sin\lambda_T\Omega'_T+(\cos\lambda_T-1)\Omega_T\times\Omega'_T$$

in which:

$\omega_T=$ instantaneous angular velocity pseudovector for time T; magnitude is equal to the instantaneous rotation rate in radians
$\lambda_T=$ total finite rotation angle for time T
$\Omega_T=$ unit vector representation of the finite pole of rotation for time T
$\times$ denotes the vector cross-product
and the prime (′) denotes the first time derivative

Given a unit vector r (at time zero), rotate the vector to r_T. r_T is related to r as:

$$r_T=\cos\lambda(\Omega_T\times r)\times\Omega_T+\sin\lambda\Omega_T\times r+(r\cdot\Omega_T)\Omega_T \tag{1.1}$$

Differentiate r_T:

$$\begin{aligned} r'_T= &-\sin\lambda\lambda'(\Omega_T\times r)\times\Omega_T\\ &+\cos\lambda(\Omega'_T\times r)\times\Omega_T\\ &+\cos\lambda(\Omega_T\times r)\times\Omega'_T\\ &+\cos\lambda\lambda'\Omega_T\times r\\ &+\sin\lambda\Omega'_T\times r\\ &+(r\cdot\Omega'_T)\Omega_T\\ &+(r\cdot\Omega_T)\Omega'_T \end{aligned} \tag{1.2}$$

Define u:

$$u=\Omega'_T||\Omega T'|| \tag{1.3}$$

Define ω_T in terms of Ω_T, with magnitudes a_1, a_2, and a_3 to be determined.

$$\omega_T=a_1\Omega_T+a_2u+a_3\Omega_T\times u \tag{1.4}$$

In addition to the form in Eq. 1.2, r'_T can be expressed in terms of ω_T and r_T:

$$r'_T=\omega_T\times r_T \tag{1.5}$$

Insert Eq. (1.4) in Eq. (1.5):

$$
\begin{aligned}
rT' = {} & a_1\cos\lambda\Omega_T\times(\Omega_T\times\Omega_T) \\
& +a_2\cos\lambda u\times(\Omega_T\times r)\times\Omega_T \\
& +a_3\cos\lambda(\Omega_T\times u)\times(\Omega_T\times r)\times\Omega_T \\
& +a_1\sin\lambda\Omega_T\times(\Omega_T\times r) \\
& +a_2\sin\lambda u\times(\Omega_T\times r) \\
& +a_3\sin\lambda(\Omega_T\times u)\times(\Omega_T\times r) \\
& +a_1(r\cdot\Omega_T)\Omega_T\times\Omega_T \\
& +a_2(r\cdot\Omega_T)u\times\Omega_T \\
& +a_3(r\cdot\Omega_T)(\Omega_T\cdot u)\times\Omega_T \\
= {} & a_1\cos\lambda\Omega_T\times(\Omega_T\times r)\times\Omega_T \\
& +a_2\cos\lambda u\times(\Omega_T\times r)\times\Omega_T \\
& +a_3\cos\lambda(\Omega_T\times u)\times(\Omega_T\times r)\times\Omega_T \\
& +a_1\sin\lambda\Omega_T\times(\Omega_T\times r) \\
& +a_2\sin\lambda u\times(\Omega_T\times r) \\
& +a_3\sin\lambda(\Omega_T\times u)\times(\Omega_T\times r) \\
& -a_2(r\cdot\Omega_T)\Omega_T\times u \\
& +a_3(r\cdot\Omega_T)u
\end{aligned}
\tag{1.6}
$$

Let r=u:

$$
\begin{aligned}
rT' = {} & a_1\cos\lambda\Omega_T\times(\Omega_T\times u)\times\Omega_T \\
& +a_2\cos\lambda u\times(\Omega_T\times u)\times\Omega_T \\
& +a_3\cos\lambda(\Omega_T\times u)\times(\Omega_T\times u)\times\Omega_T \\
& +a_1\sin\lambda\Omega_T\times(\Omega_T\times u) \\
& +a_2\sin\lambda u\times(\Omega_T\times u) \\
& +a_3\sin\lambda(\Omega_T\times u)\times(\Omega_T\times u) \\
& +a_1(r\cdot\Omega_T)\Omega_T\times\Omega_T \\
& +a_2(r\cdot\Omega_T)u\times\Omega_T \\
& +a_3(r\cdot\Omega_T)(\Omega_T\cdot u)\times\Omega_T \\
& -a_2(u\cdot\Omega_T)\Omega_T\times u \\
& +a_3(u\cdot\Omega_T)u \\
= {} & +a_1\cos\lambda\Omega_T\times u \\
& -a_3\cos\lambda\Omega_T \\
& -a_1\sin\lambda u \\
& +a_2\sin\lambda\Omega_T
\end{aligned}
\tag{1.7}
$$

Simplify:

$$
\begin{aligned}
rT' = {} & -\sin\lambda\lambda'(\Omega_T\times u)\times\Omega_T \\
& +\cos\lambda(\Omega_T'\times u)\times\Omega_T \\
& +\cos\lambda(\Omega_T\times u)\times(\Omega_T' \\
& +\cos\lambda\lambda'\Omega_T\times u \\
& +\sin\lambda\Omega_T'\times u \\
& +(u\cdot\Omega_T')\Omega_T \\
& +u(u\cdot\Omega_T)\Omega_T' \\
= {} & -\sin\lambda\lambda' u \\
& -\cos\lambda\|\Omega_T'\|\Omega_T
\end{aligned}
\tag{1.8}
$$

$$+\cos\lambda\lambda'\Omega_T\times u$$
$$+||\Omega_T'||\Omega_T$$

Equate similar terms:

$$a_1\cos\lambda\Omega_T\times u\times\cos\lambda\lambda'\Omega_T\times u \tag{1.9}$$

$$-a_1\sin\lambda u=-\sin\lambda\lambda' u \tag{1.10}$$

$$a_2\sin\lambda\Omega_T-a_3\cos\lambda\Omega_T=\cos\lambda||\Omega_T'||\Omega_T+||\Omega_T'||\Omega_T \tag{1.11}$$

Let $r=\Omega_T\times u$:

$$\begin{aligned} rT'= {} & a_1\cos\lambda\Omega_T\times\Omega_T\times(\Omega_T\times u)\times\Omega_T \\ & +a_2\cos\lambda u\times\Omega_T\times(\Omega_T\times u)\times\Omega_T \\ & +a_3\cos\lambda(\Omega_T\times u)\times\{\Omega_T\times(\Omega_T\times u)\times\Omega_T\} \\ & +a_1\sin\lambda\Omega_T\times\Omega_T\times(\Omega_T\times u) \\ & +a_2\sin\lambda u\times\Omega_T\times(\Omega_T\times u) \\ & +a_3\sin\lambda(\Omega_T\times u)\times\Omega_T\times(\Omega_T\times u) \\ & +a_2(\Omega_T\times u)\cdot\Omega_T\times u\times\Omega_T \\ & +a_3(\Omega_T\times u)\cdot\Omega_T u \\ = {} & -a_1\cos\lambda u \\ & +a_2\cos\lambda\Omega_T \\ & -a_1\sin\lambda\Omega_T\times u \\ & +a_3\sin\lambda\Omega_T \end{aligned} \tag{1.12}$$

Simplify:

$$\begin{aligned} r_T'= {} & -\sin\lambda\lambda'\Omega_T\times(\Omega_T\times u)\times\Omega_T \\ & +\cos\lambda\Omega_T'\times(\Omega_T\times u)\times\Omega_T \\ & +\cos\lambda\Omega_T\times(\Omega_T\times u)\times\Omega_T' \\ & +\cos\lambda\lambda'\Omega_T\times(\Omega_T\times u) \\ & +\sin\lambda\Omega_T'\times(\Omega_T\times u) \\ & +(\Omega_T\times u)\cdot\Omega_T'\ \Omega_T \\ & +(\Omega_T\times u)\cdot\Omega_T\ \Omega_T' \end{aligned} \tag{1.13}$$

$$\begin{aligned} r_T'= {} & -\sin\lambda\lambda'\Omega_T\times u \\ & +\cos\lambda\lambda' u \\ & +\sin\lambda||\Omega_T'||\Omega_T \end{aligned} \tag{1.14}$$

Equate similar terms:

$$-a_1\cos\lambda u=-\cos\lambda\lambda' u \tag{1.15}$$

$$-a_1\sin\lambda\Omega_T\times u=-\sin\lambda\lambda'\Omega_T\times u \tag{1.16}$$

$$a_2\cos\lambda\Omega_T+a_3\sin\lambda\Omega_T=-\sin\lambda||\Omega_T'||\Omega_T \tag{1.17}$$

Simplify and solve for a1, a2, and a3:

$$a_1\cos\lambda\Omega_T\times u=\cos\lambda\lambda'\Omega_T\times u \tag{1.18}$$

$$-a_1\sin\lambda u=-\sin\lambda\lambda' u \tag{1.19}$$

$$-a_1\sin\lambda\Omega_T\times u=-\sin\lambda\Omega_T\times u \quad (1.20)$$

$$a_1=\lambda' \quad (1.21)$$

$$a_2\sin\lambda\Omega_T-a_3\cos\lambda\Omega_T=\cos\lambda||\Omega'_T||\Omega_T+||\Omega'_T||\Omega_T \quad (1.22)$$

$$a_2\cos\lambda\Omega_T+a_3\sin\lambda\Omega_T=-\sin\lambda||\Omega'_T||\Omega_T \quad (1.23)$$

$$a_2\sin\lambda-a_3\cos\lambda=\cos\lambda||\Omega'_T||+||\Omega'_T|| \quad (1.24)$$

$$a_2\cos\lambda+a_3\sin\lambda=-\sin\lambda||\Omega'_T|| \quad (1.25)$$

$$a_2=(-\sin\lambda||\Omega'_T||-a_3\sin\lambda)/\cos\lambda \quad (1.26)$$

$$a_2=-\sin\lambda(||\Omega'_T||+a_3)/\cos\lambda \quad (1.27)$$

$$-\sin\lambda(||\Omega'_T||+a_3)/\cos\lambda\ \sin\lambda-a_3\cos\lambda=\cos\lambda||\Omega'_T||+||\Omega'_T|| \quad (1.28)$$

$$-\sin\lambda(||\Omega'_T||+a_3)\sin\lambda-a_3\cos^2\lambda=\cos^2\lambda=\cos^2\lambda||\Omega'_T||+\cos\lambda||\Omega'_T|| \quad (1.29)$$

$$-\sin^2\lambda(||\Omega'_T||+a_3)-a_3\cos^2\lambda=\cos^2\lambda||\Omega'_T||+\cos\lambda||\Omega'_T|| \quad (1.30)$$

$$-a_3\sin^2\lambda-a_3\cos^2\lambda=||\Omega'_T||\cos^2\lambda+||\Omega'_T||\sin^2\lambda+\cos^2\lambda||\Omega'_T|| \quad (1.31)$$

$$-a_3=||\Omega'_T||+\cos^2\lambda||\Omega'_T|| \quad (1.32)$$

$$a_3=-||\Omega'_T||(1+\cos\lambda) \quad (1.33)$$

$$a_2=-\sin\theta||\Omega'_T||+-||\Omega'_T||(1+\cos\lambda)/\cos\lambda \quad (1.34)$$

$$a_2=-\sin\lambda||\Omega'_T||(1-1+\cos\lambda)/\cos\lambda \quad (1.35)$$

$$a_2=-\sin\lambda \quad (1.36)$$

Substitute values of a_1, a_2, and a_3 into Eq. 1.4:

$$\omega_T=\lambda'\Omega_T+\sin\lambda u+-||\Omega'_T||(1+\cos\lambda)/\Omega_T\times\Omega'_T/||\Omega'_T|| \quad (1.37)$$

Simplify:

$$\omega_T=\lambda'\Omega_T+\sin\lambda u-(1+\cos\lambda)\Omega_T\times\Omega'_T \quad (1.38)$$

which is the desired solution.

Note that the signs of the last two terms are different from those in Smith (1984), because, for the relation between the instantaneous angular velocity pseudo-vector and the instantaneous velocity vector, he used a non-standard form ($r'_T\ni=\varrho\,(\omega_T=-\omega_T r_T)$), instead of the standard form ($r'_T=\omega_T r_T$, Eq. 1.5, above).

Appendix 2: Relative and Hotspot Plate Reconstruction Parameters

Table A2.1. Relative plate kinematic and reconstruction parameters and their sources. Instantaneous rotation rates for 0 Ma (present); finite total rotation angles for older ages

Source reconstruction parameters				
Age (Ma)	Longitude	Latitude	Angle/rate deg or deg/my	Source
North America–Eurasia				
0.0000	135.80	62.40	0.21	DeMets et al. (1990)
9.8100	133.58	65.38	2.44	Srivastava and Roest (1989)
19.5895	136.74	68.92	4.97	Srivastava and Roest (1989)
33.3015	136.95	65.64	7.51	Srivastava and Roest (1989)
47.0850	137.74	67.19	10.91	Srivastava and Roest (1989)
51.3950	140.81	62.60	12.75	Srivastava and Roest (1989)
56.1475	141.66	63.14	14.22	Srivastava and Roest (1989)
68.2360	143.96	64.84	16.95	Srivastava and Roest (1989)
79.0750	147.74	66.17	19.00	Srivastava and Roest (1989)
83.0000	148.91	66.54	19.70	Srivastava and Roest (1989)
92.0000	150.26	66.67	20.37	Srivastava and Roest (1989)
105.0000	152.34	66.85	21.49	Srivastava and Roest (1989)
118.7000	154.75	68.99	23.05	Srivastava and Roest (1989)
130.0000	154.75	68.99	23.05	Srivastava and Roest (1989)
North America–North Africa				
0.0000	38.30	78.80	0.24	DeMets et al. (1990)
9.8100	50.80	80.12	2.52	Müller et al. (1990)
19.5895	37.84	79.57	5.29	Klitgord and Schouten (1986)
33.3015	1.12	75.37	10.04	Müller et al. (1990)
47.0850	-3.88	75.30	15.25	Müller et al. (1990)
56.1475	-0.46	79.68	18.16	Müller et al. (1990)
66.5940	4.94	82.90	20.76	Müller et al. (1990)
71.2045	-9.15	81.35	22.87	Klitgord and Schouten (1986)
73.6190	-11.76	80.76	23.91	Klitgord and Schouten (1986)
79.0750	-18.35	78.30	27.06	Klitgord and Schouten (1986)
83.0000	-20.73	76.55	29.60	Klitgord and Schouten (1986)
91.0000	-19.48	73.82	34.28	Klitgord and Schouten (1986)
118.0000	-19.90	66.30	54.25	Klitgord and Schouten (1986)
126.0000	-19.00	66.13	56.39	Klitgord and Schouten (1986)
131.5000	-18.50	65.95	57.40	Klitgord and Schouten (1986)
North Africa–Central Africa				
90.0000	-172.77	-17.30	0.00	DeMets et al. (1990)
100.0000	-172.77	-17.30	0.53	Müller et al. (1991) (derived)
110.0000	-173.96	-15.69	0.86	Müller et al. (1991) (derived)
118.7000	-173.56	-18.08	1.15	Müller et al. (1991) (derived)
130.0000	15.16	29.82	1.59	Müller et al. (1991) (derived)
South America–Central Africa				
0.0000	-39.40	62.50	0.30	DeMets et al. (1990)
10.9490	-39.00	60.00	3.15	Cande et al. (1988)
19.5895	-38.00	59.50	7.05	Cande et al. (1988)
25.8870	-36.00	59.00	10.00	Cande et al. (1988)
33.3015	-34.00	57.50	13.38	Cande et al. (1988)

Age (Ma)	Longitude	Latitude	Angle/rate deg or deg/my	Source
Source reconstruction parameters				
43.1625	-32.00	57.50	17.60	Shaw and Cande (1990)
49.3755	-31.50	59.00	20.10	Cande et al. (1988)
56.1475	-32.00	61.00	22.30	Shaw and Cande (1990)
63.0665	-33.30	63.00	24.00	Cande et al. (1988)
64.3605	-33.30	63.00	24.30	Cande et al. (1988)
66.5940	-33.30	63.00	24.70	Cande et al. (1988)
68.2360	-33.50	63.00	25.40	Cande et al. (1988)
71.2045	-33.50	63.00	26.60	Cande et al. (1988)
73.6190	-33.50	63.00	27.90	Cande et al. (1988)
79.0750	-34.19	63.18	30.96	Shaw and Cande (1990)
83.0000	-34.78	62.29	33.58	Shaw and Cande (1990)
118.7000	-34.60	50.10	52.78	Cande et al. (1988)
121.0000	-34.20	50.00	53.64	Cande et al. (1988)
126.5000	-33.80	49.30	54.29	Cande et al. (1988)
Antarctica–India				
0.0000	13.10	21.90	0.48	DeMets et al. (1990)
9.8100	14.51	23.17	5.04	Müller et al. (1991) (derived)
19.5895	4.76	27.35	8.94	Müller et al. (1991) (derived)
33.3015	19.91	22.07	16.64	Müller et al. (1991) (derived)
38.9890	19.58	23.46	20.33	Müller et al. (1991) (derived)
47.0850	12.82	19.91	25.51	Müller et al. (1991) (derived)
56.1475	11.18	17.24	34.04	Müller et al. (1991) (derived)
68.2360	8.02	13.59	49.03	Müller et al. (1991) (derived)
73.6190	6.78	13.70	53.20	Müller et al. (1991) (derived)
79.0750	6.03	13.34	58.02	Müller et al. (1991) (derived)
83.0000	7.21	12.13	61.59	Müller et al. (1991) (derived)
90.0000	10.63	9.36	66.15	Müller et al. (1991) (derived)
100.0000	15.40	4.74	73.93	Müller et al. (1991) (derived)
110.0000	14.69	10.53	77.50	Müller et al. (1991) (derived)
118.7000	13.62	−0.78	85.33	Müller et al. (1991) (derived)
130.0000	19.01	−5.36	92.38	Müller et al. (1991) (derived)
Antarctic–Australia				
0.0000	38.20	13.20	0.65	DeMets et al. (1990)
10.9490	36.10	13.10	6.61	Chang (1991)
19.5895	32.70	15.40	11.97	Chang (1991)
33.3015	33.40	13.80	20.41	Chang (1991)
38.9890	29.90	16.60	23.62	Sandwell (1989)
47.0850	31.30	15.10	24.50	Sandwell (1989)
56.1475	31.70	12.50	25.24	Sandwell (1989)
68.2360	33.20	8.70	25.83	Sandwell (1989)
79.0750	35.10	6.20	26.37	Sandwell (1989)
83.0000	35.80	4.90	26.81	Sandwell (1989)
96.0000	38.00	1.00	28.30	Sandwell (1989)
130.0000	38.90	−2.00	31.50	Sandwell (1989)
Antarctic–Central Indian				
0.0000	38.20	13.20	0.65	DeMets et al. (1990)
3.1650	28.39	25.81	1.45	Müller et al. (1991) (derived)
9.8100	36.05	13.05	6.54	Müller et al. (1991) (derived)

Source reconstruction parameters				
Age (Ma)	Longitude	Latitude	Angle/rate deg or deg/my	Source
19.5895	32.68	15.38	11.97	Müller et al. (1991) (derived)
33.3015	33.22	14.46	20.28	Müller et al. (1991) (derived)
38.9890	30.83	16.56	23.80	Müller et al. (1991) (derived)
47.0850	22.82	14.40	28.48	Müller et al. (1991) (derived)
56.1475	18.76	12.75	36.81	Müller et al. (1991) (derived)
Antarctic–Central Africa				
0.0000	-39.20	5.60	0.12	DeMets et al. (1990)
9.8100	-49.48	7.91	1.51	Müller et al. (1991) (derived)
19.5895	-47.97	10.76	2.78	Müller et al. (1991) (derived)
33.3015	-50.68	14.10	5.51	Müller et al. (1991) (derived)
38.9890	-44.81	15.53	7.17	Müller et al. (1991) (derived)
47.0850	-47.19	9.03	9.52	Müller et al. (1991) (derived)
56.1475	-39.11	5.29	10.30	Müller et al. (1991) (derived)
68.2360	-41.56	1.06	11.84	Müller et al. (1991) (derived)
73.6190	-41.41	-1.85	13.47	Müller et al. (1991) (derived)
79.0750	-39.68	-4.65	16.03	Müller et al. (1991) (derived)
83.0000	-38.08	-2.55	17.91	Müller et al. (1991) (derived)
90.0000	-33.60	-4.09	22.11	Müller et al. (1991) (derived)
100.0000	-28.96	-5.63	29.30	Müller et al. (1991) (derived)
110.0000	-26.11	-6.50	36.62	Müller et al. (1991) (derived)
118.7000	-24.44	-7.08	43.05	Müller et al. (1991) (derived)
130.0000	-21.02	-11.88	50.42	Müller et al. (1991) (derived)
Indian–Central Indian				
0.0000	76.90	-8.70	0.25	DeMets et al. (1990)
9.8100	76.78	-8.70	2.72	Müller et al. (1991) (derived)
19.5895	74.22	-5.30	5.93	Müller et al. (1991) (derived)
Central Indian–Central Africa				
0.0000	-130.20	-12.40	0.63	DeMets et al. (1990)
9.8100	-130.40	-11.90	6.56	Müller et al. (1991) (derived)
19.5895	-133.46	-14.46	11.72	Müller et al. (1991) (derived)
33.3015	-130.59	-13.33	20.13	Müller et al. (1991) (derived)
38.9890	-130.80	-15.91	22.65	Müller et al. (1991) (derived)
47.0850	-136.88	-16.92	26.49	Müller et al. (1991) (derived)
56.1475	-145.74	-17.69	32.42	Müller et al. (1991) (derived)
68.2360	-155.58	-16.11	45.70	Müller et al. (1991) (derived)
73.6190	-157.38	-17.82	48.76	Müller et al. (1991) (derived)
79.0750	-158.54	-19.55	51.79	Müller et al. (1991) (derived)
83.0000	-156.95	-18.85	54.12	Müller et al. (1991) (derived)
Indian–Central African				
0.0000	-151.50	-23.60	0.43	DeMets et al. (1990)
9.8100	-146.72	-23.66	4.56	Müller et al. (1991) (derived)
19.5895	-156.21	-29.64	7.55	Müller et al. (1991) (derived)
33.3015	-139.25	-21.36	15.34	Müller et al. (1991) (derived)
38.9890	-138.05	-23.60	17.94	Müller et al. (1991) (derived)
47.0850	-144.10	-23.66	22.11	Müller et al. (1991) (derived)
56.1475	-152.66	-23.31	28.59	Müller et al. (1991) (derived)
68.2360	-160.81	-20.51	42.45	Müller et al. (1991) (derived)
73.6190	-162.28	-22.20	45.68	Müller et al. (1991) (derived)

Source reconstruction parameters				
Age (Ma)	Longitude	Latitude	Angle/rate deg or deg/my	Source
79.0750	-163.16	-23.91	48.84	Müller et al. (1991) (derived)
83.0000	-161.20	-22.90	51.01	Müller et al. (1991) (derived)
90.0000	-156.43	-22.46	52.99	Müller et al. (1991) (derived)
100.0000	-149.32	-21.38	56.80	Müller et al. (1991) (derived)
110.0000	-149.01	-33.72	56.30	Müller et al. (1991) (derived)
118.7000	-148.37	-21.74	57.02	Müller et al. (1991) (derived)
130.0000	-143.12	-22.26	61.02	Müller et al. (1991) (derived)
Atlantic-Indian Hotspots–Central Africa				
0.0000	3.60	-5.50	0.14	DeMets et al. (1990)
9.8100	-31.60	59.30	1.89	Müller et al. (1991)
19.5895	-44.50	50.90	4.36	Müller et al. (1991)
33.3015	-43.00	40.30	7.91	Müller et al. (1991)
38.9890	-41.20	37.70	9.65	Müller et al. (1991)
47.0850	-40.80	32.80	12.09	Müller et al. (1991)
56.1475	-41.70	30.10	13.89	Müller et al. (1991)
68.2360	-40.90	26.40	16.23	Müller et al. (1991)
73.6190	-39.60	22.30	17.80	Müller et al. (1991)
79.0750	-38.90	18.00	19.98	Müller et al. (1991)
83.0000	-40.90	19.00	21.53	Müller et al. (1991)
90.0000	-41.90	19.40	23.31	Müller et al. (1991)
100.0000	-41.40	18.90	25.35	Müller et al. (1991)
110.0000	-39.50	17.70	26.71	Müller et al. (1991)
118.7000	-39.70	18.70	27.37	Müller et al. (1991)
130.0000	-37.50	16.70	28.52	Müller et al. (1993)
Pacific–Nazca				
0.0000	-78.20	48.70	0.74	DeMets et al. (1990)
1.86	-93.39	53.01	3.58	Pardo-Casas and Molnar (1987)
4.235	-89.79	59.12	5.83	Pardo-Casas and Molnar (1987)
10.83	-91.44	63.68	16.57	Pardo-Casas and Molnar (1987)
16.04	-91.97	65.2	24.37	Pardo-Casas and Molnar (1987)
19.59	-93.02	62.38	31	Pardo-Casas and Molnar (1987)
24.756	-94.75	63.88	38.81	Pardo-Casas and Molnar (1987)
25.887	-94.6	64.49	40.39	Pardo-Casas and Molnar (1987)
28.398	-98.3	66.91	43.64	Pardo-Casas and Molnar (1987)
33.301	-105.82	69.74	49.24	Pardo-Casas and Molnar (1987)
38.989	-112.6	72.8	57.1	Pardo-Casas and Molnar (1987)
56.148	-125.5	75.1	68.8	Pardo-Casas and Molnar (1987)
71.204	-133	75.6	75.3	Pardo-Casas and Molnar (1987)
75	-129.82	80	80	Pardo-Casas and Molnar (1987)
Antarctic–West Antarctic (Marie Byrd Land)				
27.9720	0.00	0.00	0.00	Cande et al. (1999)
33.5450	-17.85	-18.15	0.70	Cande et al. (1999)
43.7890	-17.85	-18.15	1.70	Cande et al. (1999)
Cocos–Pacific				
0.0000	-108.60	36.80	0.00	DeMets et al. (1990)
1.0000	71.40	-36.80	1.99	
9.8100	-124.43	-72.48	6.49	Norton (1995)
19.5895	-121.38	-72.35	12.39	Norton (1995)

Source reconstruction parameters				
Age (Ma)	Longitude	Latitude	Angle/rate deg or deg/my	Source
33.3015	-100.55	-85.06	32.30	Norton (1995)
38.9890	-93.92	-86.63	43.16	Rosa and Molnar (1988)
47.0850	-56.20	-85.55	54.08	Rosa and Molnar (1988)
56.1475	-77.67	-82.14	61.87	Rosa and Molnar (1988)
71.2045	-83.02	-81.82	71.42	Rosa and Molnar (1988)
Kula–Pacific				
51.3950	100.76	14.48	0.00	Byrne (1979)
56.1475	-79.24	-14.48	1.32	Norton (1995)
57.7325	-79.24	-14.48	2.09	Norton (1995)
61.0980	-79.24	-14.48	3.23	Norton (1995)
63.0665	-79.24	-14.48	3.67	Norton (1995)
64.3605	-79.24	-14.48	4.32	Norton (1995)
66.5940	-79.24	-14.48	6.04	Norton (1995)
71.2045	-75.77	-14.48	8.66	Norton (1995)
76.3470	-75.77	-14.48	11.03	Norton (1995)
80.0000	-75.77	-14.48	11.60	Norton (1995)
West Antarctic (Marie Byrd Land)–Pacific				
0.0000	95.00	-66.00	0.90	DeMets et al. (1990)
2.5810	106.28	-67.03	2.42	Cande et al. (1995)
5.8940	102.07	-67.91	5.42	Cande et al. (1995)
8.8620	102.94	-69.68	7.95	Cande et al. (1995)
12.2425	106.23	-71.75	10.92	Cande et al. (1995)
17.4460	110.15	-73.68	15.17	Cande et al. (1995)
24.1180	112.72	-74.72	19.55	Cande et al. (1995)
28.2830	112.62	-74.55	22.95	Cande et al. (1995)
33.5450	115.26	-74.38	27.34	Cande et al. (1995)
42.5360	128.69	-74.90	34.54	Cande et al. (1995)
47.9060	129.81	-74.52	37.6[illegible]	Cande et al. (1995)
53.3470	127.50	-73.62	40.03	Cande et al. (1995)
61.0980	124.43	-71.38	44.90	Cande et al. (1995)
67.7350	126.56	-69.33	51.05	Cande et al. (1995)
72.2955	128.50	-68.38	54.54	Cande et al. (1995)
79.0750	125.27	-66.05	58.61	Cande et al. (1995)
83.0000	128.11	-65.99	62.71	Norton (1995)
90.0000	129.88	-65.23	67.64	Norton (1995)
Pacific Hotspot–Pacific				
0.0000	90.00	-60.20	0.93	Gripp and Gordon (1991)
5.0000	104.90	-56.50	4.70	Raymond et al. (2000)
10.8300	108.75	-65.98	9.24	Raymond et al. (2000)
21.1600	111.79	-70.45	17.43	Raymond et al. (2000)
25.0000	112.44	-71.20	20.49	Raymond et al. (2000)
33.3000	117.41	-69.14	25.44	Raymond et al. (2000)
40.2200	119.88	-67.88	29.60	Raymond et al. (2000)
43.0000	120.64	-67.46	31.28	Raymond et al. (2000)
43.8700	119.71	-66.87	31.43	Raymond et al. (2000)
47.8600	116.15	-64.25	32.60	Raymond et al. (2000)
53.2500	112.54	-60.86	34.14	Raymond et al. (2000)
63.3000	108.08	-55.19	37.47	Raymond et al. (2000)

Source reconstruction parameters				
Age (Ma)	Longitude	Latitude	Angle/rate deg or deg/my	Source
68.6800	106.42	-52.51	39.39	Raymond et al. (2000)
78.7800	104.10	-48.08	43.22	Raymond et al. (2000)
81.0000	101.91	-45.45	47.40	Norton (1995)
90.0000	100.86	-42.09	51.50	Norton (1995)
95.0000	99.28	-46.64	61.87	
100.0000	99.86	-46.31	64.71	
Greenland–North America				
33.3015	-92.00	59.50	0.00	Roest and Srivastava (1989)
47.0850	88.00	-59.50	2.81	Müller et al. (1990)
51.3950	69.99	-54.91	4.00	Roest and Srivastava (1989)
56.1475	42.75	-24.48	3.12	Roest and Srivastava (1989)
57.7325	31.80	-20.61	3.27	Roest and Srivastava (1989)
61.0980	30.59	-27.63	3.72	Roest and Srivastava (1989)
68.2360	34.69	-43.94	4.92	Roest and Srivastava (1989)
83.0000	57.55	-65.30	11.00	Roest and Srivastava (1989)
92.0000	60.52	-66.60	12.20	Roest and Srivastava (1989)
105.0000	61.04	-67.08	12.99	Roest and Srivastava (1989)
118.0000	61.52	-67.50	13.78	Roest and Srivastava (1989)
Farallon–Pacific				
0.0000	-124.43	-72.48	0.00	DeMets et al. (1990)
9.8100	-124.43	-72.48	6.49	Norton (1995)
19.5895	-121.38	-72.35	12.39	Norton (1995)
33.3015	-100.55	-85.06	32.30	Norton (1995)
38.9890	-93.92	-86.63	43.16	Rosa and Molnar (1988)
47.0850	-56.20	-85.55	54.08	Rosa and Molnar (1988)
56.1475	-77.67	-82.14	61.87	Rosa and Molnar (1988)
71.2045	-83.02	-81.82	71.42	Rosa and Molnar (1988)

Table A2.2. Hotspot reference frame reconstruction and finite difference parameters for the Atlantic–Indian Ocean. The total reconstruction parameters are derived from the African plate parameters of Müller et al. (1991) and extended to the other Atlantic–Indian Ocean plate by spline interpolation, as described in Chap. 3. Finite difference parameters represent the differential rotation for a given age and the next earliest age; the finite difference parameters for the youngest age correspond to the total reconstruction parameters

Atlantic–Indian Ocean plate reconstruction parameters relative to hotspots (Müller et al. 1991); time scale of Cande and Kent (1995)

Magnetic Chron	Age (Ma)	Total reconstruction parameters			Finite difference parameters		
		Longitude	Latitude	Angle	Longitude	Latitude	Angle
Eurasia:							
5	10.3445	-144.664	80.360	1.280	-144.664	80.360	1.280
6	19.5895	-97.481	63.943	3.023	-89.353	50.732	1.881
13	33.3015	-103.734	45.564	3.355	-110.179	-17.280	1.087
21	47.085	-86.360	24.786	4.648	-65.458	-11.286	2.146

Atlantic–Indian Ocean plate reconstruction parameters relative to hotspots (Müller et al. 1991); time scale of Cande and Kent (1995)

Magnetic Chron	Age (Ma)	Total reconstruction parameters			Finite difference parameters		
		Longitude	Latitude	Angle	Longitude	Latitude	Angle
25	56.1475	-90.308	26.019	5.546	-111.014	29.480	0.958
31	68.236	-77.029	17.644	7.289	-47.042	-3.642	2.398
33o	79.075	-80.702	-11.978	9.232	-84.908	-63.178	4.647
34	84	-88.738	-13.265	10.389	-136.050	-18.923	1.782
–	90	-97.483	-21.574	12.268	-142.785	-49.644	2.979
–	100	-115.275	-31.006	15.588	-170.170	-44.821	5.539
–	110	-123.449	-38.040	20.500	-152.877	-57.060	5.766
M1	118.7	-129.984	-38.647	24.804	-159.575	-41.672	4.753
M10	130	-132.915	-40.272	29.040	-149.226	-51.129	4.429
North America:							
5	10.3445	120.700	43.600	-1.390	-59.300	-43.600	1.390
6	19.5895	112.200	35.000	-3.380	-72.804	-28.964	2.031
13	33.3015	109.100	44.000	-6.480	-74.523	-53.575	3.192
18	39.278	110.200	47.700	-8.440	-63.131	-59.252	2.020
21	47.085	112.700	46.800	-11.270	-60.839	-43.386	2.849
25	56.1475	115.800	46.200	-14.510	-54.708	-42.677	3.277
31	68.236	119.500	46.000	-18.650	-48.718	-43.056	4.205
33y	73.619	119.600	49.300	-21.110	-45.694	-71.532	2.710
33o	79.075	117.000	53.400	-24.580	-82.232	-76.906	3.882
34	83	111.200	54.500	-25.810	-152.396	-47.254	1.989
–	90	104.400	57.400	-28.210	-165.913	-64.460	3.277
–	100	89.400	62.900	-31.740	160.369	-55.934	5.930
–	110	77.000	66.100	-37.270	170.817	-68.338	6.637
M1	118.7	63.900	66.500	-41.540	167.192	-53.945	5.552
M10	130	56.900	65.900	-45.420	-179.286	-56.538	4.429
North Africa:							
5	10.3445	-31.600	59.300	1.890	-31.600	59.300	1.890
6	19.5895	-44.500	50.900	4.360	-50.913	43.825	2.532
13	33.3015	-43.000	40.300	7.910	-40.779	27.902	3.713
18	39.278	-41.200	37.700	9.650	-33.382	26.395	1.797
21	47.085	-40.800	32.800	12.090	-37.629	14.590	2.609
25	56.1475	-41.700	30.100	13.890	-44.446	12.239	1.908
31	68.236	-40.900	26.400	16.230	-34.253	6.364	2.538
33y	73.619	-39.600	22.300	17.800	-24.808	-11.424	2.014
33o	79.075	-38.900	18.000	19.980	-29.241	-11.589	2.604
34	83	-40.900	19.000	21.530	-67.722	24.256	1.732
–	90	-41.900	19.400	23.310	-54.678	21.285	1.824
–	100	-40.500	18.800	25.720	-26.630	15.523	2.487
–	110	-38.200	17.500	27.340	-5.321	4.921	1.996
M1	118.7	-38.000	18.500	28.210	-40.347	47.263	0.998
M10	130	-40.100	16.200	27.520	-150.893	-63.588	1.623
South America:							
5	10.3445	-45.800	59.400	-1.800	134.200	-59.400	1.800
6	19.5895	-19.800	65.800	-3.210	-160.955	-64.841	1.518
13	33.3015	4.000	72.100	-6.370	-141.194	-73.041	3.267
18	39.278	12.900	74.200	-8.360	-126.007	-76.273	2.034
21	47.085	50.600	76.000	-10.280	-69.908	-50.126	2.476

Atlantic–Indian Ocean plate reconstruction parameters relative to hotspots (Müller et al. 1991); time scale of Cande and Kent (1995)

Magnetic Chron	Age (Ma)	Total reconstruction parameters			Finite difference parameters		
		Longitude	Latitude	Angle	Longitude	Latitude	Angle
25	56.1475	68.000	74.100	-12.740	-82.019	-58.595	2.643
31	68.236	87.500	73.200	-15.900	-55.471	-58.055	3.445
33y	73.619	93.000	72.600	-18.310	-64.285	-64.879	2.464
33o	78.781	93.700	72.300	-21.900	-84.035	-70.583	3.592
34	84	82.400	71.600	-22.870	-169.683	-33.024	1.690
–	90	68.100	71.700	-24.010	164.191	-34.239	2.151
–	100	45.300	72.200	-26.770	146.619	-44.373	4.141
–	110	24.900	71.800	-30.380	135.386	-49.316	4.755
M1	118.7	9.200	68.400	-33.260	136.662	-31.731	4.496
M10	130	10.400	67.100	-34.640	-168.734	-37.533	1.597
Antarctica:							
5	10.3445	102.000	64.900	1.560	102.000	64.900	1.560
6	19.5895	37.300	85.300	2.870	-55.804	67.364	1.567
13	33.3015	5.700	72.600	3.910	-0.199	41.650	1.308
18	39.278	-0.500	75.600	4.070	-138.074	46.750	0.288
21	47.085	17.600	71.100	5.220	42.432	51.809	1.274
25	56.1475	-39.700	73.800	6.280	-96.472	32.730	1.979
31	68.236	-23.600	67.600	7.490	1.465	35.784	1.551
33y	73.619	-17.000	64.900	7.790	21.170	13.746	0.587
33o	79.075	-16.100	65.700	8.040	71.988	80.864	0.278
34	84	-32.500	72.400	8.210	-154.806	22.166	1.263
–	90	-132.700	82.100	9.790	178.046	31.200	3.555
–	100	173.900	63.300	13.400	166.939	22.612	5.788
–	110	164.700	49.400	17.760	160.235	14.185	5.886
M1	118.7	162.200	42.000	23.450	161.936	20.165	6.294
M10	130	162.300	42.400	30.420	162.146	43.755	6.973
India:							
5	10.3445	21.900	36.100	5.810	21.900	36.100	5.810
6	19.5895	4.300	40.600	10.710	-16.770	40.648	5.288
13	33.3015	17.400	30.900	19.330	31.530	20.334	9.337
18	39.278	16.900	31.300	23.000	13.783	32.862	3.676
21	47.085	11.100	27.900	29.060	-2.728	10.906	6.642
25	56.1475	6.200	24.700	37.290	-3.983	9.455	8.787
31	68.236	3.400	19.100	53.370	3.947	5.457	16.748
33y	73.619	2.500	18.700	58.120	-3.449	10.426	4.831
33o	79.075	1.800	18.100	63.000	-1.050	8.394	4.963
34	84	12.300	17.100	65.480	91.574	28.664	10.967
–	90	5.000	16.300	67.290	-53.397	-31.963	7.911
–	100	10.300	14.500	70.310	74.821	19.906	6.823
–	110	10.100	23.700	71.680	-93.177	52.583	10.766
M1	118.7	10.300	14.200	72.090	74.149	-47.994	11.154
M10	130	15.500	13.300	75.340	78.472	29.809	6.958
Australia:							
5	10.3445	40.600	23.700	7.270	40.600	23.700	7.270
6	19.5895	31.400	27.100	13.230	19.572	29.101	6.154
13	33.3015	30.100	22.700	22.470	29.691	16.260	9.340
18	39.278	27.800	24.000	26.000	11.410	27.854	3.680

Atlantic–Indian Ocean plate reconstruction parameters relative to hotspots (Müller et al. 1991); time scale of Cande and Kent (1995)

Magnetic Chron	Age (Ma)	Total reconstruction parameters			Finite difference parameters		
		Longitude	Latitude	Angle	Longitude	Latitude	Angle
21	47.085	25.900	23.300	27.670	4.541	6.371	1.883
25	56.1475	26.200	23.000	27.860	62.949	-5.177	0.273
31	68.236	26.800	28.300	28.720	-105.661	79.335	2.741
33y	73.619	26.700	18.000	29.610	50.947	-54.390	5.256
33o	79.075	26.700	18.300	30.160	21.911	33.263	0.571
34	84	27.400	18.700	29.490	-173.837	4.372	0.780
–	90	31.300	22.400	28.400	173.521	37.652	2.816
–	100	40.700	28.600	27.410	164.241	36.656	5.153
–	110	52.600	34.600	27.500	162.851	37.125	5.575
M1	118.7	65.700	40.800	28.850	168.648	42.994	6.012
M10	130	75.500	48.900	32.300	-173.905	66.236	6.579
Central Indian Basin:							
5	10.3445	35.640	19.230	-2.166	-144.360	-19.230	2.166
6	19.5895	31.400	27.100	13.230	19.572	29.101	6.154
13	33.3015	30.100	22.700	22.470	29.691	16.260	9.340
18	39.278	27.800	24.000	26.000	11.410	27.854	3.680
21	47.085	20.400	22.100	31.640	-3.390	5.174	6.637
25	56.1475	13.700	20.000	39.460	-4.740	3.891	8.784
31	68.236	8.400	15.200	55.290	3.408	-0.350	16.765
33y	73.619	7.000	14.900	59.950	-4.276	4.258	4.847
33o	79.075	5.900	14.400	64.750	-1.399	2.712	4.952
34	84	6.200	13.600	67.280	22.009	1.496	2.695
–	100	10.300	14.500	70.310	64.196	46.668	5.513
–	110	10.100	23.700	71.680	-93.177	52.583	10.766
M1	118.7	10.300	14.200	72.090	74.149	-47.994	11.154
M10	130	15.500	13.300	75.340	78.472	29.809	6.958
Central Africa:							
5	10.3445	-31.600	59.300	1.890	-31.600	59.300	1.890
6	19.5895	-44.500	50.900	4.360	-50.913	43.825	2.532
13	33.3015	-43.000	40.300	7.910	-40.779	27.902	3.713
18	39.278	-41.200	37.700	9.650	-33.382	26.395	1.797
21	47.085	-40.800	32.800	12.090	-37.629	14.590	2.609
25	56.1475	-41.700	30.100	13.890	-44.446	12.239	1.908
31	68.236	-40.900	26.400	16.230	-34.253	6.364	2.538
33y	73.619	-39.600	22.300	17.800	-24.808	-11.424	2.014
33o	79.075	-38.900	18.000	19.980	-29.241	-11.589	2.604
34	83	-40.900	19.000	21.530	-67.722	24.256	1.732
–	90	-41.900	19.400	23.310	-54.678	21.285	1.824
–	100	-41.400	18.900	25.350	-34.772	14.450	2.060
–	110	-39.500	17.700	26.710	-7.312	3.445	1.674
M1	118.7	-39.700	18.700	27.370	-61.308	49.451	0.814
M10	130	-37.500	16.700	28.520	4.339	-9.881	1.811

Appendix 3: Calculated Reconstruction and Kinematic Parameters

Interpolated finite reconstructions and instantaneous rotation parameters calculated using the methods of Chap. 4

Age (Ma)	Total rotation:			Instantaneous rotation:		
	Longitude	Latitude	Angle	Longitude	Latitude	Rate
North America–Eurasia:						
0.000	135.800	62.400	0.000	135.800	62.400	0.209
5.000	134.082	63.741	1.161	132.917	65.037	0.253
10.000	133.595	65.461	2.492	134.738	69.338	0.274
15.000	134.882	67.742	3.837	143.222	73.600	0.263
20.000	136.881	68.924	5.064	144.152	68.429	0.227
25.000	137.835	67.755	6.095	137.913	55.574	0.192
30.000	137.591	65.926	6.967	132.305	54.966	0.168
35.000	136.488	66.075	7.798	109.999	81.014	0.181
40.000	135.217	68.425	8.784	130.016	84.156	0.237
45.000	136.230	68.710	10.158	145.045	54.167	0.342
50.000	140.068	63.761	12.161	145.296	36.583	0.491
55.000	141.558	62.705	13.918	165.434	79.428	0.285
60.000	142.135	64.135	15.172	161.506	72.958	0.242
65.000	143.108	64.635	16.289	164.289	66.617	0.214
70.000	144.505	65.003	17.294	-177.918	68.097	0.199
75.000	146.268	65.635	18.235	-162.877	69.058	0.197
80.000	148.049	66.275	19.177	-167.775	69.203	0.198
85.000	149.333	66.643	19.947	-171.012	66.438	0.106
90.000	150.029	66.700	20.279	-177.387	54.465	0.048
95.000	150.654	66.605	20.544	177.352	54.338	0.070
100.000	151.435	66.571	20.950	-174.259	63.088	0.098
105.000	152.340	66.850	21.490	-143.045	72.568	0.129
110.000	153.334	67.595	22.124	-106.135	73.439	0.149
115.000	154.261	68.489	22.728	-91.719	71.972	0.127
120.000	154.855	69.097	23.121	-90.794	72.564	0.054
125.000	154.953	69.196	23.197	95.753	-69.392	0.014
130.000	154.750	68.990	23.050	78.545	-75.677	0.048
North America–North Africa:						
0.000	38.300	78.800	0.000	38.300	78.800	0.238
5.000	46.351	79.745	1.241	53.612	80.305	0.258
10.000	50.849	80.123	2.572	53.067	80.255	0.272
15.000	47.866	79.984	3.957	29.633	78.711	0.283
20.000	36.623	79.514	5.414	-1.812	74.154	0.308
25.000	20.177	78.380	7.018	-18.621	68.166	0.350
30.000	6.668	76.540	8.795	-23.257	65.446	0.383
35.000	-0.711	74.899	10.690	-19.713	67.667	0.389
40.000	-3.433	74.224	12.608	-12.191	73.608	0.382
45.000	-3.966	74.721	14.488	6.365	83.231	0.373
50.000	-3.568	76.466	16.285	141.728	81.909	0.369
55.000	-1.504	79.049	17.859	144.247	66.661	0.326
60.000	5.081	81.644	19.005	139.607	57.558	0.263
65.000	8.289	82.977	20.206	-67.422	80.261	0.317
70.000	-6.495	81.778	22.310	-49.501	61.352	0.510

Age (Ma)	Total rotation:			Instantaneous rotation:		
	Longitude	Latitude	Angle	Longitude	Latitude	Rate
75.000	-13.381	80.281	24.595	-40.999	62.342	0.553
80.000	-19.171	77.849	27.665	-40.789	58.923	0.700
85.000	-20.843	75.818	30.797	-27.830	58.406	0.610
90.000	-19.708	74.162	33.669	-22.887	55.805	0.632
95.000	-19.045	72.354	37.010	-28.130	54.732	0.773
100.000	-19.178	70.495	40.932	-30.915	54.342	0.861
105.000	-19.568	68.828	45.131	-32.042	54.421	0.874
110.000	-19.907	67.500	49.209	-31.598	54.953	0.795
115.000	-20.014	66.600	52.684	-28.169	56.217	0.610
120.000	-19.731	66.206	55.012	-13.057	58.733	0.333
125.000	-19.125	66.146	56.211	0.378	54.087	0.190
130.000	-18.597	65.996	57.109	-13.028	50.629	0.188
North Africa–Central Africa:						
0.000	0.000	0.000	0.000	0.000	0.000	0.000
5.000	0.000	0.000	0.000	0.000	0.000	0.000
10.000	0.000	0.000	0.000	0.000	0.000	0.000
15.000	0.000	0.000	0.000	0.000	0.000	0.000
20.000	0.000	0.000	0.000	0.000	0.000	0.000
25.000	0.000	0.000	0.000	0.000	0.000	0.000
30.000	0.000	0.000	0.000	0.000	0.000	0.000
35.000	0.000	0.000	0.000	0.000	0.000	0.000
40.000	0.000	0.000	0.000	0.000	0.000	0.000
45.000	0.000	0.000	0.000	0.000	0.000	0.000
50.000	0.000	0.000	0.000	0.000	0.000	0.000
55.000	0.000	0.000	0.000	0.000	0.000	0.000
60.000	0.000	0.000	0.000	0.000	0.000	0.000
65.000	0.000	0.000	0.000	0.000	0.000	0.000
70.000	0.000	0.000	0.000	0.000	0.000	0.000
75.000	0.000	0.000	0.000	0.000	0.000	0.000
80.000	0.000	0.000	0.000	0.000	0.000	0.000
85.000	0.000	0.000	0.000	0.000	0.000	0.000
90.000	0.000	0.000	0.000	0.000	0.000	0.000
95.000	-172.173	-18.447	0.236	-172.457	-17.900	0.063
100.000	-172.770	-17.300	0.530	-174.898	-13.034	0.044
105.000	-173.851	-15.293	0.670	-179.221	-6.002	0.022
110.000	-173.960	-15.690	0.861	-171.536	-22.568	0.067
115.000	-173.000	-18.366	1.232	-171.061	-26.135	0.048
120.000	-174.593	-16.670	0.947	11.402	24.798	0.188
125.000	27.615	41.657	0.447	11.191	25.062	0.306
130.000	15.162	29.820	1.586	11.147	25.108	0.161
South America–Central Africa						
0.000	-39.400	62.500	0.000	-39.400	62.500	0.304
5.000	-39.246	61.259	1.418	-39.101	59.937	0.270
10.000	-39.043	60.150	2.837	-38.742	58.627	0.319
15.000	-38.742	59.641	4.786	-37.810	59.132	0.461
20.000	-37.896	59.487	7.254	-34.233	58.843	0.496
25.000	-36.294	59.129	9.598	-30.256	55.872	0.455
30.000	-34.811	58.159	11.880	-30.099	51.993	0.461
35.000	-33.601	57.271	14.135	-27.705	53.188	0.442

Age (Ma)	Total rotation:			Instantaneous rotation:		
	Longitude	Latitude	Angle	Longitude	Latitude	Rate
40.000	-32.530	57.141	16.282	-24.358	58.517	0.421
45.000	-31.772	57.868	18.361	-21.464	66.943	0.418
50.000	-31.494	59.169	20.332	-22.359	73.395	0.379
55.000	-31.816	60.610	21.978	-35.969	84.499	0.317
60.000	-32.846	62.390	23.260	-145.713	86.599	0.277
65.000	-33.280	62.994	24.398	-34.687	61.410	0.218
70.000	-33.542	63.004	26.102	-33.928	62.536	0.447
75.000	-33.601	63.081	28.664	-37.095	66.896	0.563
80.000	-34.338	63.055	31.546	-44.107	55.930	0.656
85.000	-35.040	61.631	34.927	-46.331	45.227	0.697
90.000	-35.564	59.723	38.121	-47.934	37.454	0.671
95.000	-35.897	57.595	41.087	-48.739	30.901	0.649
100.000	-36.034	55.428	43.857	-48.594	26.167	0.623
105.000	-35.970	53.401	46.454	-47.094	23.823	0.586
110.000	-35.691	51.700	48.894	-43.320	24.673	0.533
115.000	-35.170	50.519	51.182	-35.179	29.912	0.470
120.000	-34.365	50.049	53.307	-21.535	36.145	0.373
125.000	-33.852	49.493	54.165	-48.699	-12.269	0.150
130.000	-33.720	49.112	54.911	-33.523	42.429	0.210
East Antarctica–India:						
0.000	13.100	21.900	0.000	13.100	21.900	0.476
5.000	15.324	22.058	2.540	16.366	22.501	0.532
10.000	14.378	23.244	5.137	6.961	27.756	0.486
15.000	8.840	25.867	7.247	-16.121	34.590	0.408
20.000	4.730	27.341	9.095	4.930	26.286	0.390
25.000	9.207	25.495	11.340	34.497	9.374	0.595
30.000	16.706	22.686	14.357	39.434	10.090	0.745
35.000	20.550	22.440	17.789	26.959	29.462	0.675
40.000	18.868	23.378	20.940	-6.534	20.449	0.657
45.000	14.332	21.071	24.040	-14.201	5.876	0.787
50.000	11.744	18.810	27.852	3.633	11.193	0.878
55.000	11.297	17.570	32.735	6.773	10.614	1.115
60.000	10.355	15.838	38.882	-0.363	7.780	1.359
65.000	8.893	14.150	45.435	-3.891	9.308	1.272
70.000	7.606	13.537	50.541	-4.878	19.453	0.807
75.000	6.446	13.731	54.304	-6.509	19.914	0.858
80.000	6.189	13.100	58.899	8.766	-6.574	1.010
85.000	8.074	11.503	63.042	24.449	-26.049	0.878
90.000	10.635	9.362	66.145	21.044	-42.932	1.131
95.000	13.467	5.886	70.158	17.984	-40.691	1.267
100.000	15.398	4.742	73.929	55.228	15.214	0.707
105.000	15.636	8.441	75.671	92.162	58.571	1.127
110.000	14.692	10.532	77.503	-29.586	-5.890	0.759
115.000	13.645	4.757	81.318	-39.064	-36.676	2.323
120.000	13.869	-2.190	86.634	-22.628	-47.384	1.662
125.000	15.903	-4.953	90.240	49.234	-56.471	0.991
130.000	19.014	-5.364	92.379	83.153	-44.729	1.035

Age (Ma)	Total rotation:			Instantaneous rotation:		
	Longitude	Latitude	Angle	Longitude	Latitude	Rate
Antarctica–Australia:						
0.000	38.200	13.200	0.000	38.200	13.200	0.647
5.000	37.681	12.672	3.108	36.891	12.391	0.600
10.000	36.465	12.901	6.051	32.814	14.944	0.588
15.000	34.192	14.399	9.066	27.631	19.716	0.629
20.000	32.666	15.402	12.230	31.441	15.280	0.633
25.000	33.359	14.429	15.337	38.144	6.536	0.621
30.000	34.156	13.347	18.386	33.549	11.541	0.609
35.000	32.374	14.609	21.468	11.533	34.496	0.695
40.000	29.641	16.806	23.970	18.098	29.088	0.314
45.000	30.562	15.922	24.519	98.721	-55.956	0.220
50.000	31.824	14.110	24.620	33.530	-46.519	0.150
55.000	31.741	12.799	25.121	13.286	-30.037	0.155
60.000	31.859	11.345	25.535	37.945	-57.431	0.160
65.000	32.588	9.682	25.742	71.746	-64.599	0.169
70.000	33.527	8.254	25.879	83.216	-58.153	0.136
75.000	34.400	7.190	26.075	72.987	-50.193	0.133
80.000	35.262	5.917	26.463	55.674	-48.736	0.196
85.000	36.165	4.219	27.045	55.940	-50.051	0.213
90.000	37.064	2.629	27.615	58.104	-49.991	0.200
95.000	37.862	1.246	28.184	55.270	-46.924	0.182
100.000	38.457	0.129	28.778	48.806	-40.556	0.163
105.000	38.835	-0.714	29.387	43.105	-33.103	0.144
110.000	39.028	-1.308	29.981	38.339	-25.207	0.126
115.000	39.080	-1.688	30.528	34.423	-17.462	0.107
120.000	39.039	-1.897	30.992	31.223	-10.336	0.083
125.000	38.959	-1.983	31.332	28.582	-4.108	0.054
130.000	38.900	-2.000	31.500	25.038	0.537	0.018
East Antarctica–Central Indian Ocean:						
0.000	38.200	13.200	0.000	38.200	13.200	0.647
5.000	27.698	25.804	2.306	31.915	18.160	0.565
10.000	36.204	12.775	6.739	40.691	3.516	1.046
15.000	36.036	11.841	10.514	23.877	25.108	0.487
20.000	32.427	15.630	12.093	10.339	37.189	0.353
25.000	31.575	16.158	14.347	34.516	9.710	0.590
30.000	32.857	14.593	17.796	37.305	8.311	0.768
35.000	32.948	15.023	21.438	26.599	29.332	0.678
40.000	29.951	16.646	24.351	-6.543	20.944	0.652
45.000	24.731	15.230	27.123	-14.075	5.713	0.789
50.000	21.030	13.680	30.728	3.681	11.507	0.878
55.000	19.193	12.969	35.529	6.769	10.796	1.114
60.000	16.985	11.667	41.568	-0.446	7.509	1.364
65.000	14.554	10.250	47.984	-3.878	8.354	1.330
East Antarctica–Central Africa:						
0.000	-39.200	5.600	0.000	-39.200	5.600	0.124
5.000	-46.434	6.846	0.719	-51.107	7.815	0.162
10.000	-49.519	7.955	1.545	-51.077	10.379	0.159
15.000	-48.736	9.387	2.237	-42.604	15.001	0.122
20.000	-48.048	10.861	2.828	-52.964	16.226	0.129

Age (Ma)	Total rotation:			Instantaneous rotation:		
	Longitude	Latitude	Angle	Longitude	Latitude	Rate
25.000	-50.687	11.787	3.607	-62.571	15.012	0.189
30.000	-52.366	12.845	4.675	-50.410	18.420	0.241
35.000	-48.724	14.919	5.973	-22.491	21.028	0.312
40.000	-44.670	15.098	7.492	-45.921	2.712	0.325
45.000	-46.856	10.820	9.019	-60.476	-16.202	0.314
50.000	-45.546	7.321	9.952	7.906	-23.097	0.201
55.000	-40.100	5.613	10.249	32.559	-19.192	0.185
60.000	-38.099	4.118	10.577	-56.600	-26.973	0.111
65.000	-40.156	2.353	11.201	-73.440	-15.897	0.209
70.000	-41.830	0.298	12.280	-46.345	-19.320	0.289
75.000	-41.034	-2.952	14.061	-34.767	-26.828	0.495
80.000	-39.356	-4.294	16.471	-25.104	9.849	0.503
85.000	-36.906	-2.355	18.983	-17.121	-9.135	0.603
90.000	-33.601	-4.093	22.106	-16.472	-17.933	0.726
95.000	-31.007	-5.125	25.618	-16.735	-12.026	0.752
100.000	-28.956	-5.631	29.298	-15.555	-13.709	0.771
105.000	-27.165	-6.330	33.006	-15.723	-14.822	0.760
110.000	-26.109	-6.504	36.623	-20.084	-4.237	0.708
115.000	-25.718	-6.074	40.139	-18.227	-10.426	0.736
120.000	-23.590	-7.949	44.229	-4.855	-43.972	1.209
125.000	-20.175	-11.809	48.861	-8.146	-48.521	1.014
130.000	-21.020	-11.878	50.420	-104.699	60.345	0.724
India–Central Indian Ocean:						
0.000	76.900	-8.700	0.000	76.900	-8.700	0.251
5.000	77.220	-9.215	1.308	77.252	-9.337	0.274
10.000	76.739	-8.643	2.780	74.942	-6.202	0.321
15.000	75.334	-6.797	4.496	71.805	-1.627	0.352
20.000	74.158	-5.207	6.021	70.460	0.552	0.226
25.000	73.830	-4.724	6.577	78.603	2.207	0.006
30.000	74.085	-4.992	6.254	-110.420	-0.697	0.108
35.000	74.382	-5.293	5.875	-108.641	7.277	0.017
40.000	74.303	-5.123	5.946	70.852	3.726	0.016
45.000	74.342	-5.121	5.950	-118.770	0.295	0.008
50.000	74.407	-5.035	5.914	-116.935	50.395	0.006
55.000	74.410	-4.829	5.909	75.902	53.665	0.003
60.000	74.414	-4.890	5.924	71.379	-56.810	0.006
65.000	74.386	-5.220	5.943	29.941	-70.151	0.008
70.000	74.223	-5.525	5.943	-43.194	-35.568	0.007
75.000	74.094	-5.845	5.958	79.656	-45.197	0.013
80.000	74.348	-5.357	5.960	-138.780	65.135	0.043
85.000	74.526	-4.338	5.903	36.238	-71.336	0.006
Central Indian Ocean–Central Africa:						
0.000	-130.200	-12.400	0.000	-130.200	-12.400	0.628
5.000	-129.934	-11.734	3.310	-129.925	-11.392	0.687
10.000	-130.438	-11.932	6.678	-132.963	-13.639	0.633
15.000	-132.096	-13.346	9.486	-139.927	-18.800	0.507
20.000	-133.503	-14.488	11.933	-135.884	-15.323	0.518
25.000	-132.958	-13.850	14.846	-126.436	-9.418	0.644
30.000	-131.433	-12.990	18.141	-122.996	-12.669	0.658

Age (Ma)	Total rotation:			Instantaneous rotation:		
	Longitude	Latitude	Angle	Longitude	Latitude	Rate
35.000	-130.354	-13.980	20.978	-130.401	-34.031	0.492
40.000	-131.229	-16.242	23.080	-159.374	-23.661	0.482
45.000	-134.944	-16.833	25.421	-171.795	-7.177	0.631
50.000	-139.606	-17.255	28.066	-178.075	-10.373	0.720
55.000	-144.539	-17.734	31.432	-176.380	-6.472	0.981
60.000	-149.709	-17.038	36.505	-172.432	-1.791	1.317
65.000	-153.825	-16.096	42.483	-172.890	-4.174	1.228
70.000	-156.229	-16.470	46.932	173.979	-22.447	0.689
75.000	-157.886	-18.433	49.452	158.093	-31.889	0.697
80.000	-158.273	-19.493	52.349	-129.118	-20.101	0.688
85.000	-156.508	-18.614	55.115	-134.018	-16.379	0.519
India–Central Africa:						
0.000	-151.500	-23.600	0.000	-151.500	-23.600	0.430
5.000	-147.212	-22.809	2.317	-144.576	-22.634	0.487
10.000	-146.815	-23.747	4.640	-152.979	-28.851	0.413
15.000	-151.942	-27.319	6.286	174.804	-38.688	0.311
20.000	-156.207	-29.624	7.690	-153.825	-27.857	0.338
25.000	-150.744	-26.582	9.928	-125.023	-11.618	0.639
30.000	-142.731	-22.345	13.176	-119.142	-12.606	0.767
35.000	-138.282	-21.758	16.239	-125.224	-35.014	0.504
40.000	-138.438	-23.890	18.382	-158.818	-28.773	0.473
45.000	-142.150	-23.869	20.922	-170.227	-12.514	0.636
50.000	-146.821	-23.640	23.858	-176.904	-15.587	0.721
55.000	-151.563	-23.509	27.527	-175.611	-11.888	0.979
60.000	-156.114	-22.109	32.894	-171.796	-7.462	1.315
65.000	-159.447	-20.660	39.111	-172.186	-9.970	1.227
70.000	-161.330	-20.840	43.739	175.166	-28.535	0.690
75.000	-162.705	-22.849	46.422	157.694	-39.370	0.694
80.000	-162.875	-23.778	49.400	-131.613	-17.775	0.687
85.000	-159.949	-22.668	51.731	-92.698	-29.527	0.592
90.000	-156.433	-22.460	52.985	-85.851	-17.839	0.755
95.000	-152.302	-20.640	55.100	-92.479	-14.340	0.867
100.000	-149.317	-21.375	56.802	53.005	-75.524	0.861
105.000	-148.545	-28.441	56.604	77.721	-48.050	1.563
110.000	-149.007	-33.721	56.298	84.493	39.574	0.110
115.000	-149.021	-28.091	56.081	-100.753	46.893	1.726
120.000	-147.994	-20.315	57.514	-98.844	27.106	0.993
125.000	-145.921	-19.251	59.311	-86.557	-64.131	0.658
130.000	-143.120	-22.257	61.018	-45.537	-83.912	1.009
Tristan–Central Africa:						
0.000	3.600	-5.500	0.000	3.600	-5.500	0.143
5.000	-7.974	35.719	0.626	-30.094	63.813	0.171
10.000	-27.933	58.247	1.643	-60.340	64.711	0.265
15.000	-41.444	57.270	3.025	-52.145	44.224	0.299
20.000	-44.596	50.379	4.477	-47.834	31.444	0.301
25.000	-44.873	45.394	5.821	-44.558	27.496	0.267
30.000	-44.003	41.999	7.054	-39.229	25.721	0.258
35.000	-42.406	39.552	8.399	-35.083	26.946	0.307
40.000	-41.005	37.123	9.974	-38.423	19.058	0.337

Age (Ma)	Total rotation:			Instantaneous rotation:		
	Longitude	Latitude	Angle	Longitude	Latitude	Rate
45.000	-40.699	33.958	11.517	-43.686	11.401	0.313
50.000	-41.072	31.603	12.772	-48.636	12.699	0.225
55.000	-41.612	30.325	13.698	-49.505	15.079	0.176
60.000	-41.783	29.338	14.549	-42.485	11.964	0.185
65.000	-41.406	27.905	15.500	-37.452	1.015	0.234
70.000	-40.547	25.309	16.686	-35.853	-11.320	0.341
75.000	-39.167	20.882	18.304	-34.990	-20.467	0.509
80.000	-39.263	18.007	20.367	-60.304	25.760	0.444
85.000	-41.580	19.431	22.155	-58.172	31.057	0.297
90.000	-41.900	19.400	23.310	-42.325	13.639	0.210
95.000	-41.854	19.227	24.365	-39.373	14.761	0.210
100.000	-41.400	18.900	25.350	-24.399	-0.834	0.200
105.000	-40.262	18.147	26.144	-12.485	-13.706	0.185
110.000	-39.500	17.700	26.710	-33.329	19.844	0.088
115.000	-39.884	18.230	27.068	-63.877	65.594	0.105
120.000	-39.296	18.754	27.515	13.571	1.929	0.204
125.000	-37.128	18.249	28.173	2.070	-23.501	0.229
130.000	-37.500	16.700	28.520	-114.347	-25.574	0.282
Tristan–North Africa:						
0.000	3.600	-5.500	0.000	3.600	-5.500	0.143
5.000	-7.974	35.719	0.626	-30.094	63.813	0.171
10.000	-27.933	58.247	1.643	-60.340	64.711	0.265
15.000	-41.444	57.270	3.025	-52.145	44.224	0.299
20.000	-44.596	50.379	4.477	-47.834	31.444	0.301
25.000	-44.873	45.394	5.821	-44.558	27.496	0.267
30.000	-44.003	41.999	7.054	-39.229	25.721	0.258
35.000	-42.406	39.552	8.399	-35.083	26.946	0.307
40.000	-41.005	37.123	9.974	-38.423	19.058	0.337
45.000	-40.699	33.958	11.517	-43.686	11.401	0.313
50.000	-41.072	31.603	12.772	-48.635	12.699	0.225
55.000	-41.612	30.325	13.698	-49.508	15.080	0.176
60.000	-41.783	29.338	14.549	-42.493	11.965	0.185
65.000	-41.407	27.905	15.500	-37.449	1.015	0.234
70.000	-40.547	25.309	16.687	-35.841	-11.315	0.341
75.000	-39.169	20.882	18.303	-35.025	-20.485	0.508
80.000	-39.260	18.007	20.368	-60.107	25.733	0.445
85.000	-41.609	19.433	22.145	-59.767	31.368	0.294
90.000	-41.900	19.400	23.310	-37.862	12.806	0.228
95.000	-41.416	19.192	24.532	-29.875	12.170	0.259
100.000	-40.500	18.800	25.720	-20.640	-1.418	0.238
105.000	-39.239	17.974	26.608	-12.149	-13.523	0.204
110.000	-38.200	17.500	27.340	-15.002	13.419	0.161
115.000	-37.910	18.060	28.028	-25.564	52.490	0.130
120.000	-38.048	18.522	28.147	159.245	-19.379	0.074
125.000	-38.513	17.839	27.574	-166.550	-43.833	0.182
130.000	-40.100	16.200	27.520	-103.534	-20.856	0.280

Age (Ma)	Total rotation:			Instantaneous rotation:		
	Longitude	Latitude	Angle	Longitude	Latitude	Rate
Pacific–Nazca:						
0.000	-78.200	48.700	0.000	-78.200	48.700	0.742
5.000	-91.850	60.061	8.791	37.160	65.574	0.662
10.000	-88.398	61.539	13.082	-98.509	49.529	1.853
15.000	-92.316	60.822	22.586	-88.789	72.618	1.875
20.000	-91.398	64.604	31.580	-82.140	63.639	1.779
25.000	-90.787	63.007	39.961	-112.405	66.319	1.486
30.000	-98.806	66.419	46.122	104.922	73.522	1.419
35.000	-104.850	70.676	51.894	103.862	77.623	1.483
40.000	-114.421	73.183	58.335	121.479	76.679	1.332
45.000	-120.700	74.567	63.362	106.622	82.380	0.885
50.000	-123.437	75.295	66.629	160.947	86.557	0.511
55.000	-124.976	75.255	68.498	177.356	52.682	0.325
60.000	-128.082	74.211	69.705	167.521	26.143	0.453
65.000	-131.583	73.257	71.240	152.327	65.514	0.455
70.000	-133.186	74.645	74.202	-4.467	56.382	1.178
75.000	-129.820	80.000	80.000	-8.743	46.238	2.185
Pacific–Farallon:						
0.000	-108.600	36.800	0.000	-108.600	36.800	1.990
5.000	-104.834	59.445	4.989	59.948	50.204	0.725
10.000	56.605	71.053	6.608	64.213	17.265	1.068
15.000	61.665	59.262	9.716	60.142	84.150	0.572
20.000	58.317	73.575	12.719	-125.694	68.191	1.050
25.000	61.241	82.576	18.266	-175.728	83.464	1.399
30.000	74.765	84.604	26.163	113.447	84.516	1.771
35.000	80.090	85.490	35.617	175.444	88.057	1.970
40.000	91.084	86.814	44.868	163.606	80.389	1.660
45.000	120.914	86.261	51.792	126.711	70.790	1.211
50.000	118.451	84.471	56.904	64.080	60.881	1.010
55.000	104.613	82.518	61.018	45.577	55.406	0.884
60.000	97.901	81.406	64.548	48.225	73.844	0.676
65.000	96.260	81.407	67.683	149.691	85.988	0.607
70.000	96.766	81.777	70.680	141.466	82.363	0.612
75.000	97.436	81.683	73.855	81.969	73.390	0.667
80.000	97.603	81.028	77.223	72.269	65.876	0.705
East Antarctica–West Antarctica:						
0.000	0.000	0.000	0.000	0.000	0.000	0.000
5.000	0.000	0.000	0.000	0.000	0.000	0.000
10.000	0.000	0.000	0.000	0.000	0.000	0.000
15.000	0.000	0.000	0.000	0.000	0.000	0.000
20.000	0.000	0.000	0.000	0.000	0.000	0.000
25.000	0.000	0.000	0.000	0.000	0.000	0.000
30.000	-17.833	-18.133	0.131	-17.815	-18.117	0.105
35.000	-17.848	-18.147	0.937	-17.848	-18.147	0.158
40.000	-17.848	-18.146	1.534	-17.844	-18.143	0.080
45.000	-17.847	-18.146	1.700	-17.761	-18.069	0.000
50.000	-17.847	-18.146	1.700	-17.762	-18.070	0.000
55.000	-17.847	-18.146	1.700	-17.769	-18.077	0.000
60.000	-17.847	-18.146	1.700	-17.802	-18.106	0.000

Age (Ma)	Total rotation:			Instantaneous rotation:		
	Longitude	Latitude	Angle	Longitude	Latitude	Rate
65.000	-17.847	-18.146	1.700	-17.837	-18.137	0.000
70.000	-17.847	-18.146	1.700	-17.846	-18.145	0.000
75.000	-17.847	-18.146	1.700	-17.847	-18.146	0.000
80.000	-17.847	-18.146	1.700	-17.847	-18.146	0.000
85.000	-17.847	-18.146	1.700	-17.847	-18.146	0.000
90.000	-17.847	-18.146	1.700	-17.847	-18.146	0.000
95.000	-17.847	-18.146	1.700	-17.847	-18.146	0.000
100.000	-17.847	-18.146	1.700	-17.847	-18.146	0.000
105.000	-17.847	-18.146	1.700	-17.847	-18.146	0.000
110.000	-17.847	-18.146	1.700	162.153	18.146	0.000
115.000	-17.847	-18.146	1.700	162.153	18.146	0.000
120.000	-17.847	-18.146	1.700	162.153	18.146	0.000
125.000	-17.847	-18.146	1.700	162.153	18.146	0.000
130.000	-17.847	-18.146	1.700	162.153	18.146	0.000
Cocos–Pacific:						
0.000	71.400	-36.800	0.000	71.400	-36.800	1.990
5.000	71.567	-36.641	11.917	72.888	-35.567	2.428
10.000	73.115	-34.493	21.430	76.821	-27.159	1.447
15.000	71.700	-32.687	27.767	57.557	-27.116	1.156
20.000	67.736	-35.187	31.587	-38.366	-47.055	0.821
25.000	64.191	-49.240	29.155	-96.218	-17.431	2.548
30.000	56.023	-80.030	27.721	-100.125	-23.244	3.112
35.000	-105.643	-83.444	35.516	-112.864	-81.579	1.939
40.000	-83.806	-87.246	44.918	63.349	-74.196	1.774
45.000	-52.908	-86.473	51.848	-34.356	-68.672	1.199
50.000	-63.850	-84.312	56.873	-68.230	-57.010	1.019
55.000	-75.774	-82.461	61.012	-81.794	-55.350	0.855
60.000	-81.534	-81.522	64.556	-92.153	-70.779	0.674
65.000	-83.244	-81.516	67.690	-126.946	-83.396	0.604
70.000	-83.143	-81.796	70.682	-115.776	-86.512	0.595
Kula–Pacific:						
0.000	0.000	0.000	0.000	0.000	0.000	0.000
5.000	0.000	0.000	0.000	0.000	0.000	0.000
10.000	0.000	0.000	0.000	0.000	0.000	0.000
15.000	0.000	0.000	0.000	0.000	0.000	0.000
20.000	0.000	0.000	0.000	0.000	0.000	0.000
25.000	0.000	0.000	0.000	0.000	0.000	0.000
30.000	0.000	0.000	0.000	0.000	0.000	0.000
35.000	0.000	0.000	0.000	0.000	0.000	0.000
40.000	0.000	0.000	0.000	0.000	0.000	0.000
45.000	0.000	0.000	0.000	0.000	0.000	0.000
50.000	0.000	0.000	0.000	0.000	0.000	0.000
55.000	-79.238	-14.480	0.825	-79.244	-14.480	0.404
60.000	-79.232	-14.480	2.956	-79.174	-14.482	0.266
65.000	-79.333	-14.479	4.775	-79.324	-14.480	0.702
70.000	-76.558	-14.484	8.063	-63.867	-14.953	0.506
75.000	-75.591	-14.479	10.538	-79.209	-14.157	0.417
80.000	-75.770	-14.480	11.600	-67.715	-15.072	0.078

Age (Ma)	Total rotation:			Instantaneous rotation:		
	Longitude	Latitude	Angle	Longitude	Latitude	Rate
West Antarctica–Pacific:						
0.000	95.000	-66.000	0.000	95.000	-66.000	0.900
5.000	103.763	-67.605	4.643	93.492	-68.796	0.885
10.000	104.142	-70.435	8.939	119.958	-76.894	0.884
15.000	108.472	-72.936	13.273	123.510	-78.642	0.825
20.000	111.554	-74.258	16.889	125.855	-79.377	0.637
25.000	112.734	-74.725	20.210	112.552	-74.430	0.769
30.000	112.916	-74.436	24.403	121.845	-73.080	0.840
35.000	117.118	-74.469	28.551	167.581	-74.757	0.856
40.000	125.409	-74.862	32.655	-173.410	-72.528	0.818
45.000	130.006	-74.791	36.111	143.351	-71.317	0.586
50.000	129.115	-74.239	38.577	120.519	-58.418	0.446
55.000	126.590	-73.231	40.844	119.718	-51.796	0.562
60.000	124.556	-71.753	44.025	131.150	-53.629	0.808
65.000	125.160	-70.058	48.471	147.173	-57.057	1.002
70.000	127.975	-68.904	52.951	160.528	-59.136	0.807
75.000	127.019	-67.384	55.996	121.509	-24.514	0.698
80.000	125.670	-65.968	59.477	159.125	-66.369	1.011
85.000	129.228	-65.844	64.674	158.442	-62.238	0.918
90.000	129.880	-65.230	67.640	145.459	-49.619	0.254
95.000	129.880	-65.230	67.640	-34.620	49.514	0.070
100.000	129.880	-65.230	67.640	145.394	-49.562	0.014
105.000	129.880	-65.230	67.640	145.393	-49.563	0.001
110.000	129.880	-65.230	67.640	145.397	-49.557	0.000
115.000	129.880	-65.230	67.640	149.236	-42.336	0.000
120.000	129.880	-65.230	67.640	-50.503	65.457	0.000
125.000	129.880	-65.230	67.640	-50.122	65.231	0.000
130.000	129.880	-65.230	67.640	-50.120	65.230	0.006
Hawaiian–Pacific:						
0.000	90.000	-60.200	0.000	90.000	-60.200	0.932
5.000	104.900	-56.500	4.700	113.456	-58.331	0.906
10.000	108.538	-64.395	8.642	102.804	-86.847	0.780
15.000	110.206	-70.186	12.393	115.925	-76.061	0.792
20.000	111.625	-70.416	16.466	114.834	-70.391	0.832
25.000	112.440	-71.200	20.490	122.076	-74.638	0.740
30.000	115.482	-70.385	23.674	137.741	-55.960	0.583
35.000	117.984	-68.602	26.355	131.829	-56.278	0.566
40.000	119.740	-67.880	29.441	146.651	-68.636	0.738
45.000	118.500	-66.072	31.666	104.674	-1.453	0.512
50.000	114.613	-62.903	33.261	106.414	-7.960	0.497
55.000	111.610	-59.819	34.624	106.790	-6.936	0.472
60.000	109.422	-57.035	36.200	106.664	-10.885	0.507
65.000	107.380	-54.206	38.203	105.370	-11.165	0.586
70.000	106.421	-52.200	39.475	140.123	28.415	0.130
75.000	106.549	-51.147	39.679	108.155	-7.525	0.417
80.000	102.767	-46.516	45.586	99.405	-18.003	2.025
85.000	101.410	-43.890	49.200	160.491	52.398	0.364
90.000	100.860	-42.090	51.500	79.096	-50.378	1.546
95.000	99.280	-46.640	61.870	79.569	-59.282	1.686

Age (Ma)	Total rotation:			Instantaneous rotation:		
	Longitude	Latitude	Angle	Longitude	Latitude	Rate
100.000	99.860	-46.310	64.710	109.292	-22.488	0.186
105.000	98.410	-47.530	66.390	28.537	-40.811	0.593
110.000	96.950	-48.680	68.120	36.633	-41.817	0.450
115.000	95.490	-49.760	69.880	33.969	-41.501	0.487
120.000	94.040	-50.770	71.670	34.588	-41.619	0.478
125.000	92.580	-51.730	73.500	34.480	-41.620	0.481
130.000	91.140	-52.620	75.350	34.157	-41.447	0.478
Greenland–North America:						
0.000	0.000	0.000	0.000	0.000	0.000	0.000
5.000	0.000	0.000	0.000	0.000	0.000	0.000
10.000	0.000	0.000	0.000	0.000	0.000	0.000
15.000	0.000	0.000	0.000	0.000	0.000	0.000
20.000	0.000	0.000	0.000	0.000	0.000	0.000
25.000	0.000	0.000	0.000	0.000	0.000	0.000
30.000	0.000	0.000	0.000	0.000	0.000	0.000
35.000	102.709	-58.376	0.190	105.872	-57.900	0.125
40.000	107.344	-57.641	0.930	103.000	-58.323	0.184
45.000	96.073	-59.100	2.107	71.582	-58.590	0.310
50.000	75.795	-58.023	3.839	41.177	-27.078	0.284
55.000	51.021	-32.415	3.224	-23.382	52.510	0.574
60.000	29.118	-24.420	3.578	60.121	-75.223	0.225
65.000	33.304	-36.805	4.277	46.796	-80.884	0.225
70.000	36.465	-48.156	5.428	92.000	-78.347	0.401
75.000	44.827	-58.236	7.470	103.966	-75.019	0.528
80.000	53.765	-63.702	9.842	106.946	-73.856	0.474
85.000	59.071	-65.892	11.521	105.974	-74.561	0.221
90.000	60.444	-66.509	12.109	84.238	-78.433	0.056
95.000	60.607	-66.714	12.341	69.549	-75.530	0.052
100.000	60.802	-66.892	12.633	70.023	-73.884	0.066
105.000	61.040	-67.080	12.990	69.985	-73.995	0.078
110.000	61.287	-67.290	13.384	69.773	-74.535	0.076
115.000	61.472	-67.456	13.697	69.630	-74.804	0.045
120.000	61.520	-67.500	13.780	-110.821	74.620	0.003
125.000	61.520	-67.500	13.780	65.910	-72.259	0.002
130.000	61.520	-67.500	13.780	-117.443	68.839	0.003
Pacific – East Antarctic:						
0.000	95.000	-66.000	0.000	95.000	-66.000	0.900
5.000	103.763	-67.605	4.643	93.492	-68.796	0.885
10.000	104.142	-70.435	8.939	119.958	-76.894	0.884
15.000	108.472	-72.936	13.273	123.510	-78.642	0.825
20.000	111.554	-74.258	16.889	125.855	-79.377	0.637
25.000	112.734	-74.725	20.210	112.552	-74.430	0.769
30.000	112.916	-74.436	24.403	121.845	-73.080	0.840
35.000	117.118	-74.469	28.551	167.581	-74.757	0.856
40.000	125.409	-74.862	32.655	-173.410	-72.528	0.818
45.000	130.006	-74.791	36.111	143.351	-71.317	0.586
50.000	129.115	-74.239	38.577	120.519	-58.418	0.446
55.000	126.590	-73.231	40.844	119.718	-51.796	0.562
60.000	124.556	-71.753	44.025	131.149	-53.629	0.808

Age (Ma)	Total rotation:			Instantaneous rotation:		
	Longitude	Latitude	Angle	Longitude	Latitude	Rate
65.000	125.160	-70.058	48.471	147.170	-57.055	1.002
70.000	127.975	-68.904	52.951	160.464	-59.108	0.807
75.000	127.019	-67.384	55.996	121.135	-23.652	0.700
80.000	125.670	-65.968	59.477	146.866	-60.596	0.908
Tristan–North America:						
0.000	-10.582	-66.511	0.000	-10.582	-66.511	0.269
5.000	-33.934	-63.758	0.953	-62.504	-50.161	0.132
10.000	-58.164	-51.691	1.444	-79.552	-11.511	0.143
15.000	-66.212	-37.103	2.224	-69.240	-17.246	0.225
20.000	-66.988	-32.624	3.431	-69.223	-32.846	0.250
25.000	-68.492	-35.562	4.612	-80.139	-54.134	0.240
30.000	-70.377	-41.025	5.720	-86.547	-65.354	0.249
35.000	-70.940	-44.859	6.885	-67.877	-56.053	0.252
40.000	-69.633	-45.735	8.306	-61.505	-46.989	0.335
45.000	-68.032	-45.871	10.194	-60.277	-47.175	0.410
50.000	-66.474	-46.385	12.248	-56.484	-50.681	0.408
55.000	-64.622	-46.626	14.204	-49.132	-43.115	0.382
60.000	-62.374	-45.334	15.987	-44.381	-31.982	0.363
65.000	-60.791	-44.928	17.664	-53.991	-60.516	0.346
70.000	-60.924	-48.473	19.464	-94.160	-76.781	0.499
75.000	-60.911	-51.034	22.177	-75.011	-68.273	0.672
80.000	-64.150	-53.742	24.933	-130.004	-40.155	0.521
85.000	-71.563	-54.983	26.378	-140.988	-35.952	0.479
90.000	-77.357	-56.673	28.116	-146.813	-46.168	0.510
95.000	-83.263	-58.640	30.134	-156.584	-47.560	0.600
100.000	-89.893	-61.035	32.548	-166.503	-50.326	0.715
105.000	-97.134	-63.621	35.410	-172.111	-50.739	0.773
110.000	-104.613	-65.477	38.223	-170.749	-44.845	0.662
115.000	-111.432	-66.148	40.342	-168.338	-36.265	0.495
120.000	-116.625	-66.376	41.940	-172.162	-40.697	0.393
125.000	-120.348	-66.523	43.475	-159.357	-43.010	0.371
130.000	-122.774	-65.643	45.128	-124.209	-29.759	0.422
Tristan–Eurasia:						
0.000	53.070	-45.754	0.000	53.070	-45.754	0.086
5.000	93.576	54.136	0.230	173.535	75.039	0.133
10.000	-177.895	78.354	1.159	-120.596	70.746	0.241
15.000	-114.991	73.820	2.308	-81.551	49.866	0.243
20.000	-94.536	64.290	3.210	-84.090	27.976	0.165
25.000	-95.963	57.293	3.554	-127.376	-21.921	0.089
30.000	-102.842	50.517	3.438	-148.783	-53.679	0.106
35.000	-101.737	43.230	3.401	-60.871	-15.250	0.118
40.000	-90.102	36.161	3.868	-56.946	-2.443	0.208
45.000	-84.638	29.272	4.562	-93.262	-4.487	0.135
50.000	-93.708	27.272	4.805	-172.542	2.584	0.174
55.000	-92.608	25.202	5.309	-46.785	5.071	0.252
60.000	-82.362	23.520	6.302	-43.937	9.772	0.265
65.000	-77.266	20.684	7.087	-64.538	-38.637	0.152
70.000	-79.041	11.406	7.101	-112.261	-67.779	0.313
75.000	-78.000	-1.953	8.021	-80.349	-59.218	0.499

Age (Ma)	Total rotation:			Instantaneous rotation:		
	Longitude	Latitude	Angle	Longitude	Latitude	Rate
80.000	-82.382	-12.685	9.499	-131.268	-18.449	0.441
85.000	-92.173	-14.128	10.977	-141.215	-23.575	0.441
90.000	-98.791	-19.286	12.551	-147.071	-41.095	0.491
95.000	-105.135	-24.836	14.399	-156.454	-41.273	0.571
100.000	-111.895	-30.103	16.587	-164.628	-43.847	0.657
105.000	-118.327	-34.925	19.121	-168.090	-44.941	0.671
110.000	-123.515	-37.806	21.628	-164.551	-37.902	0.541
115.000	-127.268	-38.383	23.618	-161.208	-27.370	0.400
120.000	-129.967	-38.651	25.185	-168.471	-37.181	0.347
125.000	-132.009	-39.605	26.791	-160.666	-43.716	0.382
130.000	-132.638	-39.727	28.817	-127.850	-34.393	0.445
Tristan–South America:						
0.000	71.341	-69.942	0.000	71.341	-69.942	0.302
5.000	93.862	-67.855	0.950	130.697	-53.154	0.101
10.000	124.480	-61.273	1.215	176.194	-24.601	0.081
15.000	147.211	-63.415	1.771	-156.001	-71.747	0.196
20.000	165.565	-71.285	2.945	-132.096	-75.865	0.274
25.000	-178.590	-73.855	4.232	-143.303	-73.115	0.262
30.000	-173.550	-73.637	5.544	-167.058	-71.753	0.260
35.000	-170.118	-73.827	6.730	-117.052	-74.986	0.217
40.000	-157.317	-75.237	7.801	-74.825	-62.353	0.275
45.000	-136.473	-75.793	9.230	-73.028	-61.249	0.362
50.000	-122.288	-75.117	10.903	-80.138	-66.092	0.337
55.000	-114.226	-74.616	12.403	-59.507	-64.211	0.297
60.000	-102.892	-74.232	13.706	-40.491	-54.958	0.290
65.000	-95.665	-73.368	14.781	-55.029	-54.278	0.222
70.000	-91.474	-73.034	16.472	-72.392	-72.422	0.469
75.000	-87.391	-72.819	19.448	-58.382	-69.612	0.719
80.000	-86.962	-71.925	22.336	-132.625	-46.227	0.430
85.000	-100.664	-71.904	23.704	-177.101	-36.810	0.468
90.000	-114.813	-73.162	25.265	167.143	-36.344	0.488
95.000	-129.745	-73.548	26.601	164.470	-26.859	0.454
100.000	-143.550	-73.130	27.809	157.805	-27.245	0.439
105.000	-155.338	-72.461	29.144	155.180	-28.472	0.438
110.000	-164.948	-70.894	30.472	165.922	-17.064	0.449
115.000	-171.400	-68.144	31.648	174.179	-12.999	0.416
120.000	-174.239	-66.632	32.921	158.317	-57.160	0.230
125.000	-175.047	-67.329	33.248	30.826	-25.245	0.095
130.000	-173.794	-66.760	34.217	-143.820	-36.180	0.416
Tristan–Antarctica:						
0.000	61.858	-14.671	0.000	61.858	-14.671	0.102
5.000	88.303	31.820	0.527	111.619	54.191	0.161
10.000	105.678	55.871	1.424	147.181	77.462	0.215
15.000	107.976	73.705	2.266	-60.705	60.711	0.202
20.000	24.860	85.026	2.921	-40.451	41.709	0.181
25.000	-7.024	77.593	3.473	0.157	37.488	0.118
30.000	2.360	73.142	3.810	38.354	28.854	0.067
35.000	4.087	73.415	3.932	-140.158	27.112	0.054
40.000	1.627	75.528	4.157	51.303	54.335	0.111

Age (Ma)	Total rotation:			Instantaneous rotation:		
	Longitude	Latitude	Angle	Longitude	Latitude	Rate
45.000	17.422	71.786	4.887	34.181	51.167	0.182
50.000	7.025	72.922	5.570	-103.138	36.143	0.217
55.000	-33.090	74.477	6.132	-102.224	27.756	0.239
60.000	-44.551	71.615	6.750	-19.911	42.625	0.130
65.000	-32.145	69.443	7.237	20.808	18.631	0.173
70.000	-20.730	66.584	7.599	13.877	10.754	0.116
75.000	-15.649	64.475	7.866	27.277	25.422	0.082
80.000	-18.411	66.841	8.059	-158.928	23.294	0.258
85.000	-46.554	76.914	8.493	174.116	32.550	0.479
90.000	-132.700	82.100	9.790	160.668	30.291	0.571
95.000	-175.462	72.635	11.473	157.663	23.220	0.583
100.000	173.900	63.300	13.400	153.363	20.165	0.579
105.000	168.380	55.950	15.510	149.400	16.848	0.582
110.000	164.700	49.400	17.760	147.549	9.998	0.630
115.000	162.455	43.626	20.476	152.568	20.062	0.728
120.000	162.432	42.189	24.786	169.436	46.748	1.075
125.000	163.461	43.861	29.904	170.389	50.739	0.807
130.000	162.300	42.400	30.420	49.933	-60.521	0.653
Tristan–India:						
0.000	21.976	16.759	0.000	21.976	16.759	0.550
5.000	25.211	26.289	2.839	27.117	34.682	0.593
10.000	23.458	34.300	5.745	9.389	48.530	0.583
15.000	13.228	39.869	8.351	-27.672	44.878	0.583
20.000	3.984	40.457	10.929	-8.873	33.057	0.549
25.000	6.756	36.995	13.750	27.113	13.946	0.691
30.000	14.168	32.545	17.025	36.005	10.732	0.805
35.000	17.952	30.843	20.450	22.260	32.774	0.647
40.000	16.299	31.167	23.682	-5.015	26.775	0.725
45.000	12.625	28.962	27.397	-12.669	15.029	0.886
50.000	9.230	26.807	31.425	-13.159	19.310	0.895
55.000	6.671	25.184	36.029	-9.882	16.017	1.112
60.000	4.953	22.659	42.242	-7.449	9.665	1.464
65.000	3.909	20.114	49.345	-6.950	9.250	1.421
70.000	3.127	18.853	55.096	-7.601	17.850	0.913
75.000	2.238	18.638	59.346	-9.553	20.462	0.924
80.000	1.838	17.869	63.708	-4.635	-2.577	0.783
85.000	2.836	16.850	66.214	35.361	-20.246	0.448
90.000	5.000	16.300	67.290	46.466	-44.251	0.694
95.000	8.020	14.367	68.787	39.244	-44.436	0.832
100.000	10.300	14.500	70.310	90.234	33.115	0.769
105.000	10.703	19.795	71.096	119.677	55.194	1.451
110.000	10.100	23.700	71.680	-83.753	25.782	0.150
115.000	9.758	19.267	71.479	-53.745	-50.803	1.676
120.000	10.705	12.993	72.552	-9.816	-53.830	1.083
125.000	12.974	11.643	74.563	69.103	-14.079	0.727
130.000	15.500	13.300	75.340	107.215	5.249	0.789

Age (Ma)	Total rotation:			Instantaneous rotation:		
	Longitude	Latitude	Angle	Longitude	Latitude	Rate
Tristan–Central Indian Ocean:						
0.000	41.601	9.566	0.000	41.601	9.566	0.729
5.000	42.927	16.170	3.686	43.580	22.317	0.748
10.000	41.773	21.881	7.321	35.411	32.732	0.708
15.000	36.708	25.920	10.516	14.485	35.113	0.640
20.000	31.064	27.066	13.480	14.247	25.526	0.633
25.000	29.363	25.352	16.722	27.516	13.868	0.695
30.000	29.954	23.154	20.214	32.012	13.700	0.713
35.000	29.786	23.003	23.550	21.139	33.829	0.639
40.000	27.025	24.030	26.645	-3.735	26.338	0.730
45.000	22.357	22.793	30.099	-13.132	15.219	0.883
50.000	17.970	21.423	33.849	-13.192	19.669	0.894
55.000	14.427	20.367	38.238	-9.772	16.094	1.113
60.000	11.519	18.315	44.308	-7.553	9.447	1.466
65.000	9.399	16.089	51.317	-7.100	8.959	1.422
70.000	7.939	14.999	56.985	-7.819	17.796	0.914
75.000	6.613	14.847	61.160	-10.379	19.859	0.937
80.000	5.922	14.222	65.431	0.454	-2.952	0.741
85.000	6.216	13.326	68.394	-0.809	-0.640	0.569
90.000	7.942	11.779	71.184	28.028	-42.875	1.096
Tristan–Australia:						
0.000	41.338	9.603	0.000	41.338	9.603	0.731
5.000	43.545	16.213	3.470	45.155	23.098	0.670
10.000	42.835	22.458	6.722	36.108	33.973	0.661
15.000	37.047	26.532	10.012	16.078	33.183	0.722
20.000	31.112	27.026	13.524	17.769	23.155	0.729
25.000	30.016	24.763	17.031	30.411	11.143	0.710
30.000	30.778	22.375	20.421	30.832	12.759	0.672
35.000	29.318	22.525	23.623	5.840	38.030	0.677
40.000	26.688	24.309	26.260	21.365	36.256	0.408
45.000	27.797	24.107	27.521	64.505	-12.211	0.206
50.000	27.972	23.588	27.778	-67.814	7.754	0.137
55.000	25.654	23.542	27.714	-65.407	8.330	0.212
60.000	24.263	22.822	28.041	3.272	-17.375	0.167
65.000	24.598	21.384	28.774	33.336	-25.001	0.237
70.000	25.400	19.903	29.476	36.246	-32.315	0.183
75.000	26.094	18.800	29.997	47.053	-26.469	0.161
80.000	26.855	18.287	30.058	178.559	-20.453	0.148
85.000	28.180	19.373	29.126	156.092	14.666	0.354
90.000	31.437	22.128	28.427	145.582	16.266	0.483
95.000	35.902	25.022	27.850	143.711	11.460	0.514
100.000	41.096	27.694	27.549	139.550	11.648	0.516
105.000	46.655	30.071	27.662	136.360	11.495	0.525
110.000	52.799	32.041	28.031	136.950	7.338	0.575
115.000	59.664	34.183	28.636	145.066	20.903	0.665
120.000	65.965	39.750	30.038	166.271	49.845	1.028
125.000	70.942	47.184	32.373	168.506	53.967	0.782
130.000	72.979	46.584	32.786	47.598	-59.433	0.655

Age (Ma)	Total rotation:			Instantaneous rotation:		
	Longitude	Latitude	Angle	Longitude	Latitude	Rate
Hawaiian–Nazca:						
0.000	-9.185	-68.018	0.000	-9.185	-68.018	0.271
5.000	-113.118	60.264	4.226	90.779	-14.817	0.626
10.000	-108.542	51.086	4.692	-95.267	26.434	1.341
15.000	-101.023	46.770	10.876	-99.918	66.476	1.098
20.000	-101.035	54.724	15.668	-83.256	55.329	0.966
25.000	-96.585	51.275	20.360	-117.457	46.038	0.832
30.000	-111.475	54.879	23.634	145.976	48.452	1.155
35.000	-127.330	59.727	27.185	150.988	54.857	1.173
40.000	-140.561	59.597	31.583	168.361	40.005	0.886
45.000	-149.780	59.553	35.322	128.551	56.029	1.015
50.000	-158.221	59.701	37.942	125.195	41.646	0.627
55.000	-164.122	58.643	39.574	139.023	15.351	0.501
60.000	-167.891	55.437	40.736	150.113	-3.284	0.653
65.000	-171.847	51.895	42.270	134.843	17.709	0.686
70.000	-176.459	52.560	45.362	46.909	82.327	1.236
75.000	175.220	59.368	51.408	33.883	69.914	2.185
West Antarctica–Hawaiian:						
0.000	-107.946	-7.392	0.000	-107.946	-7.392	0.104
5.000	-70.121	-24.255	0.907	-31.492	-14.290	0.222
10.000	-49.161	-38.147	1.005	120.492	-26.720	0.181
15.000	-37.258	-71.352	1.080	-91.046	-52.141	0.050
20.000	-58.102	-38.202	1.192	-67.351	44.343	0.227
25.000	-58.838	-4.159	1.277	40.392	-28.215	0.058
30.000	-38.459	-38.971	1.865	1.402	-68.804	0.344
35.000	-39.380	-55.648	3.546	-85.894	-65.362	0.389
40.000	-55.875	-58.370	5.014	-83.609	-34.803	0.172
45.000	-65.669	-58.055	7.084	-85.312	-55.669	0.651
50.000	-68.175	-55.787	9.302	-70.262	-49.948	0.405
55.000	-68.260	-56.542	11.252	-70.972	-72.341	0.404
60.000	-69.050	-60.752	13.410	-133.502	-79.812	0.566
65.000	-74.469	-65.379	16.541	-128.404	-71.336	0.778
70.000	-82.240	-69.621	20.256	161.995	-68.156	0.824
75.000	-87.995	-75.916	22.316	105.966	-47.323	0.345
80.000	-80.030	-67.115	24.383	-63.256	-12.802	1.669
85.000	-79.836	-65.113	29.449	145.350	-76.000	1.175
90.000	-78.329	-64.872	32.024	-59.369	50.045	1.300
95.000	-80.190	-44.904	27.014	-49.048	57.209	1.754
100.000	-76.948	-39.226	27.173	-47.044	13.284	0.175
105.000	-80.195	-34.885	26.770	-121.350	59.165	0.592
110.000	-83.185	-30.339	26.578	-109.669	57.873	0.450
115.000	-85.928	-25.713	26.599	-113.321	58.450	0.487
120.000	-88.423	-21.094	26.832	-112.427	58.344	0.478
125.000	-90.738	-16.523	27.270	-112.519	58.398	0.481
130.000	-92.830	-12.130	27.905	-112.462	58.623	0.483
Tristan–West Antarctica:						
0.000	61.858	-14.671	0.000	61.858	-14.621	0.102
5.000	88.303	31.820	0.527	111.789	54.212	0.161
10.000	105.678	55.871	1.424	146.810	11.101	0.214

Age (Ma)	Total rotation:			Instantaneous rotation:		
	Longitude	Latitude	Angle	Longitude	Latitude	Rate
15.000	107.976	73.705	2.266	61.231	61.380	0.203
20.000	24.860	85.026	2.921	−35.455	38.418	0.194
25.000	−7.024	77.593	3.473	0.851	38.205	0.118
30.000	0.153	71.292	3.808	−0.333	0.148	0.139
35.000	−6.511	60.482	3.999	−40.304	−9.851	0.137
40.000	−11.039	55.347	4.318	11.181	30.292	0.131
45.000	−2.050	54.173	5.088	34.126	51.149	0.182
50.000	−6.790	56.668	5.751	−103.135	36.142	0.217
55.000	−27.127	58.929	6.273	−102.224	27.756	0.239
60.000	−34.583	57.980	6.914	−19.911	42.625	0.130
65.000	−28.070	56.439	7.486	20.808	18.631	0.173
70.000	−21.113	54.261	7.937	13.877	10.754	0.116
75.000	−17.683	52.730	8.258	27.277	25.422	0.082
80.000	−19.648	55.187	8.380	−158.928	23.294	0.258
85.000	−35.492	65.905	8.468	174.116	32.550	0.479
90.000	−70.329	79.426	9.308	160.668	30.291	0.571
95.000	−155.101	78.445	10.616	157.663	23.220	0.583
100.000	179.823	68.484	12.277	153.363	20.165	0.579
105.000	170.911	59.977	14.210	149.400	16.848	0.582
110.000	165.894	52.420	16.331	147.549	9.998	0.630
115.000	163.057	45.805	18.955	152.568	20.062	0.728
120.000	162.970	43.863	23.244	169.436	46.748	1.075
125.000	164.115	45.293	28.382	170.389	50.739	0.807
130.000	162.820	43.748	28.878	49.951	−60.521	0.653
Tristan–Hawaiian:						
0.000	−39.250	−64.506	0.000	−39.250	−64.506	0.043
5.000	−48.342	−11.917	0.452	−30.928	26.951	0.086
10.000	29.240	58.574	0.657	108.924	42.857	0.238
15.000	78.477	70.853	1.221	−77.069	47.867	0.169
20.000	−44.893	65.421	2.392	−53.433	43.087	0.410
25.000	−41.293	61.176	3.774	13.344	18.239	0.145
30.000	−22.127	44.192	3.505	−2.717	−50.552	0.415
35.000	−24.679	8.536	3.848	−68.553	−52.932	0.466
40.000	−36.216	−8.196	4.755	−46.549	−8.895	0.172
45.000	−40.381	−17.360	6.031	−73.808	−47.574	0.510
50.000	−48.706	−20.636	7.839	−86.791	−20.932	0.471
55.000	−57.754	−23.575	9.759	−94.404	−37.828	0.437
60.000	−60.767	−30.152	11.397	−93.484	−74.729	0.478
65.000	−62.048	−39.853	13.458	−106.714	−75.512	0.684
70.000	−64.118	−49.728	16.120	153.287	−73.628	0.772
75.000	−61.246	−59.060	17.309	85.436	−43.901	0.339
80.000	−66.374	−49.704	19.996	−78.947	−5.744	1.680
85.000	−76.293	−50.779	24.216	164.677	−52.413	1.114
90.000	−83.127	−52.518	25.037	−102.263	64.111	1.423
95.000	−91.465	−24.959	21.581	−89.415	68.317	1.797
100.000	−94.876	−18.319	21.251	166.530	29.217	0.468
105.000	−105.044	−12.172	20.940	179.558	42.255	0.953
110.000	−115.043	−6.619	21.432	173.802	33.206	0.811
115.000	−124.971	−1.668	22.835	177.789	36.875	0.996

Age (Ma)	Total rotation:			Instantaneous rotation:		
	Longitude	Latitude	Angle	Longitude	Latitude	Rate
120.000	-134.291	6.184	25.955	-170.351	49.568	1.449
125.000	-141.097	14.533	30.830	-163.852	50.672	1.200
130.000	-142.540	15.812	32.886	-115.362	-77.601	0.276
Eurasia–Hawaiian:						
0.000	-109.825	19.376	0.000	-109.825	19.376	0.067
5.000	-56.890	-26.789	0.621	-23.290	-39.745	0.141
10.000	18.727	-45.725	0.803	95.211	-15.645	0.243
15.000	70.949	-45.758	1.485	86.303	-54.756	0.076
20.000	41.202	-33.449	1.284	-28.086	48.159	0.277
25.000	25.957	11.215	1.748	31.297	22.236	0.222
30.000	24.118	-3.056	3.060	5.791	-34.413	0.395
35.000	12.871	-22.330	4.414	-67.237	-62.361	0.380
40.000	6.148	-36.226	4.830	88.141	-18.691	0.054
45.000	4.441	-43.072	5.691	-54.425	-56.977	0.428
50.000	-12.365	-41.173	7.237	-61.240	-18.745	0.494
55.000	-26.873	-45.000	8.396	-134.519	-52.023	0.387
60.000	-31.227	-56.702	9.613	159.517	-73.347	0.546
65.000	-31.179	-66.525	11.922	-126.730	-84.444	0.579
70.000	-30.239	-70.755	14.296	114.876	-62.746	0.519
75.000	3.136	-75.852	14.793	95.529	22.439	0.532
80.000	-28.193	-66.364	13.989	-67.825	1.403	1.455
85.000	-39.779	-65.821	17.091	112.716	-57.808	0.894
90.000	-37.690	-67.217	16.470	-116.959	76.158	1.663
95.000	-68.544	-18.191	8.122	-105.227	79.275	2.142
100.000	-67.428	17.122	7.849	102.170	66.142	0.721
105.000	-82.664	50.908	9.057	145.960	61.154	1.145
110.000	-109.969	68.415	11.850	134.661	50.488	0.841
115.000	-151.051	71.271	14.582	143.806	40.134	0.918
120.000	-175.997	65.268	19.451	157.520	46.578	1.471
125.000	175.823	60.614	26.541	155.668	51.077	1.287
130.000	171.180	60.455	29.062	30.923	-0.739	0.311
North America–Hawaiian:						
0.000	175.046	66.262	0.000	175.046	66.262	0.227
5.000	-120.906	81.554	0.769	51.812	72.682	0.147
10.000	100.764	61.032	1.939	104.704	31.597	0.368
15.000	107.873	49.829	3.270	120.807	61.529	0.217
20.000	104.191	63.797	4.496	-33.506	76.641	0.430
25.000	93.551	69.720	6.410	57.249	52.848	0.314
30.000	75.847	63.636	6.993	15.774	-14.767	0.288
35.000	58.329	58.685	6.486	-68.456	-49.067	0.214
40.000	56.084	61.122	6.162	94.764	72.539	0.232
45.000	58.800	61.681	6.445	-119.009	-42.769	0.124
50.000	58.042	69.861	6.660	-129.033	23.143	0.293
55.000	78.682	81.282	6.562	-157.834	-11.277	0.248
60.000	123.454	76.195	5.815	135.185	-53.328	0.364
65.000	129.591	59.463	4.428	153.386	-77.406	0.392
70.000	130.285	40.187	3.428	104.900	-41.892	0.380
75.000	114.387	25.719	5.581	89.016	45.062	0.622
80.000	137.958	66.265	5.206	-88.011	3.286	1.465

Age (Ma)	Total rotation:			Instantaneous rotation:		
	Longitude	Latitude	Angle	Longitude	Latitude	Rate
85.000	-167.121	58.494	3.100	91.532	-47.967	0.815
90.000	173.710	59.346	3.931	-135.089	68.044	1.708
95.000	-147.651	68.185	17.157	-128.869	71.088	2.204
100.000	-148.424	72.435	22.167	108.282	74.226	0.807
105.000	-164.099	72.535	27.671	145.701	64.558	1.261
110.000	-175.833	71.922	32.735	128.068	60.288	0.947
115.000	174.390	70.133	36.851	131.901	48.961	0.984
120.000	168.416	66.964	42.346	145.411	48.720	1.503
125.000	165.583	64.011	49.452	143.948	51.317	1.279
130.000	162.980	63.647	51.872	11.170	-3.493	0.315
Australia–Hawaiian:						
0.000	-137.192	-12.648	0.000	-137.192	-12.648	0.736
5.000	-128.972	-17.684	3.539	-127.920	-20.725	0.642
10.000	-136.158	-18.895	6.213	-162.262	-19.024	0.565
15.000	-144.181	-20.845	9.276	-152.290	-22.017	0.675
20.000	-143.170	-19.056	12.493	-130.179	-3.798	0.641
25.000	-141.815	-15.297	15.460	-141.110	-11.014	0.574
30.000	-141.680	-18.099	18.276	-143.365	-49.629	0.657
35.000	-141.970	-24.842	21.558	-166.441	-54.287	0.951
40.000	-143.323	-29.582	24.978	-140.046	-49.141	0.409
45.000	-140.547	-31.688	26.945	-88.645	-55.574	0.612
50.000	-136.936	-32.792	28.412	-96.077	-52.097	0.366
55.000	-136.088	-34.257	29.928	-160.464	-62.645	0.354
60.000	-136.762	-36.342	31.528	-179.360	-60.175	0.464
65.000	-137.696	-38.911	33.969	-165.042	-61.452	0.669
70.000	-139.405	-41.721	36.949	156.986	-45.868	0.718
75.000	-143.482	-43.744	38.165	127.449	-17.662	0.227
80.000	-135.497	-43.157	40.299	-65.935	-32.176	1.711
85.000	-132.559	-45.103	44.686	149.492	-49.421	1.031
90.000	-131.267	-45.787	46.740	-59.811	25.712	1.459
95.000	-121.640	-32.369	45.279	-52.554	31.139	1.885
100.000	-118.130	-29.435	45.665	-75.357	8.224	0.264
105.000	-118.332	-25.753	46.761	-97.058	40.359	0.722
110.000	-118.556	-22.397	48.015	-95.165	35.973	0.551
115.000	-118.775	-19.369	49.371	-96.204	36.990	0.564
120.000	-118.941	-16.632	50.771	-94.964	36.621	0.529
125.000	-119.030	-14.111	52.174	-92.302	36.993	0.508
130.000	-118.989	-11.782	53.514	-87.871	37.813	0.489
Hawaiian–Cocos:						
0.000	-115.892	18.228	0.000	-115.892	18.228	1.225
5.000	-117.747	22.124	8.179	-115.255	19.439	1.767
10.000	-111.737	15.300	15.393	-100.886	-7.963	1.289
15.000	-110.031	8.288	20.490	-123.840	-11.929	0.910
20.000	-114.480	5.350	22.263	121.070	-2.332	0.586
25.000	-124.785	7.463	15.600	79.649	7.434	2.490
30.000	165.643	33.655	8.459	81.119	20.852	3.053
35.000	121.746	41.015	16.625	146.574	70.529	1.543
40.000	137.154	53.766	21.754	-153.285	42.525	1.289
45.000	143.845	54.854	27.603	132.248	50.876	1.394

Age (Ma)	Total rotation:			Instantaneous rotation:		
	Longitude	Latitude	Angle	Longitude	Latitude	Rate
50.000	139.796	52.510	33.671	112.493	43.573	1.248
55.000	135.032	50.872	39.179	107.408	43.899	1.071
60.000	132.797	49.822	43.675	120.023	43.050	0.818
65.000	132.570	48.884	47.425	130.062	37.016	0.727
70.000	133.491	49.195	50.695	-173.305	58.660	0.682
75.000	134.794	50.717	53.309	141.128	40.511	0.501
80.000	133.916	45.308	55.943	98.412	-17.317	1.909
Hawaiian–Kula:						
0.000	90.000	-60.200	0.000	90.000	-60.200	0.932
5.000	104.900	-56.500	4.700	-113.404	-58.505	0.904
10.000	108.538	-64.395	8.642	-107.498	-85.935	0.781
15.000	110.206	-70.186	12.393	115.378	-76.856	0.788
20.000	111.625	-70.416	16.466	118.856	-69.969	0.836
25.000	112.440	-71.200	20.490	122.581	-74.586	0.741
30.000	115.482	-70.385	23.674	137.764	-55.953	0.583
35.000	117.984	-68.602	26.355	131.831	-56.278	0.566
40.000	119.740	-67.880	29.441	146.652	-68.637	0.738
45.000	118.500	-66.072	31.666	104.671	-1.454	0.512
50.000	114.613	-62.903	33.261	106.372	-7.980	0.497
55.000	111.895	-58.498	34.849	119.688	2.123	0.837
60.000	110.593	-52.681	37.222	116.608	-3.345	0.738
65.000	109.412	-47.832	40.159	122.599	0.443	1.219
70.000	110.309	-42.458	43.273	148.645	7.558	0.624
75.000	111.417	-38.993	45.017	122.340	0.749	0.812
80.000	108.648	-35.784	52.151	100.949	-17.016	2.295
North America–Pacific:						
0.000	100.936	-51.400	0.000	100.936	-51.400	0.769
5.000	106.412	-51.411	4.038	108.088	-51.757	0.802
10.000	105.520	-52.633	7.678	103.226	-57.835	0.692
15.000	106.644	-55.501	11.122	109.987	-64.289	0.641
20.000	106.126	-57.016	13.705	87.313	-59.542	0.424
25.000	103.054	-57.230	15.977	82.552	-54.019	0.577
30.000	99.423	-57.576	19.173	76.031	-60.593	0.655
35.000	98.895	-59.724	22.296	136.110	-76.627	0.667
40.000	103.380	-61.188	25.285	130.613	-59.282	0.558
45.000	103.751	-59.173	27.476	105.981	-14.732	0.457
50.000	103.861	-55.788	28.602	132.851	4.952	0.375
55.000	106.145	-52.563	29.702	129.852	-12.769	0.493
60.000	106.699	-49.650	32.460	111.036	-29.097	0.789
65.000	106.242	-47.837	36.540	107.309	-37.545	0.798
70.000	106.203	-47.121	39.338	114.657	-26.734	0.429
75.000	104.224	-43.565	41.264	96.808	24.833	0.931
80.000	101.639	-40.249	43.546	101.281	-45.692	0.691
85.000	102.549	-40.922	47.398	109.714	-25.446	0.731
90.000	101.557	-37.988	49.785	110.361	36.390	0.723
95.000	99.769	-33.670	49.654	110.151	61.572	0.801
100.000	97.680	-29.270	49.580	99.568	61.646	0.823
105.000	95.245	-24.696	49.508	90.201	63.414	0.843
110.000	92.825	-20.660	50.087	83.115	50.138	0.709

Age (Ma)	Total rotation:			Instantaneous rotation:		
	Longitude	Latitude	Angle	Longitude	Latitude	Rate
115.000	90.986	-17.643	51.675	94.379	40.313	0.706
120.000	90.122	-13.211	53.591	124.622	49.815	1.138
125.000	89.598	-7.448	55.607	115.210	54.556	0.931
130.000	87.643	-7.017	56.797	4.990	-18.654	0.773
North America–Cocos:						
0.000	-119.971	26.230	0.000	-119.971	26.230	1.337
5.000	-118.146	26.535	8.595	-115.678	24.570	1.797
10.000	-114.784	21.954	15.197	-111.273	-0.190	1.009
15.000	-115.442	15.495	19.435	-132.037	-2.317	0.842
20.000	-120.104	15.665	21.533	111.370	37.547	0.632
25.000	-133.928	29.305	15.958	72.779	12.750	2.714
30.000	139.726	53.428	13.389	71.414	18.494	3.168
35.000	108.173	47.498	21.842	150.051	69.056	1.349
40.000	123.115	57.823	26.687	-165.052	47.980	1.433
45.000	131.305	58.240	32.674	132.901	46.488	1.311
50.000	131.140	56.316	39.054	123.555	49.287	1.248
55.000	130.429	54.966	44.711	119.832	41.730	1.034
60.000	130.461	52.609	48.939	120.848	18.419	0.848
65.000	131.800	49.682	51.782	131.140	4.791	0.661
70.000	133.651	48.595	54.080	152.963	32.160	0.593
75.000	133.845	47.963	58.228	114.123	46.197	1.062
80.000	132.838	46.999	60.827	94.496	-39.107	0.306
North America–Kula:						
0.000	100.936	-51.400	0.000	100.936	-51.400	0.769
5.000	106.412	-51.411	4.038	98.811	-50.202	0.769
10.000	105.520	-52.633	7.678	111.017	-58.924	0.726
15.000	106.644	-55.501	11.122	107.083	-63.815	0.630
20.000	106.126	-57.016	13.705	92.909	-59.873	0.425
25.000	103.054	-57.230	15.977	82.861	-54.064	0.577
30.000	99.423	-57.576	19.173	76.057	-60.598	0.655
35.000	98.895	-59.724	22.296	136.114	-76.626	0.667
40.000	103.380	-61.188	25.285	130.614	-59.282	0.558
45.000	103.751	-59.173	27.476	105.979	-14.735	0.457
50.000	103.861	-55.788	28.602	132.854	4.931	0.374
55.000	106.519	-51.114	30.030	129.338	-1.728	0.876
60.000	107.810	-45.130	33.842	116.320	-19.463	0.996
65.000	108.034	-41.604	38.967	120.970	-15.828	1.342
70.000	109.658	-37.899	43.722	134.003	-11.901	0.870
75.000	108.922	-33.067	47.650	109.172	20.962	1.286
80.000	106.975	-30.101	51.078	102.856	-24.351	0.681
Eurasia–North Africa:						
0.000	-20.226	22.992	0.000	-20.226	22.992	0.122
5.000	-21.709	17.964	0.580	-22.547	12.843	0.111
10.000	-21.076	13.757	1.105	-15.974	6.536	0.102
15.000	-17.332	12.064	1.612	-5.176	12.311	0.112
20.000	-14.634	14.697	2.298	-13.176	27.251	0.182
25.000	-16.038	20.518	3.444	-22.069	34.844	0.287
30.000	-18.285	25.693	4.967	-23.813	39.301	0.327
35.000	-19.122	29.443	6.389	-16.932	46.105	0.247

Age (Ma)	Total rotation:			Instantaneous rotation:		
	Longitude	Latitude	Angle	Longitude	Latitude	Rate
40.000	−18.432	31.927	7.377	−7.960	43.829	0.182
45.000	−17.683	31.993	8.366	−17.083	20.282	0.253
50.000	−18.052	28.725	10.015	−22.205	10.863	0.350
55.000	−17.917	28.303	10.629	141.905	12.342	0.085
60.000	−16.896	29.510	10.204	134.789	−8.104	0.080
65.000	−16.147	30.260	10.337	−14.432	41.394	0.171
70.000	−16.706	31.548	12.013	−16.218	40.484	0.396
75.000	−15.757	33.128	13.750	−5.760	43.507	0.429
80.000	−14.973	34.942	16.392	−7.812	43.927	0.581
85.000	−13.749	36.728	19.153	3.031	47.223	0.551
90.000	−11.954	38.652	21.878	6.466	48.688	0.616
95.000	−10.472	40.311	25.211	1.694	47.469	0.751
100.000	−9.410	41.482	29.104	−0.008	46.567	0.815
105.000	−8.571	42.311	33.114	0.572	46.260	0.787
110.000	−7.811	42.946	36.765	3.030	46.472	0.677
115.000	−7.086	43.488	39.703	8.197	47.365	0.504
120.000	−6.381	44.003	41.602	22.690	49.399	0.288
125.000	−5.775	44.488	42.656	28.360	51.069	0.201
130.000	−5.428	44.874	43.670	14.508	54.674	0.226
Eurasia–Greenland:						
0.000	−44.200	−62.400	0.000	−44.200	−62.400	0.209
5.000	−45.918	−63.741	1.161	−47.140	−65.031	0.253
10.000	−46.405	−65.461	2.492	−45.753	−69.349	0.274
15.000	−45.118	−67.742	3.837	−38.178	−73.742	0.262
20.000	−43.119	−68.924	5.064	91.054	−64.667	2.209
25.000	−42.165	−67.755	6.095	−73.070	46.548	0.601
30.000	−42.409	−65.926	6.967	20.302	−79.329	0.308
35.000	−44.319	−64.992	7.685	−67.875	−36.085	0.119
40.000	−47.559	−63.428	8.248	−65.438	−32.649	0.143
45.000	−50.872	−58.912	8.916	−57.537	−1.251	0.289
50.000	−53.809	−48.836	10.052	−62.223	−19.017	0.433
55.000	−59.909	−55.751	12.688	177.146	−63.561	0.787
60.000	−65.962	−61.675	13.627	−49.680	−7.322	0.094
65.000	−64.003	−59.661	13.855	−26.328	15.932	0.090
70.000	−61.937	−56.943	13.613	−22.041	57.275	0.241
75.000	−59.470	−51.091	12.862	−22.170	61.260	0.364
80.000	−57.033	−43.686	12.228	−21.618	56.209	0.316
85.000	−55.229	−38.442	12.030	−16.828	51.926	0.142
90.000	−54.352	−36.769	12.012	−0.786	23.004	0.034
95.000	−53.596	−36.416	12.097	−4.932	−11.673	0.041
100.000	−52.706	−36.266	12.263	−5.641	−33.195	0.049
105.000	−51.934	−36.624	12.460	7.290	−67.395	0.055
110.000	−51.485	−37.738	12.643	110.300	−80.834	0.073
115.000	−51.270	−39.319	12.818	137.142	−75.944	0.082
120.000	−51.112	−40.847	12.993	133.272	−78.457	0.057
125.000	−51.064	−41.169	13.040	−45.010	75.233	0.016
130.000	−51.151	−40.521	12.953	−48.995	82.526	0.046

Age (Ma)	Total rotation:			Instantaneous rotation:		
	Longitude	Latitude	Angle	Longitude	Latitude	Rate
Greenland–Tristan:						
0.000	–169.418	66.511	0.000	169.418	65.511	0.269
5.000	146.066	63.758	0.953	116.961	50.334	0.131
10.000	121.836	51.691	1.444	99.933	12.217	0.144
15.000	113.788	37.103	2.224	109.593	17.120	0.224
20.000	113.012	32.624	3.431	111.060	39.936	0.251
25.000	111.508	35.562	4.612	102.020	54.854	0.239
30.000	109.623	41.025	5.720	99.770	66.426	0.249
35.000	109.061	43.309	6.844	114.927	42.492	0.238
40.000	110.725	39.356	8.141	114.669	13.915	0.342
45.000	111.145	33.974	9.832	103.272	–7.453	0.434
50.000	108.596	28.499	11.580	87.781	40.408	0.473
55.000	104.271	36.465	14.120	65.167	72.923	0.576
60.000	102.798	38.494	15.381	117.305	4.649	0.234
65.000	105.984	35.345	16.504	126.423	8.880	0.243
70.000	106.656	35.670	17.386	122.670	26.546	0.227
75.000	108.866	33.571	18.918	122.670	9.420	0.402
80.000	108.323	31.795	20.566	70.824	4.603	0.353
85.000	103.743	29.670	21.574	43.522	20.669	0.369
90.000	99.872	31.683	22.890	34.610	62.246	0.469
95.000	95.672	35.147	24.462	19.522	63.803	0.557
100.000	90.977	39.191	26.286	4.481	68.192	0.659
105.000	85.964	43.565	28.430	–5.142	68.892	0.704
110.000	80.878	46.918	30.609	–4.323	61.863	0.598
115.000	76.159	48.568	32.365	–4.217	54.162	0.463
120.000	72.417	49.709	33.811	–11.732	61.030	0.393
125.000	69.674	50.764	35.248	17.862	58.793	0.368
130.000	67.722	50.583	37.015	35.961	43.430	0.405
Greenland–Hawaiian:						
0.000	175.046	66.262	0.000	175.046	66.262	0.227
5.000	–120.906	81.554	0.769	51.812	76.682	0.147
10.000	100.764	61.032	1.939	104.704	31.598	0.368
15.000	107.873	49.829	3.270	120.804	61.521	0.217
20.000	104.191	63.797	4.496	–33.392	80.004	0.417
25.000	93.551	69.720	6.410	52.057	56.454	0.325
30.000	75.847	63.636	6.993	17.593	–22.049	0.273
35.000	59.626	57.413	6.386	–45.280	–55.873	0.195
40.000	63.456	54.501	5.675	130.543	–47.729	0.177
45.000	69.662	44.483	5.543	77.292	–57.016	0.308
50.000	68.714	35.201	5.219	–65.894	55.315	0.135
55.000	60.952	52.467	5.985	–72.584	55.800	0.321
60.000	55.332	49.774	5.405	114.277	–77.122	0.452
65.000	70.166	16.868	4.082	145.483	–81.949	0.659
70.000	75.038	–24.058	4.696	100.360	–52.365	0.765
75.000	86.849	–29.263	8.364	104.941	–32.800	0.214
80.000	81.085	–39.858	6.457	–80.840	–25.762	0.723
85.000	77.669	–65.150	8.729	74.525	–80.092	0.862
90.000	86.670	–61.161	8.906	–145.105	67.202	1.641
95.000	169.165	53.661	5.542	–128.976	69.901	2.166

Age (Ma)	Total rotation:			Instantaneous rotation:		
	Longitude	Latitude	Angle	Longitude	Latitude	Rate
100.000	170.066	68.123	10.094	125.152	66.014	0.747
105.000	164.092	65.261	15.597	156.282	58.144	1.102
110.000	159.743	63.161	20.610	140.347	52.019	0.887
115.000	156.298	59.969	24.830	145.611	42.345	0.973
120.000	156.628	56.564	30.631	157.768	43.166	1.515
125.000	158.231	54.283	38.016	154.855	47.071	1.083
130.000	156.624	54.067	40.564	83.549	51.511	0.300
Greenland–Eurasia:						
0.000	135.800	62.400	0.000	135.800	62.400	0.209
5.000	134.082	63.741	1.161	132.922	65.041	0.253
10.000	133.595	65.461	2.492	134.725	69.333	0.274
15.000	134.882	67.742	3.837	143.180	73.549	0.262
20.000	136.881	68.924	5.064	144.297	68.611	0.227
25.000	137.835	67.755	6.095	137.788	55.076	0.191
30.000	137.591	65.926	6.967	132.505	57.176	0.170
35.000	135.681	64.992	7.685	116.953	61.275	0.142
40.000	132.441	63.428	8.248	105.774	35.119	0.142
45.000	129.128	58.912	8.916	115.802	-12.976	0.352
50.000	126.191	48.836	10.052	108.352	40.645	0.443
55.000	120.091	55.751	12.688	-8.785	73.794	0.616
60.000	114.038	61.675	13.627	-19.479	57.284	0.110
65.000	115.997	59.661	13.855	137.362	-23.927	0.166
70.000	118.063	56.943	13.613	136.465	-63.668	0.218
75.000	120.530	51.091	12.862	133.476	-63.880	0.371
80.000	122.967	43.686	12.228	136.831	-60.827	0.311
85.000	124.771	38.442	12.030	143.997	-55.266	0.146
90.000	125.648	36.769	12.012	168.377	-32.298	0.036
95.000	126.404	36.416	12.097	170.067	8.226	0.040
100.000	127.294	36.266	12.263	170.740	25.271	0.049
105.000	128.066	36.624	12.460	-170.151	58.915	0.055
110.000	128.515	37.738	12.643	-112.616	74.113	0.072
115.000	128.730	39.319	12.818	-88.480	73.683	0.086
120.000	128.888	40.847	12.993	-99.548	74.709	0.053
125.000	128.936	41.169	13.040	95.758	-71.880	0.014
130.000	128.849	40.521	12.953	71.348	-77.529	0.046
Hawaiian–Farallon:						
0.000	-115.892	18.228	0.000	-115.892	18.228	1.225
5.000	179.660	14.719	1.370	90.688	-13.354	0.869
10.000	93.455	-15.580	5.548	74.125	-21.905	1.125
15.000	88.856	-20.668	8.983	116.189	-40.375	0.305
20.000	97.722	-21.138	8.819	-154.606	23.238	0.400
25.000	108.825	-8.239	8.352	177.329	59.342	0.754
30.000	118.858	21.231	10.550	160.902	64.931	1.391
35.000	126.450	43.290	16.097	179.913	70.381	1.550
40.000	135.639	53.314	21.838	-170.680	50.231	1.138
45.000	142.538	54.558	27.667	144.114	52.174	1.354
50.000	140.468	52.687	33.612	113.673	44.083	1.233
55.000	135.187	50.921	39.156	103.989	42.574	1.114
60.000	132.533	49.730	43.729	118.569	42.742	0.828

Age (Ma)	Total rotation:			Instantaneous rotation:		
	Longitude	Latitude	Angle	Longitude	Latitude	Rate
65.000	132.342	48.805	47.479	131.669	36.979	0.722
70.000	133.451	49.182	50.705	-173.212	57.660	0.700
75.000	134.716	50.546	54.051	134.531	41.143	0.819
80.000	134.208	44.933	59.592	107.764	-1.478	2.340
North America–Farallon:						
0.000	100.936	-51.4	0.000	100.936	-51.4	0.769
5.000	-176.669	38.718	1.775	95.582	-9.757	0.669
10.000	93.712	1.922	6.280	91.227	-26.265	0.668
15.000	91.317	-3.436	10.442	101.043	-12.123	0.433
20.000	97.041	4.840	10.235	119.211	42.57	0.381
25.000	103.107	24.524	11.482	144.411	66.56	0.978
30.000	106.901	38.493	15.895	161.416	70.291	1.335
35.000	111.442	49.753	21.279	173.473	67.255	1.414
40.000	121.915	57.325	26.790	173.319	54.126	1.294
45.000	130.197	57.893	32.759	146.781	47.879	1.279
50.000	131.746	56.503	38.991	124.946	48.979	1.241
55.000	130.577	55.013	44.690	115.477	41.271	1.063
60.000	130.207	52.525	48.989	118.95	17.76	0.873
65.000	131.584	49.612	51.836	136.297	5.144	0.611
70.000	133.613	48.584	54.090	146.083	34.04	0.662
75.000	133.807	47.834	58.976	121.583	40.05	0.871
80.000	133.086	46.535	64.464	105.369	32.249	0.758
Tristan–Nazca:						
0.000	-13.734	-67.895	0.000	-13.734	-67.895	0.314
5.000	-111.489	61.166	1.342	-120.434	66.725	0.741
10.000	-108.754	63.292	6.309	-104.072	62.042	1.121
15.000	-105.164	60.683	11.871	-97.166	46.435	1.204
20.000	-96.756	57.592	17.801	-73.541	52.947	1.352
25.000	-91.289	54.835	23.667	-108.226	54.347	0.814
30.000	-102.045	58.734	25.806	125.249	42.468	0.794
35.000	-111.642	62.778	27.523	164.186	50.751	0.752
40.000	-123.350	62.104	30.747	175.004	47.113	0.770
45.000	-130.758	62.409	33.242	142.374	58.582	0.527
50.000	-134.730	62.692	34.992	-171.460	46.557	0.331
55.000	-136.395	60.790	35.928	-174.154	-15.930	0.411
60.000	-139.841	56.410	35.600	166.901	-35.159	0.758
65.000	-144.544	50.557	34.742	155.391	-34.970	0.724
70.000	-148.712	48.631	34.747	78.811	49.471	0.570
75.000	-160.392	57.328	36.973	56.611	50.783	2.085
South America–Nazca:						
0.000	-57.138	-2.612	0.000	-57.138	-2.612	0.15
5.000	-102.597	64.495	2.278	-110.153	67.479	0.825
10.000	-100.511	64.546	7.434	-96.243	63.619	1.15
15.000	-98.548	63.281	13.402	-94.537	53.536	1.324
20.000	-92.481	61.836	20.292	-71.215	60.626	1.555
25.000	-88.438	60.315	27.088	-106.927	66.717	1.002
30.000	-98.692	65.282	30.297	111.550	52.978	0.962
35.000	-108.354	69.962	33.079	148.464	58.295	0.920

Age (Ma)	Total rotation:			Instantaneous rotation:		
	Longitude	Latitude	Angle	Longitude	Latitude	Rate
40.000	-124.515	70.164	36.988	153.610	52.918	0.995
45.000	-137.553	70.892	40.583	120.430	58.568	0.879
50.000	-146.442	71.440	43.693	146.897	61.175	0.611
55.000	-150.974	70.187	45.666	159.065	15.176	0.489
60.000	-156.581	66.671	46.011	147.753	-18.063	0.797
65.000	-162.466	62.157	45.370	135.086	-21.962	0.746
70.000	-168.590	60.807	46.643	79.365	60.977	1.012
75.000	175.000	65.681	53.188	54.626	61.133	2.721
Hawaiian–South America:						
0.000	80.363	-65.48	0.000	80.363	-65.48	0.269
5.000	114.557	-46.004	1.092	140.691	-41.445	0.18
10.000	153.548	-66.464	1.771	-95.43	-51.259	0.255
15.000	177.634	-74.533	2.837	135.512	-68.63	0.33
20.000	149.743	-69.469	5.306	139.906	-61.834	0.619
25.000	154.618	-69.457	7.885	-156.222	-56.369	0.363
30.000	167.553	-63.233	8.712	-176.141	10.998	0.350
35.000	164.430	-52.579	8.900	124.028	35.252	0.295
40.000	157.364	-49.600	9.103	167.880	-71.435	0.233
45.000	156.242	-49.119	9.595	105.318	22.408	0.181
50.000	148.391	-49.031	10.504	85.295	-19.625	0.336
55.000	136.786	-48.024	11.088	64.839	7.490	0.250
60.000	130.984	-46.261	10.579	-7.941	67.386	0.262
65.000	128.714	-39.042	9.064	3.528	84.337	0.502
70.000	126.764	-29.381	7.370	-66.231	43.091	0.444
75.000	142.106	-41.898	5.840	-82.913	-45.898	0.644
80.000	121.133	-42.820	9.079	94.018	-2.875	1.521
85.000	110.563	-24.424	9.806	-82.459	64.922	0.693
90.000	106.685	-26.222	10.620	60.041	-61.566	1.735
95.000	91.560	-51.120	21.726	63.647	-65.832	2.066
100.000	91.331	-54.328	25.448	7.987	-64.392	0.435
105.000	82.431	-58.615	28.368	4.169	-47.343	0.934
110.000	74.023	-61.703	31.021	-8.658	-46.108	0.632
115.000	65.425	-64.205	33.182	-12.828	-40.297	0.795
120.000	50.233	-66.252	37.309	-5.385	-39.650	1.542
125.000	33.476	-65.128	43.236	1.836	-33.821	1.285
130.000	29.125	-65.098	45.312	-141.991	3.173	0.286
Eurasia–Pacific:						
0.000	93.053	-62.807	0.000	93.053	-62.807	0.884
5.000	100.775	-63.806	4.684	103.6	-65.188	0.949
10.000	100.216	-65.975	9.182	97.966	-71.635	0.891
15.000	102.179	-68.646	13.688	107.587	-75.893	0.858
20.000	101.575	-70.285	17.282	71.716	-74.136	0.605
25.000	97.347	-70.867	20.275	72.194	-67.179	0.689
30.000	93.182	-70.858	23.996	67.814	-71.186	0.762
35.000	93.575	-72.261	27.850	167.040	-80.128	0.838
40.000	101.644	-73.233	31.794	147.633	-69.758	0.762
45.000	103.000	-72.246	34.870	94.927	-54.250	0.503
50.000	100.344	-72.198	36.610	42.010	-77.024	0.274
55.000	103.330	-71.440	38.316	143.946	-42.291	0.587

Age (Ma)	Total rotation:			Instantaneous rotation:		
	Longitude	Latitude	Angle	Longitude	Latitude	Rate
60.000	106.443	-68.991	41.566	125.796	-46.465	0.892
65.000	107.600	-66.826	46.007	122.449	-52.778	0.903
70.000	108.371	-65.978	49.331	126.490	-49.901	0.523
75.000	106.465	-63.062	51.320	107.834	8.121	0.885
80.000	103.977	-60.147	53.692	119.384	-59.425	0.837
85.000	106.047	-59.919	58.088	128.955	-36.373	0.772
90.000	105.420	-57.028	59.923	122.272	29.570	0.694
95.000	103.262	-53.739	58.956	113.517	54.987	0.744
100.000	100.802	-50.430	58.110	104.999	51.887	0.747
105.000	98.100	-47.003	57.352	99.724	50.239	0.748
110.000	95.793	-43.772	57.454	100.042	31.535	0.638
115.000	94.448	-40.963	58.694	112.490	23.752	0.668
120.000	94.107	-36.746	59.403	137.791	44.209	1.108
125.000	93.834	-31.075	59.338	128.003	50.738	0.939
130.000	91.807	-30.080	60.408	23.595	-20.620	0.755
Australia–Pacific:						
0.000	-177.313	-61.467	0.000	-177.313	-61.467	1.104
5.000	-179.059	-59.540	5.720	-178.655	-58.494	1.143
10.000	-177.428	-59.079	11.216	-174.577	-58.254	1.119
15.000	-177.187	-58.121	17.225	-176.418	-54.228	1.215
20.000	-176.753	-56.441	22.771	-169.365	-48.899	0.998
25.000	-175.823	-54.701	27.589	-174.837	-48.112	1.006
30.000	-176.967	-53.962	32.890	176.866	-51.466	1.107
35.000	-178.734	-54.064	39.126	169.007	-51.177	1.427
40.000	179.267	-54.175	45.756	170.984	-53.965	1.062
45.000	178.795	-54.457	49.201	159.828	-54.980	0.400
50.000	177.442	-54.487	50.751	145.968	-26.252	0.371
55.000	174.560	-54.046	52.455	141.083	-19.373	0.547
60.000	170.871	-53.497	54.819	142.348	-23.635	0.713
65.000	167.408	-52.763	58.298	150.245	-30.724	0.871
70.000	165.585	-52.040	62.145	159.559	-35.642	0.707
75.000	163.073	-51.217	64.359	132.067	6.728	0.612
80.000	160.365	-50.329	66.874	159.580	-39.918	0.906
85.000	160.133	-49.711	71.689	161.845	-33.434	0.809
90.000	159.410	-48.845	73.976	-174.248	3.103	0.211
95.000	160.009	-48.288	73.484	-84.750	33.809	0.228
100.000	160.536	-47.737	73.122	-116.430	29.208	0.157
105.000	160.979	-47.197	72.905	-120.817	26.413	0.144
110.000	161.336	-46.689	72.816	-128.978	22.143	0.126
115.000	161.610	-46.236	72.827	-136.626	17.560	0.107
120.000	161.804	-45.863	72.903	-143.242	13.152	0.083
125.000	161.923	-45.597	72.999	-148.561	9.511	0.054
130.000	161.974	-45.468	73.063	-142.161	21.152	0.019
Australia–Eurasia:						
0.000	-139.716	-15.482	0.000	-139.716	-15.482	0.687
5.000	-138.192	-13.500	3.340	-136.988	-11.844	0.652
10.000	-137.945	-12.569	6.549	-139.810	-12.452	0.638
15.000	-139.966	-13.488	9.793	-148.290	-17.025	0.677
20.000	-142.314	-14.723	13.310	-148.052	-16.200	0.745

Age (Ma)	Total rotation:			Instantaneous rotation:		
	Longitude	Latitude	Angle	Longitude	Latitude	Rate
25.000	-143.009	-14.751	17.170	-143.691	-13.641	0.792
30.000	-143.290	-14.864	21.081	-147.543	-18.319	0.756
35.000	-144.759	-16.896	24.461	-170.178	-35.959	0.671
40.000	-145.697	-19.900	26.875	-131.179	-46.192	0.432
45.000	-143.012	-20.342	28.354	-96.357	-13.788	0.332
50.000	-143.135	-19.458	29.318	168.660	15.958	0.246
55.000	-144.201	-19.490	29.348	20.504	-11.408	0.085
60.000	-142.865	-19.265	29.257	-69.371	0.825	0.215
65.000	-141.001	-18.582	29.960	-111.744	-16.132	0.260
70.000	-140.776	-19.123	31.404	-143.064	-45.048	0.326
75.000	-139.337	-21.170	32.847	-118.578	-65.637	0.508
80.000	-138.068	-23.174	34.721	-113.476	-32.745	0.357
85.000	-136.393	-23.609	36.254	-95.316	-58.022	0.474
90.000	-133.870	-26.308	38.152	-87.902	-68.326	0.655
95.000	-131.309	-29.007	40.458	-100.728	-65.260	0.670
100.000	-129.376	-31.352	43.243	-119.517	-63.314	0.723
105.000	-128.051	-33.507	46.464	-121.489	-60.773	0.741
110.000	-126.427	-34.876	49.578	-94.692	-52.886	0.646
115.000	-123.961	-35.482	52.156	-71.467	-58.383	0.654
120.000	-121.176	-38.021	54.898	65.357	-88.362	1.084
125.000	-118.805	-42.172	58.407	-154.909	-83.191	0.935
130.000	-118.446	-41.565	60.668	-94.900	26.068	0.767
East Antarctica–Pacific:						
0.000	97.021	-64.152	0.000	97.021	-64.152	0.870
5.000	108.824	-62.594	4.724	115.134	-66.864	0.979
10.000	111.118	-71.829	9.441	39.550	-88.792	0.990
15.000	112.918	-75.504	14.287	121.276	-72.835	0.988
20.000	115.687	-73.199	19.230	125.646	-65.581	0.991
25.000	118.701	-72.600	23.840	141.563	-71.400	0.827
30.000	123.258	-71.477	27.285	149.801	-55.908	0.618
35.000	124.791	-69.808	30.062	125.987	-56.482	0.588
40.000	125.408	-69.171	33.384	161.204	-69.653	0.827
45.000	126.540	-67.723	36.094	122.021	-16.171	0.510
50.000	122.051	-64.939	38.307	104.062	-15.211	0.680
55.000	116.040	-61.843	40.455	103.512	-13.081	0.682
60.000	113.355	-59.033	42.665	120.291	-17.498	0.590
65.000	112.916	-56.341	44.921	127.729	-15.138	0.606
70.000	113.474	-54.359	46.247	174.447	10.610	0.207
75.000	114.472	-53.287	46.521	125.043	-10.091	0.418
80.000	109.750	-49.210	52.297	99.651	-17.272	2.116
85.000	106.065	-48.007	56.313	17.822	10.207	0.197
90.000	103.808	-48.636	58.954	58.748	-55.212	1.805
95.000	100.821	-53.935	70.096	50.853	-63.696	1.912
100.000	100.122	-55.658	72.977	-2.597	-31.760	0.514
105.000	97.029	-58.706	74.958	2.772	-34.016	1.038
110.000	93.516	-61.643	76.875	-2.054	-28.904	0.902
115.000	89.241	-64.612	78.984	-2.424	-35.316	1.081
120.000	84.861	-67.837	83.142	-0.470	-53.336	1.501

Age (Ma)	Total rotation:			Instantaneous rotation:		
	Longitude	Latitude	Angle	Longitude	Latitude	Rate
125.000	81.117	–70.750	89.086	1.820	–57.664	1.243
130.000	78.530	–71.616	90.976	37.519	58.944	0.270

Appendix 4: Isotopic Ages from Inferred Hotspot Traces

Compilation of isotopic ages from inferred hotspot traces

Trace	Island/ seamount	Lati- tude	Longi- tude	Age (Ma)	Reference
Austral-Cook	Atiu	-20.00	-158.20	7.40	Duncan and Clague (1985)
Austral-Cook	Macdonald	-29.00	-140.20	0.00	Duncan and Clague (1985)
Austral-Cook	Mangaia	-21.90	-157.90	13.70	Duncan and Clague (1985)
Austral-Cook	Mangaia	-21.90	-157.90	19.60	Duncan and Clague (1985)
Austral-Cook	Marotiri	-27.60	-143.80	3.50	Duncan and Clague (1985)
Austral-Cook	Marotiri	-27.60	-143.80	4.00	Duncan and Clague (1985)
Austral-Cook	Mauke	-20.10	-157.50	4.80	Duncan and Clague (1985)
Austral-Cook	Mauke	-20.10	-157.50	6.10	Duncan and Clague (1985)
Austral-Cook	Mitiaro	-19.80	-157.80	12.30	Duncan and Clague (1985)
Austral-Cook	Raivavae	-23.90	-147.70	5.60	Duncan and Clague (1985)
Austral-Cook	Raivavae	-23.90	-147.70	7.60	Duncan and Clague (1985)
Austral-Cook	Rapa	-27.00	-144.30	5.20	Duncan and Clague (1985)
Austral-Cook	Rapa	-27.00	-144.30	5.90	Duncan and Clague (1985)
Austral-Cook	Rarotonga	-21.20	-159.80	1.10	Duncan and Clague (1985)
Austral-Cook	Rarotonga	-21.20	-159.80	2.30	Duncan and Clague (1985)
Austral-Cook	Rimatara	-22.80	-152.70	4.80	Duncan and Clague (1985)
Austral-Cook	Rimatara	-22.80	-152.70	28.60	Duncan and Clague (1985)
Austral-Cook	Rurutu	-22.40	-151.30	0.60	Duncan and Clague (1985)
Austral-Cook	Rurutu	-22.40	-151.30	12.30	Duncan and Clague (1985)
Austral-Cook	Tubuai	-23.30	-149.50	8.50	Duncan and Clague (1985)
Austral-Cook	Tubuai	-23.30	-149.50	10.60	Duncan and Clague (1985)
Bouvet	Bouvet	-49.00	8.00	0.00	Crough et al. (1980)
Bouvet	South Africa	-30.70	23.00	78.30	Crough et al. (1980)
Bouvet	South Africa	-29.50	24.00	94.10	Crough et al. (1980)
Bouvet	South Africa	-28.50	24.30	110.00	Crough et al. (1980)
Bouvet	South Africa	-30.50	22.70	115.00	Crough et al. (1980)
Bouvet	South Africa	-25.50	26.20	147.00	Crough et al. (1980)
Bouvet	South Africa	-29.50	29.00	150.00	Crough et al. (1980)
Canary	Nova Scotia	49.50	-55.40	140.00	Crough et al. (1980)
Caroline	Hotspot	3.10	164.00	0.00	Harada and Hamano (2000)
Caroline Islands	Kusaie	5.40	-162.90	1.20	Duncan and Clague (1985)s
Caroline Islands	Kusaie	5.40	-162.90	2.60	Duncan and Clague (1985)
Caroline Islands	Ponape	6.90	-158.30	3.00	Duncan and Clague (1985)
Caroline Islands	Ponape	6.90	-158.30	8.60	Duncan and Clague (1985)
Caroline Islands	Truk	7.30	-151.80	4.80	Duncan and Clague (1985)
Caroline Islands	Truk	7.30	-151.80	13.90	Duncan and Clague (1985)
Cobb	Cobb	46.00	-130.00	0.00	Harada and Hamano (2000)
Cobb	Cobb	46.80	-130.80	1.60	Duncan and Clague (1985)
Cobb	Cobb	46.77	-130.83	3.27	Harada and Hamano (2000)
Cobb	Cobb	55.20	-140.30	14.90	Harada and Hamano (2000)
Cobb	Cobb	54.00	-137.40	19.70	Harada and Hamano (2000)
Cobb	Cobb	56.50	-146.60	21.90	Harada and Hamano (2000)
Cobb	Cobb	56.90	-149.20	24.80	Harada and Hamano (2000)
Cobb	Eickelberg	48.53	-133.10	7.05	Harada and Hamano (2000)
Cobb	Eickelberg	48.29	-133.09	9.03	Harada and Hamano (2000)
Cobb	Gluttony	47.14	-131.46	1.55	Harada and Hamano (2000)
Cobb	Horton	50.30	-142.60	19.40	Duncan and Clague (1985)

Trace	Island/ seamount	Lati- tude	Longi- tude	Age (Ma)	Reference
Cobb	Horton	50.30	−142.60	23.20	Duncan and Clague (1985)
Cobb	Lust	47.50	−131.51	4.40	Harada and Hamano (2000)
Cobb	Miller	53.32	−144.20	23.20	Harada and Hamano (2000)
Cobb	Miller	53.50	−144.30	25.20	Duncan and Clague (1985)
Cobb	Miller	53.32	−144.20	27.00	Harada and Hamano (2000)
Cobb	Miller	53.32	−144.20	27.80	Harada and Hamano (2000)
Cobb	Murray	53.60	−148.30	22.50	Harada and Hamano (2000)
Cobb	Murray	53.60	−148.30	24.80	Harada and Hamano (2000)
Cobb	Murray	53.60	−148.30	25.10	Harada and Hamano (2000)
Cobb	Murray	53.60	−148.30	25.60	Harada and Hamano (2000)
Cobb	Murray	53.60	−148.30	25.60	Harada and Hamano (2000)
Cobb	Murray	53.90	−148.50	25.70	Duncan and Clague (1985)
Cobb	Pathfinder	50.60	−143.20	22.90	Harada and Hamano (2000)
Cobb	Pathfinder	50.60	−143.20	23.50	Harada and Hamano (2000)
Cobb	Pathfinder	50.60	−143.20	23.80	Harada and Hamano (2000)
Cobb	Patton	54.30	−150.30	26.10	Harada and Hamano (2000)
Cobb	Patton	54.30	−150.30	27.20	Harada and Hamano (2000)
Cobb	Patton	54.30	−150.30	30.20	Harada and Hamano (2000)
Cobb	Patton	54.30	−150.30	32.00	Harada and Hamano (2000)
Cobb	Sloth	47.62	−131.70	5.20	Harada and Hamano (2000)
Cobb	Unnamed	48.28	−133.22	7.73	Harada and Hamano (2000)
Cobb	Warwick	48.03	−132.45	6.91	Harada and Hamano (2000)
Comores	Comores	−12.00	44.00	0.00	Emeric and Duncan (1982)
Comores	Comores	−11.67	43.37	0.13	Emeric and Duncan (1982)
Comores	Comores	−12.25	45.08	5.41	Emeric and Duncan (1982)
Comores	Comores	−13.40	48.30	10.38	Emeric and Duncan (1982)
Easter	Easter	−27.10	−106.30	0.00	Emeric and Duncan (1982)
Easter	Tuamotu	−24.76	−121.67	7.70	O'Connor et al. (1995)
Easter	Tuamotu	−24.85	−121.89	8.30	O'Connor et al. (1995)
Easter	Tuamotu	−18.20	−134.20	33.30	This volume
Easter	Tuamotu	−17.50	−137.40	39.00	This volume
Easter	Tuamotu	−15.70	−139.90	43.20	This volume
Easter–Nazca	Sala y Gomez	−27.00	−109.00	0.13	O'Connor et al. (1995)
Easter–Nazca	Sala y Gomez	−27.12	−109.66	0.22	O'Connor et al. (1995)
Easter–Nazca	Sala y Gomez	−26.93	−110.30	0.25	O'Connor et al. (1995)
Easter–Nazca	Sala y Gomez	−26.92	−110.97	3.00	O'Connor et al. (1995)
Easter–Nazca	Sala y Gomez	−25.70	−93.24	11.70	O'Connor et al. (1995)
Easter–Nazca	Sala y Gomez	−24.93	−88.34	14.90	O'Connor et al. (1995)
Easter–Nazca	Sala y Gomez	−25.64	−82.37	21.60	O'Connor et al. (1995)
Easter–Nazca	Sala y Gomez	−23.28	−83.00	25.60	O'Connor et al. (1995)
Easter	Nazca	−22.00	−82.00	30.70	This volume
Easter	Nazca	−19.00	−79.00	33.30	This volume
Easter	Tuamotu	−17.00	−77.50	39.00	This volume
Erebus	Erebus	−65.60	161.00	0.00	Duncan and Richards (1991)
Erebus	Erebus	−47.00	151.00	36.00	Duncan and Richards (1991)
Foundation	Hotspot	−36.60	−110.80	0.00	Harada and Hamano (2000)
Foundation	Foundation	−36.688	−113.461	2	O'Connor et al. (1998)
Foundation	Foundation	−36.348	−113.929	3.7	O'Connor et al. (1998)
Foundation	Foundation	−35.791	−115.645	4.8	O'Connor et al. (1998)
Foundation	Foundation	−36.561	−115.274	5.1	O'Connor et al. (1998)

Trace	Island/ seamount	Lati- tude	Longi- tude	Age (Ma)	Reference
Foundation	Foundation	−36.029	−115.988	6.2	O'Connor et al. (1998)
Foundation	Foundation	−35.442	−116.651	7.2	O'Connor et al. (1998)
Foundation	Foundation	−35.448	−117.199	7.6	O'Connor et al. (1998)
Foundation	Foundation	−35.383	−118.088	7.7	O'Connor et al. (1998)
Foundation	Foundation	−35.797	−117.438	7.7	O'Connor et al. (1998)
Foundation	Foundation	−35.110	−118.554	8.4	O'Connor et al. (1998)
Foundation	Foundation	−33.697	−124.909	8.8	O'Connor et al. (1998)
Foundation	Foundation	−35.361	−119.113	9.3	O'Connor et al. (1998)
Foundation	Foundation	−34.957	−120.400	10.4	O'Connor et al. (1998)
Foundation	Foundation	−34.329	−121.977	11.6	O'Connor et al. (1998)
Foundation	Foundation	−34.869	−121.556	11.6	O'Connor et al. (1998)
Foundation	Foundation	−34.135	−122.364	12.5	O'Connor et al. (1998)
Foundation	Foundation	−33.528	−124.100	12.9	O'Connor et al. (1998)
Foundation	Foundation	−33.439	−123.986	13.1	O'Connor et al. (1998)
Foundation	Foundation	−35.053	−120.720	13.4	O'Connor et al. (1998)
Foundation	Foundation	−32.474	−126.068	16.1	O'Connor et al. (1998)
Foundation	Foundation	−32.481	−126.012	16.3	O'Connor et al. (1998)
Foundation	Foundation	−32.506	−127.502	18.2	O'Connor et al. (1998)
Foundation	Foundation	−32.849	−130.762	21.2	O'Connor et al. (1998)
Foundation	Foundation	−27.6833	−139.417	24.2	McNutt et al. (1997)
Foundation	Foundation	−27.6833	−139.417	26.1	McNutt et al. (1997)
Foundation	Foundation	−27.4667	−140	23.4	McNutt et al. (1997)
Foundation	Foundation	−28.5333	−140.217	27.7	McNutt et al. (1997)
Foundation	Foundation	−28.7667	−141.117	27.5	McNutt et al. (1997)
Foundation	Foundation	−28.2	−141.217	28.8	McNutt et al. (1997)
Foundation	Foundation	−28.2	−141.217	31.3	McNutt et al. (1997)
Foundation	Foundation	−27.0333	−143.15	31.7	McNutt et al. (1997)
Foundation	Foundation	−27.0333	−143.15	39.1	McNutt et al. (1997)
Foundation	Foundation	−27.95	−143.6	3.75	McNutt et al. (1997)
Foundation	Foundation	−27.95	−143.6	39.6	McNutt et al. (1997)
Galapagos	Hotspot	−0.50	−91.50	0.00	This volume
Galapagos	Galapagos	0.25	−89.90	0.00	White et al. (1993)
Galapagos	Galapagos	−0.20	−91.30	0.00	White et al. (1993)
Galapagos	Galapagos	−0.45	−91.10	0.00	White et al. (1993)
Galapagos	Galapagos	0.90	−90.75	0.00	White et al. (1993)
Galapagos	Galapagos	−0.80	−90.30	0.02	White et al. (1993)
Galapagos	Galapagos	−0.80	−90.30	0.03	White et al. (1993)
Galapagos	Galapagos	0.90	−90.75	0.04	White et al. (1993)
Galapagos	Galapagos	−0.90	−89.60	0.05	White et al. (1993)
Galapagos	Galapagos	0.30	−91.60	0.05	White et al. (1993)
Galapagos	Galapagos	−0.45	−91.10	0.07	White et al. (1993)
Galapagos	Galapagos	−1.25	−90.40	0.08	White et al. (1993)
Galapagos	Galapagos	0.25	−90.50	0.10	White et al. (1993)
Galapagos	Galapagos	0.25	−90.50	0.10	White et al. (1993)
Galapagos	Galapagos	0.25	−90.50	0.11	White et al. (1993)
Galapagos	Galapagos	−0.45	−91.10	0.12	White et al. (1993)
Galapagos	Galapagos	0.25	−90.50	0.12	White et al. (1993)
Galapagos	Galapagos	−0.45	−91.10	0.15	White et al. (1993)
Galapagos	Galapagos	−0.45	−91.10	0.15	White et al. (1993)
Galapagos	Galapagos	−1.25	−90.40	0.35	White et al. (1993)

Trace	Island/ seamount	Lati- tude	Longi- tude	Age (Ma)	Reference
Galapagos	Galapagos	1.70	-91.95	0.39	White et al. (1993)
Galapagos	Galapagos	1.70	-91.95	0.41	White et al. (1993)
Galapagos	Galapagos	0.25	-90.50	0.56	White et al. (1993)
Galapagos	Galapagos	-0.90	-90.10	0.72	White et al. (1993)
Galapagos	Galapagos	-0.90	-89.60	0.77	White et al. (1993)
Galapagos	Galapagos	1.40	-91.75	0.88	White et al. (1993)
Galapagos	Galapagos	0.90	-90.75	0.89	White et al. (1993)
Galapagos	Galapagos	-0.90	-89.60	0.89	White et al. (1993)
Galapagos	Galapagos	-0.60	-90.75	1.04	White et al. (1993)
Galapagos	Galapagos	-0.70	-90.30	1.12	White et al. (1993)
Galapagos	Galapagos	-0.70	-90.30	1.31	White et al. (1993)
Galapagos	Galapagos	-0.90	-89.60	1.33	White et al. (1993)
Galapagos	Galapagos	-0.60	-90.75	1.40	White et al. (1993)
Galapagos	Galapagos	-1.25	-90.40	1.52	White et al. (1993)
Galapagos	Galapagos	1.40	-91.75	1.60	White et al. (1993)
Galapagos	Galapagos	-0.90	-89.60	2.33	White et al. (1993)
Galapagos	Galapagos	-0.90	-89.60	2.35	White et al. (1993)
Galapagos	Galapagos	-0.90	-90.10	2.50	White et al. (1993)
Galapagos	Galapagos	-1.30	-89.70	2.61	White et al. (1993)
Galapagos	Galapagos	-0.90	-90.10	2.76	White et al. (1993)
Galapagos	Galapagos	-1.30	-89.70	2.77	White et al. (1993)
Galapagos	Galapagos	-1.19	-89.11	5.30	Christie et al. (1992)
Galapagos	Galapagos	-1.19	-89.11	5.80	Christie et al. (1992)
Galapagos	Galapagos	-1.12	-88.26	7.30	Christie et al. (1992)
Galapagos	Galapagos	-0.93	-87.12	7.80	Christie et al. (1992)
Galapagos	Galapagos	-2.01	-85.66	8.70	Christie et al. (1992)
Galapagos	Galapagos	-2.01	-85.66	9.10	Christie et al. (1992)
Gilbert–Marshall	Hotspot	-14.10	-155.50	0.00	Harada and Hamano (2000)
Hawaii	Hotspot	18.92	-155.25	0.00	Harada and Hamano (2000)
Hawaii	Hawaii	20.00	-155.00	0.00	Compiled
Hawaii	Hotspot	19.60	204.50	0.00	Raymond et al. (2000)
Hawaii	Hawaii	20.10	-155.70	0.46	Duncan and Clague (1985)
Hawaii	Hawaii	20.70	-156.10	0.86	Duncan and Clague (1985)
Hawaiian–Emperor	Lanai	20.80	-156.30	1.28	Harada and Hamano (2000)
Hawaii	Hawaii	20.90	-156.60	1.34	Duncan and Clague (1985)
Hawaii	Hawaii	21.20	-156.90	1.52	Duncan and Clague (1985)
Hawaii	Hawaii	21.20	-157.20	1.90	Duncan and Clague (1985)
Hawaii	Hawaii	21.50	-157.80	2.70	Duncan and Clague (1985)
Hawaii	Hawaii	21.50	-158.20	3.90	Duncan and Clague (1985)
Hawaii	Hawaii	22.00	-159.50	5.80	Duncan and Clague (1985)
Hawaii	Nihoa	23.10	-161.90	7.20	Duncan and Clague (1985)
Hawaii	Necker	23.60	-164.70	10.30	Duncan and Clague (1985)
Hawaii	LaPerouse	23.80	-166.30	12.00	Duncan and Clague (1985)
Hawaiian–Emperor	Laysan	25.80	-171.80	20.30	Harada and Hamano (2000)
Hawaii	Pearl and Hermes Bank	27.90	-175.80	20.60	Duncan and Clague (1985)
Hawaiian–Emperor	Laysan	25.80	-171.80	20.60	Harada and Hamano (2000)

Trace	Island/ seamount	Latitude	Longitude	Age (Ma)	Reference
Hawaiian–Emperor	Laysan	25.80	−171.80	21.10	Harada and Hamano (2000)
Hawaiian–Emperor	Laysan	25.80	−171.80	21.40	Harada and Hamano (2000)
Hawaiian–Emperor	Necker	25.30	−172.00	23.40	Harada and Hamano (2000)
Hawaii	Midway	28.30	−177.30	27.70	Duncan and Clague (1985)
Hawaii	Hawaii	27.40	−178.80	28.00	Duncan and Clague (1985)
Hawaii	Hawaii	33.70	171.60	39.90	Duncan and Clague (1985)
Hawaii	Hawaii	32.10	172.30	42.40	Duncan and Clague (1985)
Hawaiian–Emperor	Diakakuji	32.70	172.00	43.40	Harada and Hamano (2000)
Hawaiian–Emperor	Koko	34.40	171.40	48.10	Harada and Hamano (2000)
Hawaii	Emperor	37.50	170.70	55.20	Duncan and Clague (1985)
Hawaii	Emperor	39.00	170.00	55.40	Duncan and Clague (1985)
Hawaiian–Emperor	Jingu	38.40	171.20	55.40	Harada and Hamano (2000)
Hawaiian–Emperor	Nintoku	41.20	170.20	55.60	Harada and Hamano (2000)
Hawaiian–Emperor	Ojin	37.50	170.30	56.20	Harada and Hamano (2000)
Hawaii	Emperor	41.30	170.50	56.20	Duncan and Clague (1985)
Hawaiian–Emperor	Nintoku	41.20	170.20	56.30	Harada and Hamano (2000)
Hawaiian–Emperor	Ojin	37.50	170.30	56.60	Harada and Hamano (2000)
Hawaiian–Emperor	Ojin	37.50	170.30	57.20	Harada and Hamano (2000)
Hawaiian–Emperor	Suiko	44.40	170.00	64.30	Harada and Hamano (2000)
Hawaii	Emperor	44.80	170.00	64.70	Duncan and Clague (1985)
Hawaiian–Emperor	Suiko	44.40	170.00	65.80	Harada and Hamano (2000)
Hawaii	Emperor	53.00	165.00	71.00	Duncan and Clague (1985)
Hawaiian–Emperor	Suiko	44.40	170.00	81.50	Harada and Hamano (2000)
Islas Revillagigedos	Clarion	18.30	−114.70	1.10	Duncan and Clague (1985)
Islas Revillagigedos	Clarion	18.30	−114.70	2.40	Duncan and Clague (1985)
Islas Revillagigedos	San Benedicto	19.30	−110.80	0.00	Duncan and Clague (1985)
Islas Revillagigedos	Socorro	18.70	−111.10	0.00	Duncan and Clague (1985)
Islas Revillagigedos	Socorro	18.70	−111.10	0.30	Duncan and Clague (1985)
Juan Fernandez	Hotspot	−34.00	−82.00	0.00	This volume
Juan Fernandez	Juan Fernandez	−33.78	−80.78	0.87	Baker et al. (1987)
Juan Fernandez	Juan Fernandez	−33.78	−80.78	1.03	Baker et al. (1987)

Trace	Island/ seamount	Lati- tude	Longi- tude	Age (Ma)	Reference
Juan Fernandez	Juan Fernandez	-33.78	-80.78	1.33	Baker et al. (1987)
Juan Fernandez	Juan Fernandez	-33.78	-79.00	3.18	Baker et al. (1987)
Juan Fernandez	Juan Fernandez	-33.78	-79.00	3.59	Baker et al. (1987)
Juan Fernandez	Juan Fernandez	-33.78	-79.00	4.00	Baker et al. (1987)
Kergueulen	Kergueulen	-49.00	63.00	0.00	Raymond et al. (2000)
Kergueulen	Kergueulen	-49.00	69.30	27.00	Compiled
Line Islands	LI-119	2.70	-165.00	68.00	Duncan and Clague (1985)
Line Islands	Line Islands	-15.00	-149.00	41.80	Harada and Hamano (2000)
Line Islands	Line Islands	-15.00	-149.00	47.40	Harada and Hamano (2000)
Line Islands	Line Islands	-9.00	-150.40	70.50	Harada and Hamano (2000)
Line Islands	Line Islands	-7.30	-151.30	71.90	Harada and Hamano (2000)
Line Islands	LI-123	5.80	-160.70	76.40	Duncan and Clague (1985)
Line Islands	LI-128	9.20	-160.70	78.70	Duncan and Clague (1985)
Line Islands	LI-130	8.30	-164.30	72.00	Duncan and Clague (1985)
Line Islands	LI-133	12.00	-165.80	83.00	Duncan and Clague (1985)
Line Islands	LI-134	10.30	-168.00	48.00	Duncan and Clague (1985)
Line Islands	LI-137	14.50	-169.00	56.00	Duncan and Clague (1985)
Line Islands	LI-142	18.00	-169.10	93.40	Duncan and Clague (1985)
Line Islands	LI-142	18.00	-169.10	128.00	Duncan and Clague (1985)
Line Islands	LI-143	19.50	-169.00	88.10	Duncan and Clague (1985)
Line Islands	LI-33	8.20	-161.90	39.30	Duncan and Clague (1985)
Line Islands	LI-41	2.10	-157.30	35.50	Duncan and Clague (1985)
Line Islands	LI-43	-0.70	-155.30	59.00	Duncan and Clague (1985)
Line Islands	LI-44	-7.60	-151.50	71.90	Duncan and Clague (1985)
Line Islands	LI-45	-9.10	-150.70	70.50	Duncan and Clague (1985)
Line Islands	LI-52	-15.00	-149.00	44.60	Duncan and Clague (1985)
Line Islands	LI-59	12.50	-167.00	85.00	Duncan and Clague (1985)
Line Islands	LI-62	15.00	-167.50	82.00	Duncan and Clague (1985)
Line Islands	LI-63	16.50	-168.20	86.00	Duncan and Clague (1985)
Line Islands	LI-PC6	2.50	-158.50	69.80	Duncan and Clague (1985)
Louisville	Hotspot	-53.50	-141.12	0.00	Wessel and Kroenke (1997)
Louisville	Hotspot	-51.50	-141.00	0.00	Raymond et al. (2000)
Louisville	Hotspot	-50.90	-138.10	0.00	Harada and Hamano (2000)
Louisville	Louisville	-50.90	-138.10	0.10	Compiled
Louisville	Louisville	-50.50	-139.20	0.50	Compiled
Louisville	Unnamed	-50.43	-139.15	0.50	Harada and Hamano (2000)
Louisville	Unnamed	-48.20	-148.80	12.50	Harada and Hamano (2000)
Louisville	Osborn	-26.00	-175.00	30.00	Duncan and Clague (1985)
Louisville	Valerie	-41.61	-164.20	34.40	Harada and Hamano (2000)
Louisville	Osborn	-26.00	-175.00	36.00	Duncan and Clague (1985)
Louisville	Louisville	-41.50	-164.30	36.60	Compiled
Louisville	Valerie	-41.61	-164.20	36.60	Harada and Hamano (2000)
Louisville	Unnamed	-36.95	-169.83	44.60	Harada and Hamano (2000)
Louisville	LV-VM36-04	-38.30	-167.70	44.60	Duncan and Clague (1985)
Louisville	Louisville	-38.40	-168.00	45.50	Compiled
Louisville	Unnamed	-38.33	-167.73	45.50	Harada and Hamano (2000)
Louisville	LV-VM36-02	-40.80	-165.30	45.50	Duncan and Clague (1985)
Louisville	LV-VM36-05	-33.90	-171.20	53.30	Duncan and Clague (1985)
Louisville	Louisville	-33.90	-171.20	53.40	Compiled
Louisville	Unnamed	-39.95	-171.19	53.40	Harada and Hamano (2000)

Trace	Island/ seamount	Lati- tude	Longi- tude	Age (Ma)	Reference
Louisville	Louisville	-28.20	-173.60	61.20	Compiled
Louisville	Unnamed	-30.01	-173.25	61.20	Harada and Hamano (2000)
Louisville	Unnamed	-25.53	-175.04	65.90	Harada and Hamano (2000)
Louisville	Unnamed	-25.53	-175.04	66.60	Harada and Hamano (2000)
Louisville	Unnamed	-27.28	-174.21	66.70	Harada and Hamano (2000)
Louisville	Unnamed	-27.28	-174.21	70.10	Harada and Hamano (2000)
Louisville	Louisville	-27.60	-174.20	70.10	Compiled
Louisville	Louisville	-25.70	-175.30	75.00	Compiled
Marquesas	Eliao	-8.00	-140.70	5.00	Duncan and Clague (1985)
Marquesas	Eliao	-8.00	-140.70	5.20	Duncan and Clague (1985)
Marquesas	Eliao	-8.00	-140.70	8.80	Duncan and Clague (1985)
Marquesas	Fatu Hiva	-10.50	-138.60	1.30	Duncan and Clague (1985)
Marquesas	Fatu Hiva	-10.50	-138.60	1.40	Duncan and Clague (1985)
Marquesas	Hiva Oa	-9.80	-139.00	1.60	Duncan and Clague (1985)
Marquesas	Hiva Oa	-9.80	-139.00	2.50	Duncan and Clague (1985)
Marquesas	Nuku Hiva	-8.90	-140.10	3.00	Duncan and Clague (1985)
Marquesas	Nuku Hiva	-8.90	-140.10	4.30	Duncan and Clague (1985)
Marquesas	Tahuata	-10.00	-139.10	1.80	Duncan and Clague (1985)
Marquesas	Tahuata	-10.00	-139.10	2.10	Duncan and Clague (1985)
Marquesas	Ua Huka	-8.90	-139.50	2.70	Duncan and Clague (1985)
Marquesas	Ua Huka	-8.90	-139.50	2.80	Duncan and Clague (1985)
Meteor	Meteor	30.00	-28.50	0.00	Duncan (1984)
Meteor	Montergian Hills	45.30	-72.60	131.00	Duncan (1984)
Meteor	Montergian Hills	44.65	-73.35	136.00	Duncan (1984)
Meteor	Montergian Hills	42.90	-74.50	146.00	Duncan (1984)
Meteor	Montergian Hills	48.00	-79.50	151.00	Duncan (1984)
Meteor	New England Seamounts	34.70	-50.50	74.00	Duncan (1984)
Meteor	New England Seamounts	35.40	-51.50	81.00	Duncan (1984)
Meteor	New England Seamounts	36.32	-57.30	83.50	Duncan (1984)
Meteor	New England Seamounts	36.83	-58.82	83.80	Duncan (1984)
Meteor	New England Seamounts	38.08	-62.20	90.80	Duncan (1984)
Meteor	New England Seamounts	38.42	-63.25	99.50	Duncan (1984)
Meteor	New England Seamounts	39.82	-67.43	104.30	Duncan (1984)
Meteor	White Mountain	43.37	-70.56	124.00	Duncan (1984)
Meteor	White Mountain	43.40	-70.80	125.00	Duncan (1984)
Musician Seamounts	Khatchaturian	27.10	-162.20	66.90	Duncan and Clague (1985)
Musician Seamounts	Rachmaninoff	28.60	-163.50	88.80	Duncan and Clague (1985)

Trace	Island/ seamount	Lati- tude	Longi- tude	Age (Ma)	Reference
Musician Seamounts	Unnamed	33.50	-166.50	95.60	Duncan and Clague (1985)
New Hebrides–Samoa Lineament	Combe	-12.70	-177.60	13.50	Duncan and Clague (1985)
New Hebrides–Samoa Lineament	Lalla Rookh	-13.00	-175.60	10.00	Duncan and Clague (1985)
New Hebrides–Samoa Lineament	Taviuni	-12.30	-174.60	5.40	Duncan and Clague (1985)
Kergueulen	Ninetyeast	-31.00	87.90	38.00	Duncan (1991)
Kergueulen	Ninetyeast	-27.35	88.30	43.00	Duncan (1991)
Kergueulen	Ninetyeast	-25.50	88.50	47.00	Duncan (1991)
Kergueulen	Ninetyeast	-22.80	88.50	50.00	Duncan (1991)
Kergueulen	Ninetyeast	-20.10	88.50	53.00	Duncan (1991)
Kergueulen	Ninetyeast	-17.00	88.50	58.00	Duncan (1991)
Kergueulen	Ninetyeast	-12.00	88.50	60.00	Duncan (1991)
Kergueulen	Ninetyeast	-11.30	88.70	62.00	Duncan (1991)
Phoenix	Hotspot	-21.50	-142.80	0.00	Harada and Hamano (2000)
Pitcairn–Gambier	Gambier	-23.20	-135.00	5.20	Duncan and Clague (1985)
Pitcairn–Gambier	Gambier	-23.20	-135.00	7.20	Duncan and Clague (1985)
Pitcairn–Gambier	Mururoa	-22.00	-139.00	8.00	Duncan and Clague (1985)
Pitcairn–Gambier	Pitcairn	-24.10	-130.10	0.50	Duncan and Clague (1985)
Pitcairn–Gambier	Pitcairn	-24.10	-130.10	0.90	Duncan and Clague (1985)
Pratt-Welker	Bowie	53.30	-135.60	0.70	Duncan and Clague (1985)
Pratt-Welker	Bowie	53.30	-135.60	0.75	Duncan and Clague (1985)
Pratt-Welker	Davidson	53.70	-136.50	17.40	Duncan and Clague (1985)
Pratt-Welker	Dickens	54.60	-136.90	3.80	Duncan and Clague (1985)
Pratt-Welker	Dickens	54.60	-136.90	4.10	Duncan and Clague (1985)
Pratt-Welker	Giacomini	56.50	-146.60	20.60	Duncan and Clague (1985)
Pratt-Welker	Giacomini	56.50	-146.60	21.40	Duncan and Clague (1985)
Pratt-Welker	Hodgkins	53.50	-136.00	2.50	Duncan and Clague (1985)
Pratt-Welker	Hodgkins	53.50	-136.00	14.40	Duncan and Clague (1985)
Pratt-Welker	Kodiak	56.90	-149.20	23.40	Duncan and Clague (1985)
Pratt-Welker	Kodiak	56.90	-149.20	24.80	Duncan and Clague (1985)
Pratt-Welker	Welker	55.20	-140.30	14.90	Duncan and Clague (1985)
Reunion	Chagos	-13.50	62.00	31.00	Duncan and Hargraves (1990)
Reunion	Chagos	-13.12	61.37	35.50	Duncan and Hargraves (1990)
Reunion	Laccadive	-4.19	72.50	49.00	Duncan and Hargraves (1990)
Reunion	Laccadive	5.08	73.00	57.00	Duncan and Hargraves (1990)
Reunion	Laccadive	20.00	73.50	66.00	Duncan and Hargraves (1990)
Reunion	Mauritius	-20.00	58.00	7.00	Duncan and Hargraves (1990)
Reunion	Reunion	-21.00	55.50	0.00	This volume
Reunion	Reunion	-21.20	55.60	0.00	Raymond et al. (2000)

Trace	Island/ seamount	Lati- tude	Longi- tude	Age (Ma)	Reference
Reunion	Reunion	-21.00	56.00	2.00	Duncan and Hargraves (1990)
Society	Bora Bora	-16.50	-151.80	3.10	Duncan and Clague (1985)
Society	Bora Bora	-16.50	-151.80	3.40	Duncan and Clague (1985)
Society	Huahine	-16.70	-151.00	2.00	Duncan and Clague (1985)
Society	Huahine	-16.70	-151.00	2.60	Duncan and Clague (1985)
Society	Maupiti	-16.40	-152.20	4.00	Duncan and Clague (1985)
Society	Maupiti	-16.40	-152.20	4.50	Duncan and Clague (1985)
Society	Mehetia	-17.90	-148.30	0.00	Duncan and Clague (1985)
Society	Moorea	-17.50	-149.80	1.50	Duncan and Clague (1985)
Society	Moorea	-17.50	-149.80	1.70	Duncan and Clague (1985)
Society	Raiatea	-16.80	-151.50	2.40	Duncan and Clague (1985)
Society	Raiatea	-16.80	-151.50	2.60	Duncan and Clague (1985)
Society	Tahaa	-16.60	-151.50	2.60	Duncan and Clague (1985)
Society	Tahaa	-16.60	-151.50	3.20	Duncan and Clague (1985)
Society	Tahiti	-17.60	-149.50	0.50	Duncan and Clague (1985)
Society	Tahiti	-17.60	-149.50	1.20	Duncan and Clague (1985)
Society	Tahiti-iti	-17.80	-149.20	0.40	Duncan and Clague (1985)
Society	Tahiti-iti	-17.80	-149.20	0.50	Duncan and Clague (1985)
St. Paul	Ninetyeast	2.50	91.00	82.00	Duncan (1991)
St. Paul	Ninetyeast	23.00	88.00	117.00	Duncan (1991)
St. Paul	St. Paul	-38.72	77.50	0.00	This volume
Tasman	Tasman	-40.40	155.00	0.00	This Volume
Tasman	Tasman	-36.60	156.20	7.00	McDougall and Duncan ('88)
Tasman	Tasman	-33.10	156.30	14.80	McDougall and Duncan ('88)
Tasman	Tasman	-30.90	155.20	16.50	McDougall and Duncan ('88)
Tasman	Tasman	-28.60	155.50	21.00	McDougall and Duncan ('88)
Trinidade	Brazil	-22.60	-42.10	51.00	Herz (1977)
Trinidade	Brazil	-22.00	-44.70	66.00	Herz (1977)
Trinidade	Brazil	-18.50	-46.00	68.00	Herz (1977)
Trinidade	Brazil	-19.50	-47.00	70.00	Herz (1977)
Trinidade	Brazil	-17.90	-47.80	83.00	Herz (1977)
Trinidade	Brazil	-17.50	-50.50	85.00	Herz (1977)
Trinidade	Brazil	-15.60	-51.30	89.00	Herz (1977)
Trinidade	Brazil	-13.90	-54.10	123.00	Herz (1977)
Trinidade	Trinidade	-21.00	-28.50	0.00	Duncan and Richards (1991)
Tristan	Tristan	-37.00	-12.00	0.00	Duncan and Richards (1991)
Tristan	Walvis	-37.09	-7.88	31.00	O'Connor and Duncan ('90)
Tristan	Walvis	-34.50	-3.48	37.20	O'Connor and Duncan ('90)
Tristan	Walvis	-34.51	-3.63	38.50	O'Connor and Duncan ('90)
Tristan	Walvis	-35.00	-4.50	39.00	O'Connor and Duncan ('90)
Tristan	Walvis	-35.00	-4.50	40.00	O'Connor and Duncan ('90)
Tristan	Walvis	-34.33	-1.57	46.20	O'Connor and Duncan ('90)
Tristan	Walvis	-32.97	-0.02	64.10	O'Connor and Duncan ('90)
Tristan	Walvis	-28.53	2.32	77.60	O'Connor and Duncan ('90)
Tristan	Walvis	-29.07	2.98	79.40	O'Connor and Duncan ('90)
Tristan	Walvis	-21.00	8.00	113.00	O'Connor and Duncan ('90)

Appendix 5: Paleostresses

Compilation of paleostresses [maximum horizontal principal component (MHPC)] from published sources. Sources for the paleostress measurements are included in the Supplementary References (for North America) and References (for other continents)

Plate	Latitude	Longitude	Age (Ma)	Stress azimuth
0	45.3	5	2.5	-10
0	43.833	5.333	4	8
0	44.333	4.167	6	107
0	45.3	5	6	74
0	47.167	5	6	115
0	47.333	3.5	6	114
0	47.667	6	6	132
0	48.333	9.5	6	110
0	49.533	11.75	6	92
0	43.833	5.333	20	43
0	44	4.833	20	40
0	44	5.75	20	56
0	44.333	4.167	20	49
0	48.5	7.5	20	44
0	49.533	11.75	20	44
0	38.8	-9.2	30	45
0	45.3	5	30	0
0	43.833	5.333	32	7
0	44	4.833	32	-50
0	44	5.75	32	-66
0	44.333	4.167	32	-66
0	47.167	5	32	-64
0	47.333	3.5	32	-9
0	47.667	6	32	-76
0	48.333	9.5	32	4
0	48.5	7.5	32	1
0	49.533	11.75	32	-6
0	52	0	35	0
0	43.833	5.333	38	3
0	44	4.833	38	-2
0	44	5.75	38	3
0	44.333	4.167	38	-8
0	47.167	5	38	3
0	47.333	3.5	38	3
0	47.667	6	38	1
0	48.333	9.5	38	-2
0	48.5	7.5	38	-1
0	49.533	11.75	38	1
0	59	5	45.1	70
0	52	3	46	90
0	52	3	49	-45
0	54.2	-5.8	56	-45
0	56.4	-5.7	56.6	-45
0	57.3	-6.2	59.3	-40
0	56	-6	60	-30
0	56.8	-6.1	60.5	-30

Plate	Latitude	Longitude	Age (Ma)	Stress azimuth
0	57	-6.4	61	-20
0	52	3	66	-45
0	52	3	70	45
0	56	4	70.1	45
1	32.12	-113.5	0.1	-180
1	31.45	-109.3	0.2	-152
1	37.667	-118.883	0.72	-180
1	35.25	-111.42	1	-60
1	35.55	-111.42	1	-145
1	34.215	-109.448	1.04	-135
1	40.5	-120.953	1.8	-30
1	34.14	-109.68	3	-75
1	41.833	-109	3.1	-32.5
1	40.487	-120.782	3.4	-170
1	40.487	-120.782	3.9	-170
1	43.167	-114.5	4.15	-60
1	41	-122	4.5	-10
1	42	-122	4.5	-20
1	44	-121.5	4.5	-5
1	46	-121.5	4.5	-180
1	37.709	-108.1	5.6	-95
1	38.517	-118.267	5.8	-80
1	35.42	-110.17	6	-55
1	40.75	-118.62	6.1	-180
1	45.75	-119.45	6.6	-20
1	35.9	-104.53	8	-50
1	38.2	-119	8	-135
1	40.2	-106.95	8	-50
1	46.5	-119	8.5	-25
1	43.833	-115.167	9	-50
1	35.8	-105.21	9.4	-174
1	32.2	-111.38	9.7	-28
1	32.209	-111.371	9.7	-180
1	36.11	-106.22	9.7	-170
1	36.25	-106.28	9.8	-180
1	40.8	-107.1	9.8	-50
1	36.875	-116.467	10	-20
1	44.35	-116.908	10.3	-10
1	36.5	-116.592	10.5	-120
1	36.1	-106.223	10.67	-112.5
1	37.917	-117.208	10.8	-45
1	36.933	-116.483	11.3	-170
1	36.933	-116.5	11.6	-175
1	35.15	-104.52	12.6	-162
1	33.2	-111.5	13	-45
1	36.279	-114.431	13	-30
1	39.325	-119.65	13	-180
1	39.9	-107.25	13.4	-50
1	34.42	-113.92	13.5	-50
1	40.667	-11.2	13.93	-40
1	46	-117.3	14.2	-134
1	37.25	-114.75	14.4	-180

Plate	Latitude	Longitude	Age (Ma)	Stress azimuth
1	34.94	−105.75	14.5	−176
1	39.017	−118.432	14.6	−134
1	34.42	−113.92	15	−140
1	35.592	−114.63	15	−10
1	42.5	−118.5	15.3	−170
1	32.583	−111.5	15.4	−60
1	39.0167	−118.432	15.4	−25
1	46	−117.1	15.4	−22.5
1	37.5	−114.333	15.5	−180
1	35.97	−104.95	15.6	−170
1	41.833	−117.567	15.6	−15
1	45.3	−117.4	15.6	−5
1	37.29	−106.24	15.9	−170
1	43	−117	16	−32.5
1	33.89	−112.817	16.2	−180
1	37.71	−107.89	16.2	−30
1	40.214	−116.549	16.2	−10
1	37.3	−114.333	16.4	−40
1	40.22	−116.53	16.4	−15
1	40.217	−116.533	16.7	−20
1	34.25	−114.75	17	−115
1	41	−116.8	17	−22
1	33.897	−112.807	17.1	−20
1	39.933	−112.533	17.1	−135
1	33.8	−112.9	17.8	−95
1	37.5	−115.2	17.8	−160
1	41.833	−117.567	17.8	−180
1	30.595	−104.725	17.9	−6
1	30.583	−104.8	18	−110
1	30.583	−104.8	18	−155
1	30.583	−104.8	18	−110
1	30.597	−104.71	18	−173
1	34.333	−114.833	18	−30
1	37.5	−116.25	18	−180
1	39	−119.25	18	−180
1	36.91	−104.44	18.1	−75
1	30.56	−104.79	18.2	−6
1	33.947	−112.84	18.2	−170
1	34.667	−115.667	18.2	−180
1	33.8	−112.8	18.3	−10
1	38.125	−113.567	18.3	−170
1	38.95	−114.17	18.5	−15
1	39.9	−116.333	18.6	−20
1	33.5	−115.55	18.8	−90
1	33.333	−112.083	19	−115
1	34.283	−114.233	19	−120
1	34.283	−114.233	19	−120
1	39.27	−118.352	19.25	−90
1	30.597	−104.71	19.4	−173
1	38.6	−105.683	19.4	−15
1	30.623	−104.763	19.7	−30

Plate	Latitude	Longitude	Age (Ma)	Stress azimuth
1	30.53	−104.767	19.8	−28
1	39.27	−118.352	20.01	−90
1	38.417	−113.2	20.1	−20
1	29.385	−103.652	20.2	−20
1	30.527	−104.772	20.2	−164
1	32.583	−111.5	20.5	−60
1	30.6	−104.773	20.8	−7
1	30.623	−104.763	20.8	−30
1	30.623	−104.763	20.9	−30
1	30.583	−104.8	21	−110
1	34.25	−114.75	21	−115
1	34.367	−114.837	21	−30
1	38.15	−117.208	21	−55
1	30.53	−104.767	21.1	−28
1	30.55	−104.758	21.1	−21
1	30.5	−104.767	21.4	−10
1	30.585	−104.773	21.4	−18
1	30.527	−104.772	21.5	−164
1	30.55	−104.758	21.6	−21
1	38.375	−112.333	21.6	−175
1	30.588	−104.773	21.8	−44
1	30.61	−104.788	21.9	−65
1	30.595	−104.725	22	−65
1	38.167	−113.867	22	−90
1	38.208	−113.867	22	−30
1	29.375	−103.6	22.1	−30
1	29.79	−102.98	22.3	−18
1	38.35	−113.6	22.5	−170
1	30.608	−104.787	22.7	−49
1	38.05	−117.458	22.7	−20
1	32.833	−110.758	22.8	−68
1	33.417	−18.783	23	−20
1	38.133	−112.4	23	−25
1	38.308	−113.6	23	−17.5
1	37.967	−107.5	23.1	−5
1	30.547	−104.79	23.2	−40
1	39.517	−111.733	23.3	−180
1	36.83	−107.02	23.5	−173
1	37.62	−104.88	23.5	−98
1	30.61	−104.788	23.7	−58
1	35.06	−106.48	23.8	−180
1	30.602	−104.78	23.9	−64
1	33.19	−107.14	23.9	−10
1	36.14	−104.68	23.9	−85
1	36.14	−104.68	23.9	−85
1	30.583	−104.8	24	−155
1	33	−109.667	24	−20
1	34.283	−114.233	24	−120
1	38.283	−113.2	24	−180
1	29.37	−103.595	24.1	−30
1	30.647	−104.738	24.2	−89

Plate	Latitude	Longitude	Age (Ma)	Stress azimuth
1	36.67	−104.49	24.2	−75
1	34.35	−107.22	24.3	−180
1	30.602	−104.784	24.4	−65
1	36.6	−105.5	24.5	−50
1	30.608	−104.787	24.6	−49
1	32	−110	24.6	−45
1	32.3	−111.6	24.6	−20
1	34.33	−107.26	24.8	−180
1	33.333	−112.083	25	−115
1	33.417	−18.783	25	−20
1	38.2	−118.18	25	−105
1	33.333	−112.083	25.01	−115
1	38.2	−118.18	25.01	−105
1	30.622	−104.782	25.4	−57
1	36.55	−107.2	25.4	−175
1	33.056	−110.746	25.5	−180
1	33.333	−112.042	25.5	−30
1	37.351	−105.058	25.7	−90
1	40.7	−118.28	25.7	−135
1	40.7	−118.28	25.7	−135
1	38.7	−117.067	25.9	−45
1	32.28	−107.16	26	−120
1	32.28	−107.16	26	−120
1	32.6	−107.41	26	−32
1	36.75	−105.5	26	−135
1	38.633	−111.975	26	−115
1	38.9	−118.18	26	−180
1	37.8	−106.95	26.4	−20
1	33.5	−106	26.5	−125
1	31.6	−110.7	26.7	−125
1	37.933	−114.833	26.7	−135
1	37.933	−115.167	26.7	−135
1	37.95	−112.35	26.7	−80
1	33	−109.667	27	−20
1	37.233	−112.583	27	−180
1	37.88	−112.6	27	−180
1	37.983	−112.583	27	−140
1	47.317	−107.05	27	−150
1	34.459	−108.308	27.2	−21
1	35.41	−105.95	27.2	−125
1	36.93	−106.96	27.2	−175
1	34.31	−108.17	27.7	−35
1	33.67	−104.04	27.9	−93
1	33.9	−106.2	27.9	−65
1	34	−107.13	28	−25
1	34.04	−107.13	28	−180
1	34.11	−107.2	28	−10
1	34.46	−108.87	28	−30
1	34.11	−107.2	28.3	−10
1	34.38	−107.98	28.3	−33
1	33.72	−107.38	28.6	−25

Plate	Latitude	Longitude	Age (Ma)	Stress azimuth
1	31.6	-110.7	28.7	-55
1	32.15	-110.5	28.7	-25
1	40.458	-105.892	28.7	-95
1	34.45	-107.18	28.8	-180
1	32.12	-107.62	29.1	-100
1	35.32	-106.33	29.3	-107
1	38.767	-117.233	29.4	-135
1	38.125	-114.125	29.5	-135
1	33.48	-107.64	30.1	-180
1	34.28	-106.34	30.2	-117
1	34.28	-106.34	30.2	-117
1	38.783	-117.233	30.2	-135
1	38.783	-117.233	30.2	-135
1	35.03	-107.62	30.7	-30
1	39.767	-167.6	30.8	-180
1	40.617	-111.4	31.2	-120
1	34.3	-106.8	31.3	-150
1	32.947	-108.197	31.5	-70
1	29.925	-104.545	32	-90
1	38.05	-113.471	32	-30
1	38.283	-113.9	32	-20
1	40.567	-111.7	32	-25
1	32.44	-106.59	32.5	-48
1	39.9	-114.85	32.9	-180
1	29.052	-103.4	33.4	-124
1	29.052	-103.4	33.4	-128
1	39.81	-116.153	33.5	-130
1	39.81	-116.153	33.5	-130
1	40.617	-111.55	34	-10
1	38.717	-105.55	34.1	-135
1	33.57	-105.87	34.4	-74
1	38.667	-106.583	34.5	-135
1	32.2	-109.5	34.8	-120
1	39.34	-106.13	34.9	-157.5
1	40.667	-111.2	34.9	-60
1	39.383	-116.167	35	-120
1	39.5	-114.3	35	-175
1	33.69	-105.68	35.1	-11
1	38.7	-117.05	35.2	-140
1	40.191	-116.614	35.2	-10
1	33.08	-107.15	35.4	-45
1	38.783	-113.533	35.4	-160
1	39.554	-105.835	35.5	-22.5
1	31.188	-105.5	35.9	-112.5
1	30.328	-103.743	36	-101
1	39.667	-114.917	36	-170
1	32.2	-109.5	36.1	-65
1	33.57	-105.91	36.5	-81
1	30.842	-105.247	36.7	-101
1	31.428	-104.897	36.7	-112.5
1	29.057	-103.433	36.8	-124

Plate	Latitude	Longitude	Age (Ma)	Stress azimuth
1	29.057	−103.433	36.8	−118
1	29.057	−103.433	36.8	−124
1	29.057	−103.433	36.8	−118
1	39.75	−111.733	36.9	−40
1	32.43	−107.73	37.8	−22
1	40.533	−112.167	38	−130
1	32.35	−107.4	38.1	−139
1	40.55	−117.1	38.9	−30
1	40.417	−116.917	39	−180
1	40.533	−112.167	39	−130
1	39.767	−112.625	39.5	−5
1	39.367	−106.25	40	−100
1	29.415	−104.533	40.3	−109
1	29.437	−103.477	41	−92
1	39.697	−105.732	41.5	−45
1	29.435	−103.357	41.7	−88
1	29.432	−102.56	42.6	−133
1	39.671	−105.777	42.8	−135
1	29.337	−103.353	43.5	−43.5
1	39.327	−106.121	43.9	−157.5
1	32.84	−105.88	44.2	−135
1	44.417	−114.667	45	−120
1	47.125	−112.125	45.5	−120
1	47.125	−112.167	46.3	−120
1	46.985	−110.696	46.9	−120
1	29.32	−103.435	47.2	−2
1	47.125	−112.125	47.5	−120
1	31.768	−106.508	48.4	−135
1	44.457	−114.485	48.5	−120
1	54.1	−126.1	49.9	−5
1	43.042	−111.292	50	−140
1	48.168	−109.703	50	−120
1	53.817	−127.367	50	−135
1	47.878	−108.986	50.6	−58
1	46.967	−108.617	51.1	−95
1	48.78	−117.9	51.2	−140
1	48.78	−117.9	51.2	−150
1	47.808	−110.069	51.7	−55
1	32.95	−109.667	52.2	−135
1	47.808	−110.069	53.1	−160
1	49.9	−117.6	53.1	−160
1	32.9	−113.5	54.4	−125
1	33.1	−109.4	56.5	−130
1	45	−115	58.9	−90
1	32	−94.3	60	−105
1	31.6	−110.7	61.6	−30
1	32	−111.1	61.6	−120
1	33.3	−111	62.6	−105
1	39.337	−106.127	63.6	−112.5
1	32.2	−109.5	63.7	−153
1	32.9	−110.7	63.7	−95

Plate	Latitude	Longitude	Age (Ma)	Stress azimuth
1	33.05	−109.4	63.7	−130
1	33.5	−110.5	63.7	−120
1	32.4	−112.9	64.7	−95
1	40.107	−105.352	65.9	−164
1	47.183	−109.533	66.6	−130
1	38.179	−107.612	66.9	−58
1	40.022	−105.304	67.1	−173
1	32.3	−111.6	67.3	−105
1	47.308	−109.433	67.5	−110
1	47.308	−109.433	67.5	−110
1	32.967	−109.633	67.6	−20
1	39.25	−118.37	67.8	−60
1	33.875	−112.842	68.4	−112.5
1	31.6	−110.7	68.8	−90
1	33.8	−112.9	68.8	−120
1	32.3	−111.6	68.9	−105
1	38.667	−106.583	69	−150
1	38.15	−117.217	69.6	−60
1	39.75	−117	69.6	−180
1	38.133	−116.783	69.7	−160
1	45.583	−115.667	71	−170
1	33.2	−111.1	71.9	−105
1	34.7	−112.3	71.9	−145
1	34.7	−112.3	71.9	−113
1	34.7	−112.3	71.9	−145
1	34.7	−112.3	71.9	−113
1	34.5	−113.2	72.8	−115
1	38.031	−116.53	72.9	−40
1	39.717	−117.267	73.2	−135
1	35.3	−114.2	73.9	−45
1	34.7	−50.5	74	−80
1	35	−114	74.8	−30
1	33.051	−110.721	76.2	−90
1	37.692	−108.067	78.4	−90
1	35.4	−51.5	81	−99
1	29.2	−98.6	86	−75
1	30.68	−97.72	86	−85
1	39.608	−117.467	87.8	−157.5
1	37.667	−95.75	88	−60
1	40.6	−117.163	89.2	−120
1	38.265	−118.238	89.5	−80
1	40.385	−118.508	89.6	−93
1	34.62	−92.63	90	−90
1	40.017	−118.717	90	−92
1	37.667	−95.75	91	−60
1	40.783	−118.25	91.1	−45
1	41.217	−117.267	92.1	−180
1	40.02	−119.065	92.5	−60
1	30.685	−97.71	93	−75
1	37.233	−116.217	94.1	−135
1	41.2	−117.267	94.5	−85

Plate	Latitude	Longitude	Age (Ma)	Stress azimuth
1	43.366	−71.918	96	−76
1	43.516	−73.056	105	−165
1	39.492	−115.983	105.55	−90
1	40.733	−117.383	106.5	−45
1	59.3	−129.883	110	−120
1	38.069	−117.449	115.8	−45
1	39.267	−114.883	117.8	−90
1	45.328	−70.647	121	−85
1	43.369	−70.557	124	−78
1	43.369	−70.557	124	−78
1	39.467	−118.08	126	−90
1	41.75	−114.667	127	−10
1	44.641	−72.758	130	−156
2	32.7	−17.17	2.5	100
2	32.6	−16.65	4.63	100
2	32.7	−17.17	14.31	40
8	11.5	40.3	3.286	−10
8	−21	58	8	82
8	11.22	39.75	11.089	−40
8	9.3	40.6	11.294	45
8	11.21	39.725	11.397	−30
8	11.24	39.775	13.861	−45
8	7.2	39.2	15.401	45
8	7.3	39.1	15.401	70
8	9.38	40.64	15.401	50
8	7	39.7	15.915	55
8	11.3	39.9	15.915	−60
8	7.15	39.325	16.428	60
8	9.62	41.76	18.481	25
8	11.3	39.5	18.995	3
8	7.1	39.5	19.508	50
8	7.5	39.2	19.508	60
8	11.438	40	19.508	0
8	9.8	38.67	19.816	10
8	11.45	40	19.919	10
8	11.23	39.64	21.048	50
8	7.05	39.625	21.562	50
8	11.22	39.65	21.562	60
8	11.33	39.95	21.767	10
8	11.21	39.67	22.178	−10
8	9.1	37.9	23.102	10
8	11.26	39.59	23.204	8
8	11.29	39.52	24.128	0
8	11.375	40	25.155	−10
8	11.4	40	25.669	55
8	11.413	40	25.669	59
8	11.475	40	25.669	−60
8	11.2	39.7	26.695	−15
8	14	43.9	28	−74
8	9.3	42	28.749	70
8	9.7	41.8	28.749	−50

Plate	Latitude	Longitude	Age (Ma)	Stress azimuth
8	16.1	44	29	86
8	9.41	41.94	30.802	30
8	11.34	39.975	30.802	-10
8	-17	62	31	29
8	10.5	43.2	31	10
8	9.1	38.1	31.829	-25
8	11.488	40	31.829	-20
8	11.2	44.2	32	32
8	9.7	38.4	32.342	0
8	9.9	38.33	32.856	0
8	11.32	39.925	32.856	30
8	-13.115	61.372	35.5	12.5
8	9.65	41.8	35.936	40
8	11.26	39.825	35.936	20
8	14.3	42.4	36	11
8	11.29	39.875	36.963	-40
8	-35	-4.5	39	66
8	11.25	39.8	49.797	50
8	11.363	40	56.471	-15
8	11.425	40	58.524	0
8	9.1	37.5	59.551	0
8	-31.99	2.39	61.5	13.5
8	11.25	39.62	67.765	20
8	-28.53	2.32	77.6	15
6	-18.5	144.7	0.2	45
6	-20	145	4.5	48
6	-36.6	156.2	7	8
6	-33.1	156.3	14.8	12
6	-30.8	155.3	15.9	0
6	-30.9	155.2	16.5	-2
6	-28.6	155.5	21	-28
6	-26	148.3	23	5
6	-20.5	148.5	32	-40
6	-29.4	152.8	58	-30
7	-27.35	89.7	43	6
7	-4.193	73.395	49	18.5
7	5.082	73.832	57	1
7	-17.02	89.7	58	3
7	12	72	60	-10
7	-11.3	88.7	62	2
7	1.5	90.2	77	2
7	5.38	89.9	82	2
9	-50.9	-138.1	0.01	-63
9	18.7	-111	0.3	-105
9	20.1	-155.7	0.46	-57
9	-50.4	-139.1	0.5	-63
9	-17.8	-149.2	0.5	-69
9	53.3	-135.6	0.7	-60
9	-14.3	-170.7	0.72	-75
9	-18.7	-159.8	0.79	-66
9	20.7	-156.1	0.86	-57

Plate	Latitude	Longitude	Age (Ma)	Stress azimuth
9	-24	-130.1	0.96	-69
9	-17.6	-149.5	1.2	-69
9	20.9	-156.6	1.34	-57
9	21.2	-156.9	1.52	-63
9	-17.5	-149.8	1.7	-69
9	-21.2	-159.8	1.88	-70
9	21.2	-157.2	1.9	-63
9	18.3	-114.7	2.4	-105
9	-16.8	-151.5	2.6	-66
9	-16.7	-151	2.6	-66
9	5.4	162.9	2.6	-70
9	21.5	-157.8	2.7	-68
9	-16.6	-151.5	3.2	-66
9	-16.5	-151.8	3.4	-66
9	21.5	-158.2	3.9	-68
9	-27.6	-143.8	4	-75
9	54.6	-136.9	4.1	-65
9	-16.4	-152.2	4.5	-68
9	22	-159.5	5.8	-69
9	-27	-144.3	5.9	-75
9	29	-118.3	7.1	-67
9	-23.2	-135	7.2	-69
9	23.1	-161.9	7.2	-72
9	23.1	-161.9	7.2	-86
9	-23.9	-147.7	7.6	-69
9	-22	-139	7.8	-69
9	6.9	158.3	8.6	-70
9	23.6	-164.7	10.3	-82
9	-23.3	-149.5	10.6	-70
9	23.8	-166.3	12	-64
9	-22.4	-151.3	12.3	-61
9	-48.2	-148.8	12.5	-65
9	7.3	151.8	13.9	-70
9	53.5	-136	14.4	-60
9	55.2	-140.3	14.9	-67
9	55.2	-140.3	14.9	-50
9	-21.9	-157.9	19.4	-61
9	54	-137.4	19.7	-76
9	27.9	-175.8	20.6	-66
9	56.5	-146.6	21.9	-69
9	56.9	-149.2	24.8	-75
9	29.5	179	27.4	-65
9	28.3	-177.3	27.7	-66
9	27.4	-178.8	28	-64
9	2.1	-157.4	35.5	-66
9	-41.5	-164.3	36.6	-47
9	8.2	-161.9	39.3	-82
9	33.7	171.6	39.9	-65
9	32.1	172.3	42.4	-29
9	32.7	172.2	43.4	-29
9	-36.9	-169.8	44.6	-22

Plate	Latitude	Longitude	Age (Ma)	Stress azimuth
9	-38.4	-168	45.5	-44
9	-15	-149	47.4	-60
9	35.2	171.8	48.1	-6
9	-33.9	-171.2	53.4	-27
9	37.5	170.7	55.2	-19
9	39	170	55.4	-2
9	41.3	170.5	56.2	-5
9	21.5	159.5	56.5	-12
9	0.7	-155.3	59	-43
9	11.5	162.3	60.6	-12
9	-28.2	-173.6	61.2	-43
9	44.8	170	64.7	-5
9	-26	-175	66.6	-28
9	28.1	-162.3	66.9	-42
9	2.6	-158.5	69.8	-39
9	-27.6	-174.2	70.1	-28
9	-9.1	-150.7	70.5	-22
9	53	165	71	-40
9	-7.6	-151.6	71.9	-34
9	23.9	159.5	76.3	0
9	5.8	-160.8	76.4	-32
9	9.2	-160.8	78.7	-36
9	15	-166.4	82.6	-47
9	12	194.2	82.7	-32
9	12.5	-167.1	85	-33
9	16.4	-168.2	86	-39
9	29.6	-163.4	90.4	-42
9	18	-169.1	93.4	-32
10	-33	-81	0	90
10	-0.5	-92	0	95
10	-25	-91	20	85
10	-22	-82	30.7	40
10	-19	-79	33.3	40
10	-17	-77.5	39	40

0 Eurasia, 1 North America, 2 North Africa, 3 South America, 4 Antarctica, 5 India, 6 Australia, 7 Central Indian Ocean, 8 Central Africa, 9 Pacific; 10 Nazca.

Subject Index

Printing: Mercedes-Druck, Berlin
Binding: Stein + Lehmann, Berlin